D1703394

Expected Effects of Climatic Change on Marine Coastal Ecosystems

Developments in Hydrobiology 57

Series editor

H. J. Dumont

Expected Effects of Climatic Change on Marine Coastal Ecosystems

Edited by

Jan J. Beukema, Wim J. Wolff and Joop J.W.M. Brouns

KLUWER ACADEMIC PUBLISHERS
DORDRECHT / BOSTON / LONDON

Library of Congress Cataloging in Publication Data

Expected effects of climatic change on marine coastal ecosystems /
edited by Jan J. Beukema, Wim J. Wolff, and Joop J.W.M. Brouns.
p. cm. -- (Developments in hydrobiology ; 57)
Papers from an international workshop held on Texel, the
Netherlands, Nov. 11-15, 1988.
ISBN 0-7923-0697-X (alk. paper)
1. Marine ecology--Congresses. 2. Coastal ecology--Congresses.
3. Climatic changes--Environmental aspects--Congresses.
I. Beukema, Jan J., 1935- . II. Wolff, W. J. (Wim J.)
III. Brouns, Joop J. W. M., 1945- . IV. Series.
QH541.5.S3E97 1990
574.5'2636--dc20 90-4171

ISBN 0-7923-0697-X

Published by Kluwer Academic Publishers,
P.O. Box 17, 3300 AA Dordrecht, The Netherlands.

Kluwer Academic Publishers incorporates
the publishing programmes of
D. Reidel, Martinus Nijhoff, Dr W. Junk and MTP Press.

Sold and distributed in the U.S.A. and Canada
by Kluwer Academic Publishers,
101 Philip Drive, Norwell, MA 02061, U.S.A.

In all other countries, sold and distributed
by Kluwer Academic Publishers Group,
P.O. Box 322, 3300 AH Dordrecht, The Netherlands.

Printed on acid-free paper

Printed in the Netherlands

CONTENTS

J.J. BEUKEMA, W.J. WOLFF & J.J.W.M. BROUNS. *Introduction.* 1

CLIMATIC CHANGES.

G.P. HEKSTRA. *Man's impact on atmosphere and climate: a global threat? Strategies to combat global warming.* 5
H. DE BOOIS. *Ecological interpretation of climate projections.* 17
C.M. GOODESS & J.P. PALUTIKOF. *Western European regional climate scenarios in a high greenhouse gas world and agricultural impacts.* 23

ELEVATED CARBON-DIOXIDE CONCENTRATIONS.

S.P. LONG. *The primary productivity of Puccinellia maritima and Spartina anglica: a simple predictive model of response to climatic change.* 33
D. OVERDIECK. *Direct effects of elevated CO_2 concentration levels on grass and clover in 'model-ecosystems'.* 41
J. ROZEMA, G.M. LENSSEN, R.A. BROEKMAN & W.P. ARP. *Effects of atmospheric carbon dioxide enrichment on salt-marsh plants.* 49

TEMPERATURE CHANGES.

C. VAN DEN HOEK, A.M. BREEMAN & W.T. STAM. *The geographic distribution of seaweed species in relation to temperature: present and past.* 55
A.M. BREEMAN. *Expected effects of changing seawater temperatures on the geographic distribution of seaweed species.* 69
C.G.N. DE VOOYS. *Expected biological effects of long-term changes in temperatures on benthic ecosystems in coastal waters around The Netherlands.* 77
J.J. BEUKEMA. *Expected effects of changes in winter temperatures on benthic animals living in soft sediments in coastal North Sea areas.* 83
J.G. WILSON. *Effects of temperature changes on infaunal circalittoral bivalves, particularly T. tenuis and T. fabula.* 93
M.J. COSTA. *Expected efffects of temperature changes on estuarine fish populations.* 99

SEA-LEVEL RISES.

W. SIEFERT. *Sea-level changes and tidal-flat characteristics.* 105
C. CHRISTIANSEN & D. BOWMAN. *Long-term beach and shoreface changes, NW Jutland, Denmark: effects of a change in wind direction.* 113
R. MISDORP, F. STEYAERT, F. HALLIE & J. DE RONDE. *Climate change, sea level rise and morphological developments in the Dutch Wadden Sea, a marine wetland.* 123
W.E. WESTERHOFF & P. CLEVERINGA. *Sea-level rise and coastal sedimentation in central Noord-Holland (The Netherlands) around 5000 BP: a case study of changes in sedimentation dynamics and sediment distribution patterns.* 133
J.C. LEFEUVRE. *Ecological impact of sea level rise on coastal ecosystems of Mont-Saint-Michel Bay (France).* 139
J.W. DAY, JR & P.H. TEMPLET. *Consequences of sea level rise: implications from the Mississippi Delta.* 155
A.H.L. HUISKES. *Possible effects of sea level changes on salt-marsh vegetation.* 167
K.S. DIJKEMA, J.H. BOSSINADE, P. BOUWSEMA & R.J. DE GLOPPER. *Salt marshes in The Netherlands Wadden Sea: rising high-tide levels and accretion enhancement.* 173
J.D. GOSS-CUSTARD, S. MCGRORTY & R. KIRBY. *Inshore birds of the soft coasts and sea-level rise.* 189

UV-B RADIATION.

K.J.M. KRAMER. *Effects of increased solar UV-B radiation on coastal marine ecosystems: an overview.* 195
J. VAN DE STAAY, J. ROZEMA & M. STROETENGA. *Expected changes in Dutch coastal vegetation resulting from enhanced levels of solar UV-B.* 211

Index 219

INTRODUCTION

J.J. Beukema, W.J. Wolff & J.J.W.M. Brouns

Man is changing the biosphere at an ever increasing rate. Several of these man-made changes are on a worldwide scale, such as the increase in atmospheric concentrations of several gases. In particular the ongoing increase of the concentration of atmospheric carbon dioxide, by excessive burning of fossil fuels and forest destruction, is well-documented. By the year 2050, CO_2 levels will almost certainly be twice the pre-industrial concentrations and this is expected to have far-reaching consequences. Direct effects include higher rates of plant production (also in agriculture). Indirect effects might be less favourable: by the intensified 'greenhouse process' (to which several other gases contribute as well), a global warming is generally expected, in its turn probably resulting in hardly predictable local changes in climate and an accelerated rise of sea level. Other gases (particularly freons) are depleting the stratospheric ozone layer, which shields the biosphere from excessive UV-B radiation. This is another threat to life on earth.

All of these threats are expected to affect ecosystem functioning and human well-being in the near future. None of these threatening changes will be easy to reverse. We are now at a stage where the beginning of the effects will become detectable. It is the scientific community's responsibility to assess the first signals of changes in environmental factors (temperature, sea level, UV radiation), to predict the effects of these changes on living systems, to advise on countermeasures and to communicate the findings to policy makers and the general public. Because our present knowledge, both on the expected climatic change and on its consequences, is far from complete, the first steps to be taken are: making an inventory of present knowledge, an identification of uncertainties and a formulation of research needs.

In this spirit, a workshop assembled to evaluate the expected effects of climatic change on marine coastal ecosystems, particularly in soft-sediment areas at temperate latitudes. The vulnerability of these systems to rises in CO_2 concentrations, temperature, sea level and UV-B radiation was discussed by a group of 30 scientists from 13 (mostly west-European) countries, representing the fields of physics, chemistry, geology and biology. The workshop was held on Texel, the Netherlands, from 11 to 15 November 1988. It was sponsored by the Netherlands ministery of Housing, Physical Planning and Health (represented by Dr. G.P. Hekstra), chaired by Dr. W.J. Wolff (Research Institute for Nature Management) and housed by the Netherlands Institute for Sea Research (N.I.O.Z., represented by Dr. J.J. Beukema).

The written versions of the presentations by 23 participants have been brought together in these proceedings of the Workshop.

The first paper, by G.P. HEKSTRA, explains how trace gases affect UV-B radiation, alkalinity of the sea, rate of photosynthesis, and greenhouse warming. It expresses the expectation that global warming and sea-level rise will take place and warns against the impacts, which will be particularly serious in vulnerable coastal areas, such as most of the Netherlands.

Regional scenarios of climatic change are discussed both in the paper by H. DE BOOIS and the one by C.M. GOODESS & J.P. PALUTIKOF. Global rises of CO_2 concentration and temperature could affect local climates in quite different ways with totally different ecological consequences. Current scenarios for regional climate changes, however, result in highly different conclusions. As yet, they should be considered insufficiently reliable for local impact predictions.

All following papers deal with effects to be expected from specific large-scale changes, in the first place rises of atmospheric CO_2 concentration and temperature. S.P. LONG develops a model in which primary production is predicted for two species of salt-marsh plants. After satisfactory validation, this model is used to predict production at a doubled atmospheric CO_2 concentration and an increased air temperature. D. OVERDIECK presents results of experiments on the direct effects of elevated CO_2 concentrations on single plants and small vegetation units in model-ecosystems. J. ROZEMA *et al.*, working with two species of salt-marsh plants, also observed enhancements of growth at elevated as compared to natural CO_2 concentrations.

Expected effects of temperature changes are discussed in 6 papers. C. VAN DEN HOEK *et al.* and A.M. BREEMAN show temperature-related distribution patterns in seaweed species. Major changes in these patterns appear to have taken place during former periods of climatic changes. Therefore, these authors

J.J. Beukema et al. (eds.), Expected Effects of Climatic Change on Marine Coastal Ecosystems, 1–3.

predict profound alterations of the species composition of seaweed floras if seawater temperature changes by only a few degrees centigrade. Latitudinal displacements and regional extinction would change local community structure. C.G.N. DE VOOYS discusses this topic for bottom animals living in coastal waters. He concludes that a warming of a few degrees centigrade would change the benthic marine ecosystems of the Netherlands to those now occurring in similar areas in France (and formerly in the Eemian interglacial in the Netherlands). J.J. BEUKEMA deals in more detail with the response of benthic animals of the Wadden Sea to high and low winter temperatures. A high proportion of the tidal-flat fauna of the Dutch Wadden Sea is sensitive to low winter temperatures. This fauna would become richer in species number and more stable in biomass if winter temperature increases. J.G. WILSON examines the effects of temperature on two closely related species of benthic animals, one of them being more thermally sensitive than the other. M.J. COSTA compares the fish population of the Wadden Sea with those of a couple of more southerly (and thus warmer) estuaries in Portugal. It is expected that higher temperatures in the Wadden Sea would promote growth in some flatfish species in this nursery.

Effects of rising sea levels are discussed in the next part, made up of 8 papers. W. SIEFERT discusses sea-level records from a high number of stations along the Dutch, German and Danish coasts. During the last decade, no acceleration could be observed of the present long-term rate of mean-sea-level rise in this area (10 to 20 cm during the last century). Only the tidal range has increased rapidly over the last 20 years as a consequence of higher mean high-tide levels. This change is expected to enhance erosion and might counteract the sediment deposition necessary to keep tidal-flat sedimentation in pace with sea-level rise. The cause of the acceleration in the rise of high-tide levels is not clear, but dredging and changed directions of prevailing winds might contribute to the phenomenon. The paper by C. CHRISTIANSEN & D. BOWMAN discusses shore-line changes in Denmark as a consequence of long-term changes in wind regime: during the last decades prevailing winds have moved from W towards NW. This change turned certain sedimentation areas into stretches with significant erosion and a retreating shore line. R. MISDORP *et al.* work out a hypothesis on the future developments of the tidal flats and channels in the western Wadden Sea if mean sea level should rise more rapidly. Even at a relatively slow rate of 0.5 m per century, intertidal areas would tend to disappear in the tidal basins with relatively small intertidal areas (such as the Texelstroom basin), whereas it is yet uncertain whether in the other basins (with extended tidal flats) sedimentation could keep pace with a rise of 0.5 or 1.0 m per century.

Three papers deal with historical changes in specific geographic areas, both in Europe and in North America. W.E. WESTERHOFF & P. CLEVERINGA describe sea-level rise and sedimentation dynamics in a part of the Netherlands' coast around 5000 B.P. Besides the rising sea level, such factors as the pre-existing relief, the supply of sediments and the hydrodynamic setting appear to have been important for the sedimentation pattern. J.C. LEFEUVRE describes the history of the area of Mont-Saint-Michel Bay (France) and develops two scenarios for the fate of this area, *viz.* for a continuation of the present low rate of sea-level rise (0.2 cm per year) and for a significantly higher rate (more than 1 cm per year). J.W. DAY & P.H. TEMPLET discuss in detail the history of the morphology and the management of the Mississippi River Delta (USA), where relative sea-level rise has been exceptionally rapid. In this area, loss rates of wetland are now high and saline water moves inland. Experience acquired in the management of this area will be indispensable wherever areas become affected by rapid sea-level rise.

The fate of salt marshes along the Dutch coast under the stress of changing sea levels and tidal regimes is discussed in two papers. A.H.L. HUISKES describes the zonation of the vegetation patterns of salt marshes as caused by the inundation regime, sedimentation/erosion balance and interaction between species. Changes in such factors are bound to affect the vegetation patterns. K.S. DIJKEMA *et al.* compare the balances of sedimentation and erosion in the marsh zone itself and in the pioneer zone in front of the marsh. In the Netherlands, this accretion balance is mostly positive in the salt-marsh zone itself and in this zone it could keep up with even higher rates of sea-level rise. In the pioneer zone, however, the balance is negative in most areas. The present accelerated rising of high-tide levels will lead to accelerated cliff formation and increased marsh erosion from the seaward edge. Thus, management techniques should be directed to the protection of these pioneer zones.

Possible effects of an accelerated rise in sea level on the birds that utilise the intertidal zone are discussed by J.D. GOSS-CUSTARD *et al.* The effects are difficult to predict, but erosion or insufficient sedimentation would surely reduce both the quality and the size of areas suitable for the feeding and breeding of shore birds.

The final section deals with the possible effects of increased UV-B irradiation. The overview paper by K.J.M. KRAMER evaluates the possible biological effects of increased UV-B radiation on coastal marine ecosystems in the Netherlands. Coastal waters in this area show relatively high concentrations of such transmission-limiting material as particulate matter and 'yellow substance'. Only in special conditions (*e.g.* shallow tidal pools with clear water), can detrimental effects on living organisms be expected. Salt-marsh plants are exposed more directly to UV-B radiation. J. VAN DE STAAY *et al.* report on the response of two species of such plants to enhanced UV-B levels. In both species, photosynthesis and growth were inhibited, though to a different extent.

The coastal environment is and will be affected in a variety of ways by the large-scale changes in the concentrations of atmospheric gases and probably resulting climatic changes. Some of these effects might be mainly beneficial (*e.g.* the increased productivity due to higher availability of carbon dioxide), others could be exclusively detrimental (*e.g.* increased levels of UV-B irradiation). Consistent changes in temperature will cause shifts in areas of distribution of great numbers of species of plants and animals, changing the composition of ecosystems with risks of imbalance and extinction of rare and sensitive species. Global temperature rises are bound to be followed by rises in sea level, even from thermal expansion alone. The precise effects of sea-level rise in coastal environments will depend largely on geomorphological processes, in particular on the net result of changing rates of sedimentation and erosion. Such decisive but indirect effects make predictions highly uncertain. Nevertheless, it is a worthwhile exercise to consider possible effects of climatic changes, in particular to realize in time the gaps in our knowledge and to be prepared when climatic changes become greater.

MAN'S IMPACT ON ATMOSPHERE AND CLIMATE: A GLOBAL THREAT? STRATEGIES TO COMBAT GLOBAL WARMING

G.P. HEKSTRA

Ministry of Housing, Physical Planning and Environment, P.O. Box 450, NL-2260 MB Leidschendam, The Netherlands

ABSTRACT

Climate is in constant interaction with solar irradiation and emissions from the biosphere, including those induced by man. In particular trace gases with more than two atoms in a molecule can influence valued properties of the atmosphere, such as absorption of UV radiation, constancy of the radiation budget, and proper photochemistry. Only the long-lived trace gases with a global distribution significantly affect climate and the ozone layer; the others are much more involved in the regional and continental scale problems of acidification and other forms of air pollution.

Discussed are increased UV-B irradiation, alkalinity of sea water, enhancement of photosynthesis and the phenomenon of greenhouse warming due to increasing emissions of CO_2, CH_4, N_2O and CFC's, giving for each the proportional contribution to greenhouse warming. The impact of a warmer world is discussed with special reference to the Northern Hemisphere terrestrial biomes.

An indication is given of required reductions of (expected) emissions of greenhouse gases, following a scenario approach regarding world economic and population growth. Discussed are accelerated emissions in case of no measures for reduction, the measures to be taken to reach stabilization in warming commitment and the measures that can additionally decrease the warming commitment. As these requirements are unlikely to be met, global warming and sea level rise is undoubtedly going to take place over the next decades to century. Sea-level rise will not only affect coastal lowlands, but salt water intrusion as far inland as to the 5-metre contourline is not unrealistic. Shown are vulnerable areas in the world at large and in Europe in particular.

The societal costs are discussed, assuming different strategies to prevention (limiting the emissions) and adaptation to the impacts. Different countries (coastal versus continental; developed versus developing) may have different perceptions and attitudes to either prevention, adaptation or both. The Netherlands government has taken various initiatives to enhance public awareness and international cooperation, in the framework of the Intergovernmental Panel on Climatic Change.

1. COMPONENTS OF THE CLIMATE SYSTEM IN RELATION WITH THE BIOSPHERE

Climate is a composite. It comprises daily and weekly weather events on the meso- (local), synoptic- (regional) and global scale that constantly couple back with memory functions from soil moisture, atmospheric heat and moisture content, snow and sea ice, sea currents and oceanic up- and downwelling, causing storage and release of heat, and sequestration (photosynthesis) and release (respiration) of solar energy through ecosytems.

2. TRACE GASES AND THEIR BIOSPHERIC ORIGINS

The major components of the atmosphere have fairly constant concentrations (N_2 about 79%; O_2 about 19%; Argon just over 1%). The remaining 1% comprises water vapour and many trace gases of varying concentrations and most of them with great potential for affecting valued properties of the atmosphere. The trace-gas concentrations are highly influenced by human industrious activities, or changes in landuse, agriculture, animal husbandry *etc.* (Table 1).

3. TRACE GASES AND VALUED PROPERTIES OF THE ATMOSPHERE

Table 2 indicates for the listed trace gases their capacity to influence various valued components of the atmosphere, and thus affect general human interests: UV-B enhancement, global warming, photochemical oxidant formation, acidification of precipitation, corrosion of materials, degradation of visibility and bad smell.

4. LIFETIMES AND SPREADING OF ENVIRONMENTALLY RELEVANT TRACE GASES

Each trace gas has its characteristic lifetime and

J.J. Beukema et al. (eds.), Expected Effects of Climatic Change on Marine Coastal Ecosystems, 5–16.

TABLE 1

Sources of major disturbance to atmospheric chemistry. After Clark (in CLARK & MUNN, 1986).

	Oceans and Estuaries	Vegetation and Soils	Wild Animals	Wetlands	Biomass Burning	Crop Production	Domestic Animals	Oil combustion	Coal Combustion	Industrial Processes
C-soot					x			x	x	
CO_2	x	x			x	x		x	x	
CO	x	x			x			x	x	x
CH_4		x	x	x	x	x	x			
C_xH_y	x	x			x	x				
NO_x	x				x	x		x	x	x
N_2O	x	x			x	x		x	x	
NH_3/NH_4		x	x	x	x	x	x		x	
SO_x								x	x	x
H_2S/methyl S	x	x		x						
COS	x	x		x						
Organic S	x	x		x						
Halocarbons										x
Other halogens	x							x	x	x
Trace elements	x				x			x	x	x
O_3 groundlevel								x	x	x
O_3 stratosph.										x

TABLE 2

Major impacts of atmospheric chemistry on valued properties of the atmosphere. After Clark (based on data from CRUTZEN & GRAEDEL, in CLARK & MUNN, 1986). Last column with bad smell added.

	Ultraviolet energy absorption	Alteration thermal radiation budget	Formation of photo-chemical oxidants	Acidification of precipitation	Degradation of visibility	Corrosion of materials	Contribution to bad smell
C-soot		x			x		x
CO_2		x					
CO			x				
CH_4		x					
C_xH_y			x			x	x
NO_x			x	x	x		x
N_2O		x					
NH_3/NH_4				x			x
SO_x		x		x	x	x	x
H_2S/methyl S						x	x
COS							x
Organic S							x
Halocarbons		x					
Other halogens						x	x
Trace elements							x
O_3 groundlevel			x				x
O_3 stratosph.	x	x					

spatial spreading as shown in Fig. 1. Only those with a global spreading and lifetimes of many years (CH_4, CO_2, CFC's and N_2O) are significantly affecting climate and the stratospheric ozone layer. The others, clustered at left in the diagram, collectively cause the more commonly known local and regional air pollution and derived soil and water pollution problems.

5. CLIMATE RELATED EFFECTS OF CHANGES IN ATMOSPHERIC CHEMISTRY

5.1. INCREASED UV-B IRRADIATION

When CFC's escape into the stratosphere they reduce the ozone concentration, that filters out UV light. A 10-percent reduction of the ozone shield would result in a 1% increase of the total UV component, mostly UV-B, some UV-A and almost no UV-C (the most lethal component). Increased UV-B can cause serious biotic damages to chlorophyl, eye pigment, skins and DNA. The increase in UV-B is not linear but progressive and so is the damage. WORREST (1986) mentions calculations that a 25% reduction of the ozone layer could lead to 35% reduction of oceanic primary production, but KRAMER (1987, this issue) argues that marine organisms can shelter themselves against increased UV-B by floating or swimming deeper and that the effects are likely negligeable. However, organisms on tidal flats, marshes or on the land that cannot creep away will get more damaged. In fact far too little has been investigated about UV-B effects in outdoor marine environments to warrent conclusions.

5.2. DECREASE OF ALKALINITY OF SEA WATER

Average pH of ocean water is between 8.0 and 8.2.

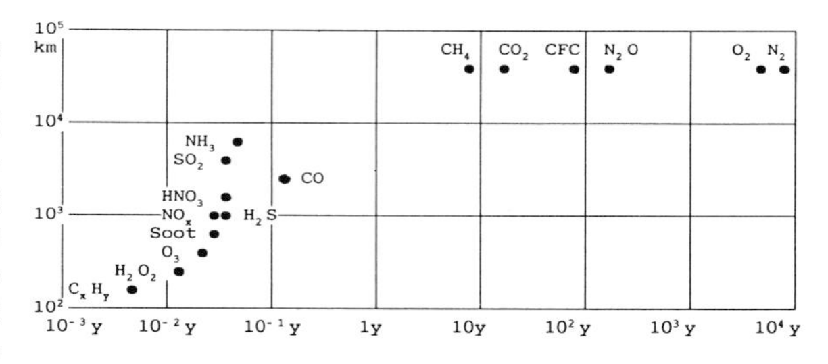

Fig. 1. Characteristic scales of atmospheric constituents: half-life times and global spreading. Both axes exponential scales. After data from CRUTZEN & GRAEDEL and an idea of Clark (in CLARK & MUNN, 1986).

According to BROUNS (1988) there is general agreement on a 0.3 lowering of pH at a doubling of atmospheric CO_2. This is of the same order of magnitude as the daily amplitude that reflect photosythesis and respiration (7.8 - 8.3). Brouns doesnot mention calculations of pH shifts due to acid deposition derived from SO_2, NO_x and sulphides. He indicates a strong enhancement of bio-availability of Cu^{--} and hence toxic effects on some algae by a lowering of pH of that magnitude. The marine environment is most toxic during periods of maximum respiration (warm and dark). The higher, eukaryotic algae have mechanisms to cope with these differences in pH, but the more primitive, prokaryotic Bluegreen Algae face more difficulties and for them pH 7.6 is a limit to growth.

5.3. CO_2 ENHANCEMENT OF PHOTOSYNTHESIS

At pre-industrial levels of approximately 270 ppmv, CO_2 almost certainly was a limiting factor in terrestrial photosynthesis, but not in marine photosynthesis, as the marine bicarbonate solution is buffered much more to the great surplus of carbonate sediment than to the atmospheric CO_2. Systematic experiments with CO_2 enhancement of photosynthesis in land plants are mainly done with crops in greenhouses. CO_2 was never identified as an environmental variable during the ten years of the International Biological Programme (1964-1974). The number of experiments with wild plants is now rapidly increasing. These include plants from salt marshes, both with C3 and C4 carbon cycles, and experiments with mixed species communities (TERAMURA, 1986). It appears that CO_2 enhancement and UV-B damage in most respects are counteracting processes, as is shown in Table 3.

TABLE 3

Effects of UV-B and CO_2 on plants (from TERAMURA, 1986).

Characteristic	Enhanced UV-B	Enhanced CO_2 (doubling)
Photosynthesis	decreases in many C3 and C4 plants	in C3 plants increase up to 100 % but in C4 plants only a small increase
Leaf conductance	no effects in many plants	decreases in both C3 and C4 plants
Wateruse efficiency	decreases in most plants	increases in both C3 and C4 plants
Dry matter production and yield	decreases in many plants	almost a doubling in C3 plants and only small increase in C4 plants
Leaf area	decreases in many plants	increases more in C3 than in C4 plants
Specif.leaf weight	increases	increases
Drought tolerance	no influence	greater tolerance
Crop maturation	no influence	accelerated
Flowering	some plants faster, others slower	accelerated
Intraspecific differences	response varies among cultivars	response may vary among cultivars
Interspecific differences	response may vary per community	C3 plants get more competitive over C4 plants

5.4. GREENHOUSE WARMING

We now confine ourselves to the climatically relevant gases CO_2, CH_4, N_2O and CFC's. The world is already for long alerted about global warming or the greenhouse effect of carbon dioxide. Next to water vapor this is the most important gas for keeping the globe at acceptable temperature conditions to support life. But too much CO_2 may cause an unbearable heating or climatic change. Only more recently has it become known that the so-called other greenhouse gases are potentially an even greater threat. RAMANATHAN *et al.* (1985) have computed the capacity of several gases to trap heat and transmit it to other air molecules, in particular to water vapor. The 'heat trapping efficiency' can be expressed relative to CO_2, either per unit mass or per molecule as in Table 4. From this table it can be concluded that per emitted molecule methane is thirty times, dinitrous-oxide two hundred times and the freons about twenty thousand times as efficient in causing global warming! Fortunate enough the latter have still extremely low atmospheric concentrations relative to carbon dioxide.

TABLE 4

Relative efficiency of greenhouse gases to increase global mean temperatures (after RAMANATHAN *et al.*, 1985).

	rel. ΔT per unit mass	rel. ΔT per mol
CO_2	1	1
CH_4	90	30
N_2O	200	200
$CFCl_3$ (= F11)	5800	19000
CF_2Cl_2 (= F12)	7700	21000

6. PAST AND PROSPECTED INCREASE RATES OF GREENHOUSE GASES

6.1. CARBON DIOXIDE

The rapid recent increase of atmospheric CO_2 concentrations as caused by particularly industrialisation and deforestation are well known (Fig. 2). It has become more and more clear that up to one quarter of the increased emissions originates from the biosphere, by direct burning of biomass and by slow oxidation of organic matter and soil carbon in connection with deforestation, drainage and erosion of soils. More than half of the emitted CO_2 is taken up in the oceans. The proportion being absorbed is decreasing with increasing emission rates. Hence, the airborne fraction is growing and, with the ocean water becoming warmer, will increase even faster (negative feedback by less rapid uptake of CO_2 in warmer water; BOLIN *et al.*, 1986).

Table 5 shows a low and high scenario for the contribution of the terrestrial biosphere. The low one,

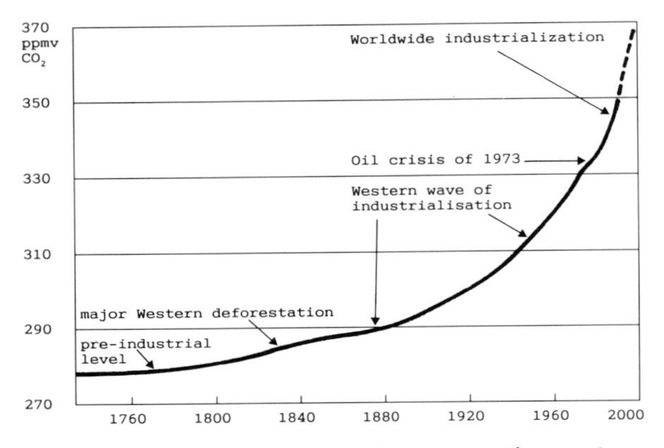

Fig. 2. Atmospheric carbon dioxide concentrations as trapped in ice cores from Sipple Station, Antarctica. After SIEGENTHALER & OESCHER (1987).

corresponding with an airborne fraction close to 50%, is given most credibility by scientists. It means about equal emissions from fossil fuels and the terrestrial biosphere over the two-century period 1760 to 1957. Over the last 30 years fossil fuels by far outpace the biospheric emission, in spite of the present speed of deforestation, erosion and desertification. The fossil emissions simply accelerated much faster. BOLIN (1987) calculated that the annual growth of fossil-CO_2 emissions dropped from 4% per year over the period 1945-1973 to 1% per year over the period 1973 to 1986. This has brought 20 Pg C (= 20×10^{15} g C) less in the atmosphere than if the pre-1973 trend had continued (about 4 to 5 ppm less increase in atmospheric CO_2).

6.2. METHANE

Methane concentrations in the air have been very stable over at least 10 000 years time till the 18th century, and have risen exponentially over the last 200 years, with a doubling per 50 years (Fig. 3). The present global concentration is only about 0.5% of the concentration of CO_2, but the rate of increase is much faster. Per molecule, CH_4 is 30 times more radiatively active than CO_2 (1.75 ppmv CH_4 equals 52.5 ppmv CO_2).

Two major causes of increasing CH_4 emission stem from agriculture, *i.e.* methanogenesis in wetland farming (irrigation) and in the digestive system of ruminantia (intensive cattle raising). The third cause is from fermentation in domestic waste heaps. About 20% of the CH_4 emissions is energy related (leaks in the production and use of fossil fuels).

TABLE 5

Estimated changes of atmospheric components of the global carbon cycle for the periods 1760-1957 and 1958-1985 (after BOLIN, 1987).

	1760-1957	1958-1985	1760-1985
increase of CO_2 (in ppm)	34	33	67
(in Pg C)	72	70	142
Pg C emissions: fossil fuel	80	112	192
terrestrial-low	80	20	100
(airborne fraction-high)	(0.45)	(0.53)	(0.49)
terrestrial-high	130	70	200
(airborne fraction-low)	(0.34)	(0.38)	(0.36)

6.3. DINITROUS OXIDE

N_2O is a by-product in the biotic processes of nitrification and denitrification. It easily escapes into the air from aerated soils, but in anaerobic soils N_2O is reduced to N_2. Once in the air N_2O can easily become oxidized to NO and NO_2, two gases without absorbtion bands in the infrared part of the spectrum. About 30% of the emissions is energy related. Emission estimates widely vary from 11.1 to 54.7×10^{12} g N per year. The greatest range of uncertainty stems from cultivated land. The emissions from the application of nitrogen fertilizers is estimated between 0.7 and 3.0×10^{12} g N per year (BOLLE *et al.*, 1987). This number is certainly going to increase with greater use of fertilizers around the world. On the other hand the amounts naturally emitted from rainforest and other tropical forest will likely decrease due to deforestation. The trend is shown in Fig. 4.

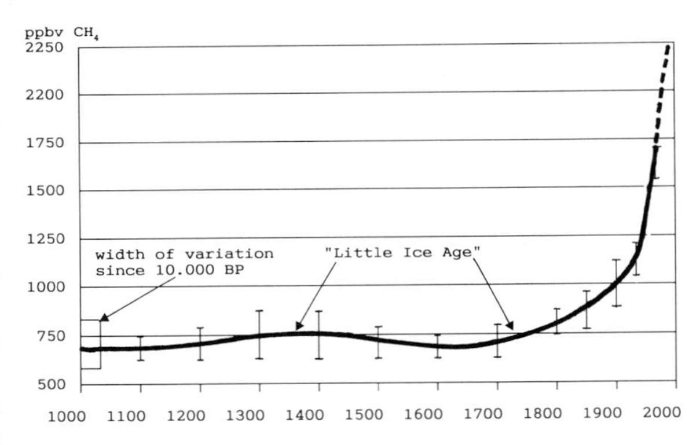

Fig. 3. Increase in atmospheric methane trapped in ice cores at Greenland and Antarctica. After RASMUSSEN & KHALIL (1984).

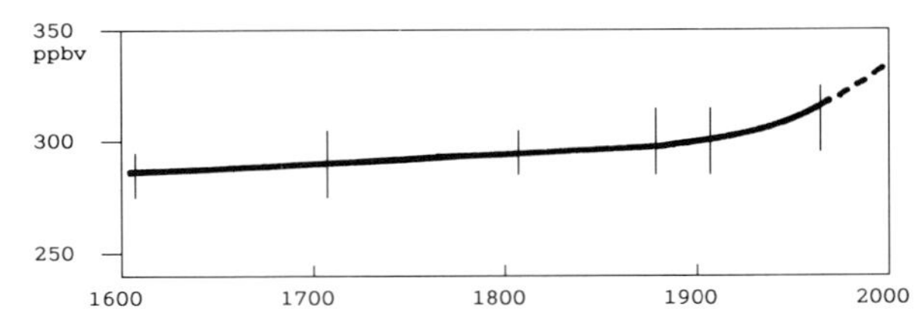

Fig. 4. Atmospheric N_2O concentrations from ice cores bubbles 1600 - 1980 (after TORRENS, 1989).

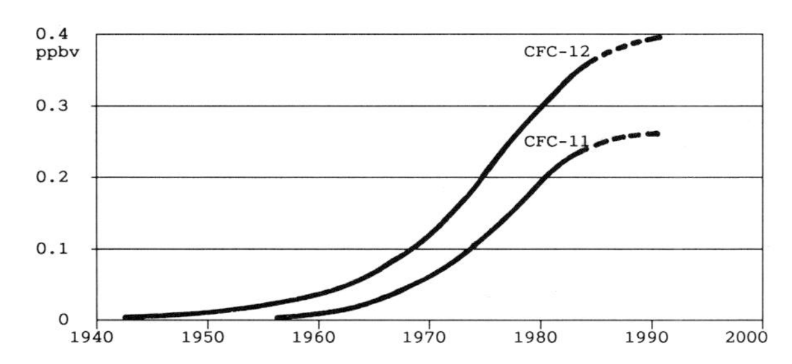

Fig. 5. Computed atmospheric concentrations of CFC-11 and CFC-12 based on data by CMA 1984 and with residence times determined to fit observed concentrations in 1980 (after WMO; through BOLIN, 1987).

6.4. CFC'S 11 AND 12, AND OTHER HALOCARBONS

Rapid exponential increases took place of atmospheric concentrations of the two most radiatively active industrial gases CFC-11 and CFC-12 (Fig. 5), used for spray cans, extrusion of foams and cooling liquids. These 'new' chemicals exist only about 50 years. Their concentrations in the air are still extremely low: about 0.0001% of CO_2, but they are 20 000 x more radiatively active (the present level of 0.7 ppbv CFC's 11 + 12 equals the effect of 14 ppmv CO_2). Another radiatively active substance is CFC-22; much less active are CFC's 13 and 14. These minor CFC's contribute less to global warming than CFC-11 alone. Brominated hydrocarbons ('halons', used as flame retardants) are extremely radiatively active. They should be entirely replaced and their production completely banned.

7. TREND IN MEAN ATMOSPHERIC TEMPERATURE RISE

Fig. 6 shows the trend in global mean annual temperature: a rise of ~0.7°C since 1860. From 1940 to 1960 a stagnation in the global trend to a warmer planet occurred, but only in the Northern Hemisphere. The

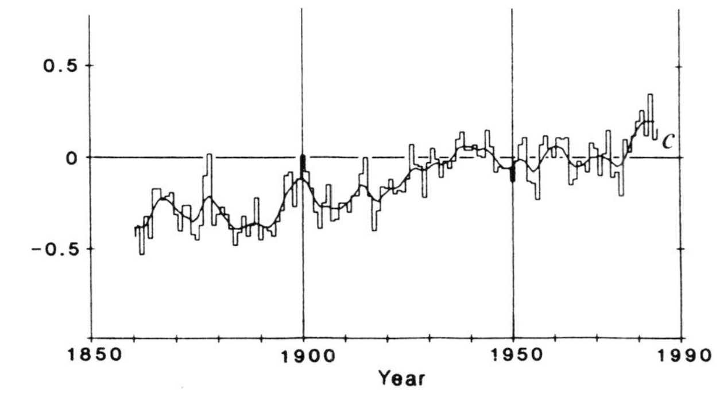

Fig. 6. Global annual mean surface temperature changes with 10-year Gaussian filtered values (after JONES *et al.*, 1986).

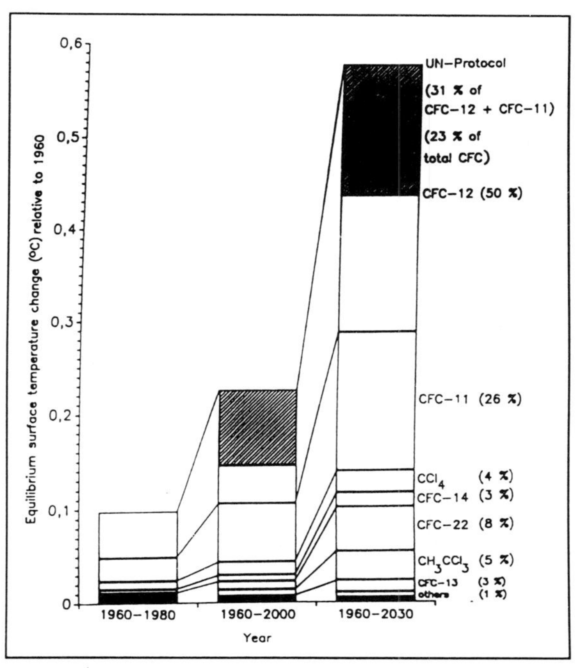

Fig. 7. CFC contributions to global warming relative to 1960 (from BACH, 1988; with permission).

warming trend was uninterrupted in the Southern Hemisphere.

8. RELATIVE CONTRIBUTIONS TO PRESENT AND FUTURE GLOBAL WARMING

Of the realized global warming (Fig. 6) about 61% is attributed to CO_2, 24% to CH_4, 4% to N_2O and 11% to CFC's (BOLIN *et al.*, 1987). BACH (1988) has calculated the historic and future contribution of trace gases to global warming till 2030, without or with UN programme for CFC control (Montreal protocol). The Montreal Protocol of 1986 aims at 31% reduction of atmospheric concentrations of CFC's 11 + 12 and 23% of total CFC's by 2030.

The potential impact of CFC's alone is shown in Fig. 7 and the total potential warming with high, medium and low scenarios for CO_2, CH_4 and N_2O is shown in Fig. 8. In each of these scenarios CO_2 alone is responsible for half of the warming and the needed 'efforts' to reduce emissions other than CFC's are distributed about proportionally over all greenhouse gases. This shows a sort of ideal, but not very realistic case, as there are many legally permitted exemptions, particularly for developing countries.

9. IMPACT OF TEMPERATURE RISE ON TERRESTRIAL BIOMES

According to ecologists, a further global temperature

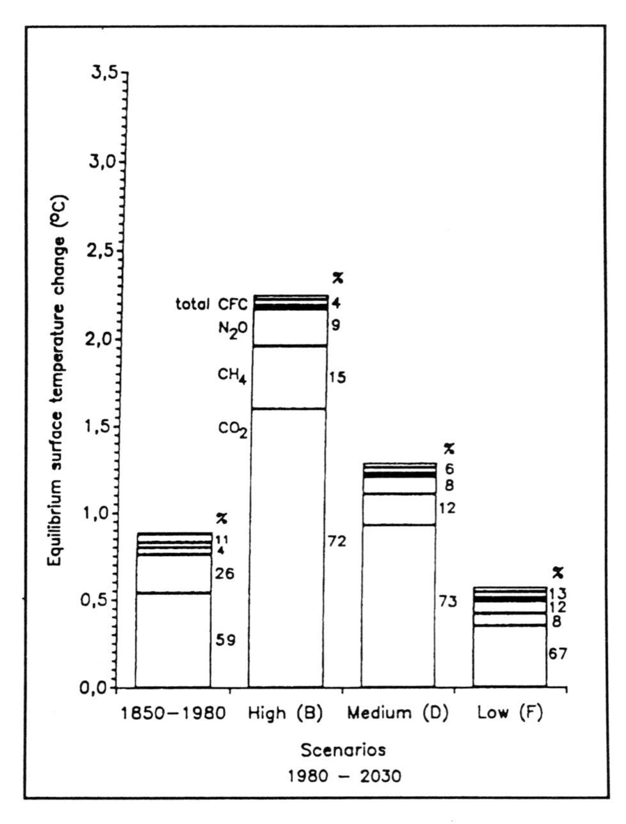

Fig. 8. Historic and future contribution of trace gases to global warming for three scenarios: high, medium and low; with Montreal CFC control (from BACH, 1988; with permission).

rise of about 1.2 K from now till 2100 seems to be the very maximum that the world could tolerate - and even that is not desirable as it could cause major regional problems. This can best be illustrated in Fig. 9, (adapted after BOLIN *et al.*, 1979) which shows the north to south transsect of biomes in eastern Europe at about the middle of last century before the major landuse changes had taken place. A global warming of 2 K would conform with about 1.5 K in the south and about 4 K in the north. This would bring a steppe climate into the southern forests and a taiga climate into the tundra and make the permafrost disappear.

The soils formed under entirely different climate regimes, when exposed to warmer conditions, will perform more rapid metabolic transformations. When more inundated, the wetter lands will become a sink for nitrate and sulfate and produce more CH_4. But lands that get warmer and dryer tend to mobilize nitrate and sulphate, leading to acidification, eutrophication and eventually salinification and to increased oxidation of organic matter in the soil. Sequestered carbon then becomes a source of CO_2. Thus in both cases, wetter or dryer, there will be positive feedback to the greenhouse problem.

10. PREVENTIVE STRATEGIES TO REDUCE GREENHOUSE WARMING

During the 1980s global energy production and use contributed 57%, CFC's 17%, agricultural practices 14%, other landuse changes 9% and other industrial activities 3% to global warming (USEPA, 1989). Regardless of the scenario chosen for the future

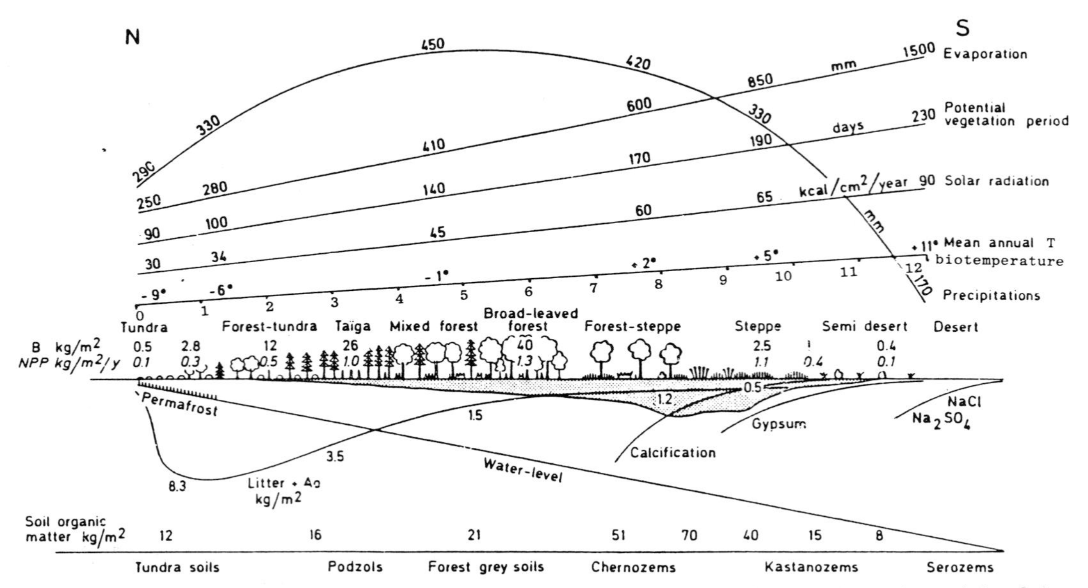

Fig. 9. North-to-south cross-section of biomes and soils through eastern Europe from tundra to desert (after Schennikov, in BOLIN *et al.*, 1979; with adaptation of annual mean bio-temperatures, *i.e.* no-growth season being excluded).

TABLE 6

Reductions of trace gas emissions required to limits ΔTs (the global mean transient surface temperature change in 2100 relative to 1860) to about 1 or 2 K for a climate sensitivity Te of respectively 1.5 or 4.5 K (after BACH, 1988).

Gas	ΔTs(K) in 2100* for Te=1.5K		ΔTs(K) in 2100* for Te=4.5K	Emission rate ca. 1980	Reduction required by 2100 to	Required annual reduction starting from 1990
CO_2	0.50	-	1.18	16-20 Gt**	6-9 Mt	1.1-1.4 Gt***
CH_4	0.18	-	0.44	135-395 Mt	37-224 Mt	1.6-2.0 Mt
N_2O	0.09	-	0.19	16-28 Mt	4-9 Mt	200-284 kt
CFC-11	0.04	-	0,08	330 kt	63 kt	5 kt
CFC-12	0.08	-	0.15	440 kt	89 kt	6.4 kt
TOTAL	0.90	-	2.05			

* ΔTs values calculated for the Efficiency Scenario for Energy by Lovins et al., 1981 and Low Trace Gas Scenario of Bach, 1988. ** Gt = 10^9 tons; Mt = 10^6 tons; kt = 10^3 tons. *** Slower reduction initially and faster reduction toward 2100 hereby achieving a ΔTs very similar to a ₃.5 % annual reduction rate for CO_2.

economic growth and world population development, the commitment to global warming can only be limited within a range of 1.5 to 2 K in case of extremely rigorous reductions of emissions (Fig. 10).

Assuming a climate sensitivity of 1.5 K (low) to 4.5 K (high) for a doubling of atmospheric CO_2 or equivalent other trace gases one can calculate how far emissions have to be reduced annually to keep within about 2 K warming by 2100 (see Table 6).

So far it does not seem likely that the reductions proposed by Bach will be effectuated. More likely will be that CFC's get phased-out slowly and no significant reductions in emissions of CO_2, CH_4 and N_2O will occur. Global mean temperature will then, by the middle of next century, be 4 K higher than in 1860 (from 2.5 K in the tropics to 7 K at the poles). This corresponds with an increase of about 0.3 K per decade.

There is even a chance that emissions will accelerate in view of justified claims in developing countries for rapid economic growth and adoption of wasteful western lifestyles (spreading of the Cola and Hamburger culture). Cheap coal and synfuels will be widely available, solar and nuclear energy remain expensive and/or unaccepted, and the efficiency improvements in heating and cooling and in industrial production remain slow because of reluctance against the discarding of obsolete but widely accepted technologies. These factors would jointly increase world emissions by 60 (2050) to 70 (2100) percent over the next century (USEPA, 1989).

Several strategies have been developed to decrease the commitment to global warming, based on the relative effect of individual measures (USEPA, 1989). Efficiency gains in heating, cooling, transportation and industrial production are among the top priorities, together with CFC phaseout. Commercialized biomass, rapid reforestation and control of emissions from landfills and agriculture also rank high. Solar energy has more promise than nuclear, which is facing growing opposition. Curious enough the promotion of natural gas will not make a great impact, if the use of coal and synfuel is not drastically curbed. Among the additional measures coal phaseout and a taxation on carbon rank high. Very advanced innovations in cars, power plants and buildings beyond technical improvements that are already feasible now do not seem much rewarding. Prolonged reforestation is necessary, coupled with increased use of biomass.

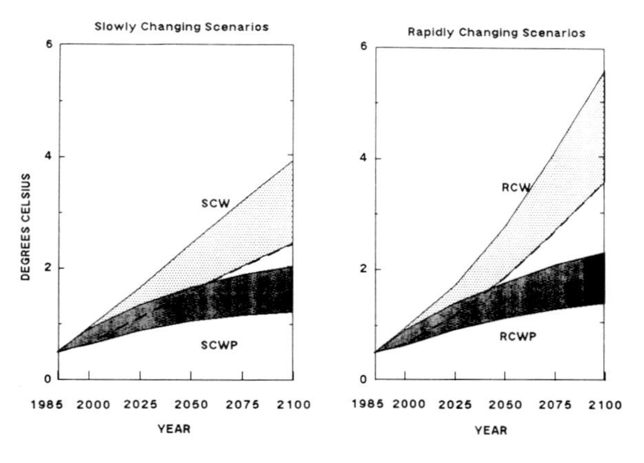

Fig. 10. Realized warming till 2100 at no response and at stabilizing policy scenarios. (A) Slow economic growth and rapid population growth. (B) Rapid economic growth and slow population growth (in °C with 2-4 degrees climate sensitivity in the models; shaded areas represent the range based on an equilibrium climate sensitivity to doubling; from USEPA, 1989).

In a world faced with cash scarcities, investments should first go to measures yielding the most rapid results and that are relatively cheap, such as CFC phaseout, reforestation and insulation that improves the efficiency of heating and cooling of buildings. These pay-off immediately, contrary to investments in nuclear and in high-tech solar energy (not to consider the acceptability). The collective potential of the measures to stabilize and additionally decrease the worldwide threat of global warming is, according to this USEPA study, 63% in 2050 and 87% in 2100. That still means a slight global warming! But ecologically that is more acceptable while less disruption to the major ecosystems (forests, tundras, grasslands, wetlands), water balance and crop production.

11. ADAPTATION TO INEVITABLE SEA LEVEL RISE (THE NETHERLANDS CASE)

The sea level rise discussion is still confuse (OERLEMANS, 1987). If there is no sudden break-off of major parts of the West Antarctic ice shelves (which then would drift away and melt in warmer parts of the ocean) it seems unlikely that the global sea level rise would be more than one meter; between 40 and 80

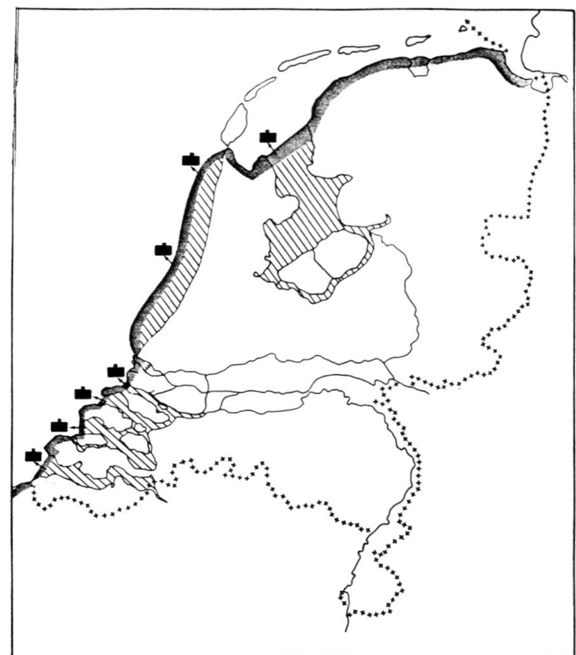

Fig. 11. Strategic options for the Netherlands at an extreme relative sea level rise of 5 m. (left) Option of a strategic retreat in which only the segment with major metropolitan area defended. (right) Option of maximal defence but with the Wadden Islands given up. (From DE RONDE, 1987; with permission).

cm seems more plausible. Yet that means a doubling to quadrupling of the rates measured over the past half century.

Locally land subsidence (*e.g.* due to oil and gas mining and groudwater extraction) or rising (due to geostatic rebound processes) may influence the relative rate of sea level rise and also other local and regional features (*e.g.* surges in estuaries) may modify the global trend (HEKSTRA, 1986, 1988).

Along most of the southern North Sea coastline from Esbjerg to Dunkerk major floodings have occurred throughout history. Rijkswaterstaat has made flooding risk assessments for the three segments of the Netherlands' coast at different sea level scenarios, including an extreme one of 5 m relative sea level rise (*i.e.* including regional and local effects). That extreme situation would require either a strategic retreat or a new dike conception (Fig. 11a, b), but both options are yet to be regarded as unrealistic.

Another, more realistic risk would be more upwelling of brackish or saline ground water, that would much affect agriculture and drinking water production. Lake and canal levels have to be put up artificially to counteract salt intrusion.

12. ENDANGERED COASTS OF THE WORLD

European areas below the 5-m contour line, that is the area that is likely to be affected by 1 to 1.5 m sea level rise, either by flooding or by saltwater intrusion and drinking water problems, are shown in Fig. 12. The sites at the Atlantic coasts have in common wide V-shaped estuaries with broad inland coastal plains, salt marshes, wide tidal flats and deep tidal gullies. At the southern Baltic Sea coasts tidal differences are much less important and estuaries are insignificant; some rivers even form deltas.

This is even more so in the Mediterranean, where tidal differences hardly exist and all rivers form deltas. Protruding deltas and incised estuaries and lagoonal areas are very unevenly scattered over the world (see fig. 2 of WIND (1987). What they have in common is a delicate balance of sedimentation and erosion, which can easily become upset by changes in sea level and river runoff.

13. ASSESSING COASTAL VULNERABILITY AND COSTS FOR THE SOCIETY

Coastal countries get increasingly aware of their vulnerability to climatic change and sea level rise. If they undertake nothing they might well become faced with substantial losses over the next four to five decades. The loss of harbours, coastal cities,industries and fertile farmland may amount to several hundred percents of present costs of coastal protection. This is the scenario 'no limitation - no adapta-

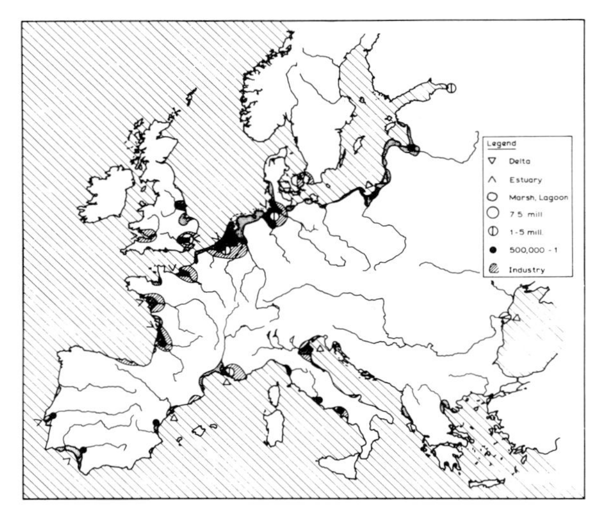

Fig. 12. European areas vulnerable to an extreme sea level rise of 5 m (from JELGERSMA, 1986; with permission).

tion' in Fig. 13 (after VELLINGA, 1987). An extremely concerned country may already now start with maximal adaptation to sea level rise and also maximal limitation of its contribution to the causes of climatic change. They will start to make investments already

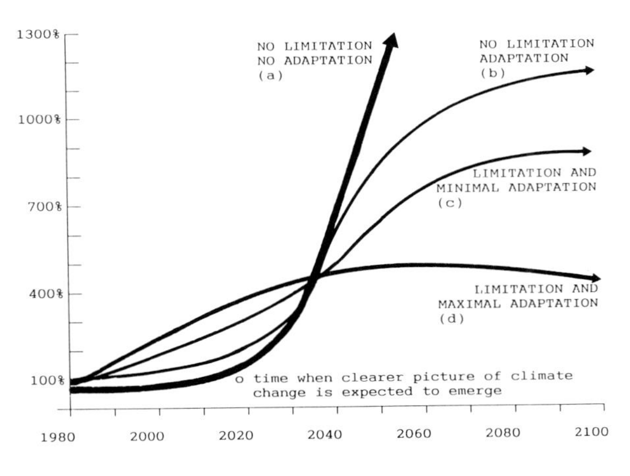

Fig. 13. Cost issues for management regarding a rising sea level. The graphs are only meant as an indication; not as real estimates of costs, as for each coastal country the situation is specific. The very approximate costs are shown here relative to current investments made over the last few decades (= 100% for each specific region). The 4 scenarios are: (a) do nothing, neither limiting emissions nor making adaptations to rising sea level; within 4-5 decades major losses, worth a multitude of recently made expenses. (b) no limitation of emissions, but some adaptations for defence. (c) limitation of emissions of greenhouse gases and minimal adaptation after disasters. (d) maximal efforts to reduce emissions and strenuous efforts to adapt vulnerable coastal regions.

now and have to continue to do so for may be more than a century, but in the long run would be better off than coastal countries that do less. Their particular problem is to persuade other countries also to reduce emissions that affect the global climate.

The graphs are only meant as an indication; not as real estimates of costs, as for each coastal country the situation is specific. The very approximate costs are shown here relative to current investments made over the last few decades (=100% for each specific region). The 4 scenarios are: (a) do nothing, neither limiting emissions nor making adaptations to rising sea level; within 4 - 5 decades major losses, worth a multitude of recently made expenses. (b) no limitation of emissions, but some adaptions for defense. (c) limitation of emissions of greenhouse gases and minimal adaptation after disasters. (d) maximal efforts to reduce emissions and strenuous efforts to adapt vulnerable coastal regions.

14. INITIATIVES IN THE NETHERLANDS

The Netherlands have over the last 35 years invested about 16 billion guilders (8×10^9 US$) in coastal defense works and would have to spend the same or less over the period up to 2100 to cope with a sea level rise of one meter. It seems as if the country is in a comparatively advanced position relative to other vulnerable countries. To find an outlet for its present surplus engineering capacity the country is getting more and more involved at foreign coasts. In cooperation with UNEP the Delft Hydraulics Laboratory has developed a computerized management framework for assisting developing countries in cost-benefit analyses and policy measures regarding the Impact of Sealevel rise On Society (ISOS project).

In the Netherlands itself much effort is now being put in policies to reduce emissions. This has, however, much less standing tradition than coastal engineering and waterworks. Besides, it requires much more international cooperation and control on a truly global scale and on a sound scientific basis, for which the political will has to be built up.

That is why the Netherlands' government has organized and hosted conferences and workshops like the European Workshop on Interrelated Bioclimatic and Landuse Changes held at Noordwijkerhout (Oct. 1987) and the International Workshop on the Impact of Climatic Change on Marine Environments, held at Texel (Nov. 1988). Three more scientific conferences of this kind followed:

- Aug. 1989 at Wageningen: World conference on the impact of soil and landuse changes on greenhouse gases and global climate.
- Dec. 1989 at Lunteren: European workshop on landscape ecological impact of climatic changes.

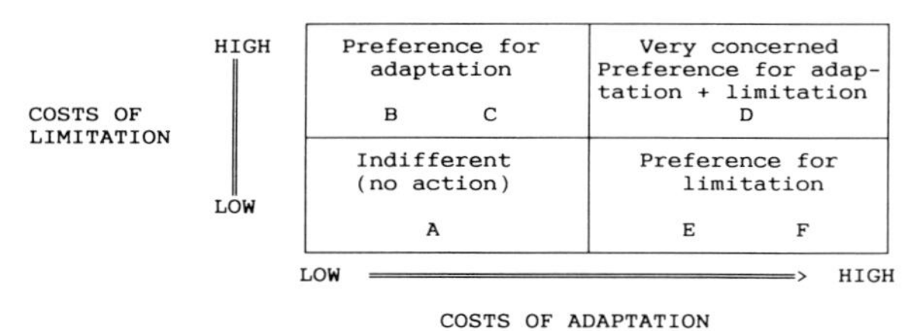

Fig. 14. National response options to climatic change. (After a working paper at an expert meeting on policy responses at Bellagio (Nov. 1987; JÄGER, 1988).

- April 1990 at Wageningen: International Workshop on the impact of climatic change on plant productivity and food production.

Furthermore the Netherlands government has invited other governments to a high-level meeting of ministers of the environment (in Nov. 1989 in The Hague) to discuss further political action with regard to the global atmosphere and climatic change.

15. DIFFERENTIAL NATIONAL RESPONSES TO CLIMATIC CHANGE

Country A (Fig. 14) in the humid tropics with no coastline and using very little fossil fuels will be less affected than temperate countries and care less about internationally proposed policies.

Country B in the humid tropics with no coastline is industrializing rapidly. Being a creator but not a major receiver of the problem it may prefer to adapt to some climatic inconveniences.

Country C is a rich industrial northern country with an Arctic coastline, rich in minerals and short of good agricultural land. It may prefer adaptation as it believes that it can better exploit the ice-free coastline and that the added revenues will more than compensate for costs of adaptation, but which is not proven.

Country D is a rich, highly industrialized and urbananized temperate country with important low-lying coastal areas. It will be very concerned to do limitation as well as adaptation.

Country E is a poor, weakly industrialized semi-arid coastal developing country that cannot afford the costs of adaptation measures and in international fora will stress limitation strategies.

Country F is a poor, non-industrial coastal state with harsh cold climate, living from mining and agriculture. It perceives to benefit from a milder climate for agriculture and longer open ports and will formally opt for limitation as it has little costs to bear for it.

Many more and intermediate options could be envisaged. In fact more intangible motivations such as national prestige and international competition will also influence attitudes and options.

16. THE DECISION-MAKERS PERSPECTIVE

Decision-makers may have thoughts in the back of their minds such as mentioned in one of the Bellagio working papers:

- what is the latest date I can leave a decision without seriously foreclosing any of my options for mangement in the future?
- do I have a good case which is credible enough to convince the general public and/or senior colleagues and that will silence the anti-lobby groups, so that I can gain legitimacy to devote funding? And when do I obtain such a strong case?
- what are the interactions of sea level rise and climatic change with other problems in my jurisdiction, making the issue more manageable or more complex?
- what are the cost of doing minimal management versus doing nothing and when would these cost incur?
- will negative consequences of doing minimal or nothing fall within my period of office?
- am I being made to look 'backward' by more advanced actions in other countries, and if so, can such be easily discredited?
- how long can I afford to do nothing without causing a serious public outcry or what events or observations might trigger public demand to act?
- might my political opposition make a big issue of the problem and thus endanger my position?
- can I play the card of insufficient international agreement?

17. RATIONAL RISK MANAGEMENT

The latter question of course seemingly appears to be crucial. Although most scientist agree on a sea level rise between half to one metre over the next century, they do not exclude that in case of increased snow fall, in particular over Antarctica, the sea level might not rise at all, or even fall. Besides, how reliable judgement can be given about changes in storm surges?

Thus the policy dilemmas seemingly look awkward indeed, and it is quite understandable that decision-makers try to postpone action. But the questions mostly are wrongly posed, as what matters is not the 'likely' events or trend but the thinkable risk of sudden extreme events, which are calculable. It is really important not to get trapped in the usual national and international swings and flaws of day to day policies, but to prepare for a solid scientific core of risk assessment and applied systems analysis by internationally credited institutes.

The International Institute on Applied Systems Analysis (IIASA) at Laxenburg, Austria, has started in 1985 a broad project on Sustainable Development of the Biosphere, involving climatic change, impact of acid precipitation, water and land management and options of demographic, agricultural and industrial development. IIASA and related international institutes could jointly coordinate the best professional judgement and modelling of climatic change and regional impacts. IIASA has succesfully organized policy workshops that involve senior advisers to governments and environmental, technical and business specialist from different parts of Europe, and could facilitate similar workshops elsewhere (*e.g.* Zambesi Basin). IIASA has now developed an action plan for the development of a strategy for reducing greenhouse gas emissions (IIASA, 1989). Parts of it will be included in the action plan of the Intergovernmental Panel on Climate Change (IPCC) and discussed at the Ministers Conference at Noordwijk, The Netherlands, in November 1989.

The workshop at Texel on the Impact of Climatic Change on Marine and Coastal Ecosystems (Nov. 1988) brought together scientist with special knowledge of the West European Atlantic coasts. Their recommendations for further cooperation regarding policy oriented research for decisionmakers were well at time. They have been submitted for implementation to the appropriate governments and the European Community.

The Intergovernmental Panel on Climatic Change, set up by WMO and UNEP, may well develop porposals for a United Nations Authority with far reaching mandate to coordinate control measures. A Fund for climatic change could be established on the basis of levies on CO_2 emissions and other greenhouse gases. An International Climate Convention, to be established in 1990, should pave the way. The Netherlands have already committed themselves to the aims of the Convention and paymant to the Climate Fund, when established.

18. REFERENCES

BACH, W., 1988. The endangered climate. Report no. 15. In: F. KRAUSE & W. BACH. Energy and climate change: what can Western Europe do? Report for the Netherlands' Ministry of Housing, Physical Planning and the Environment and the European Community. (Arbeitsunterlage 11/45 der Enquete Kommission Vorsorge zum Schutz der Erdatmosphäre Deutsche Bundestag).

——, 1989. An effective greenhouse gas reduction strategy for the protection of global climate. Paper for the IEA/OECD expert seminar on energy technologies for reducing emissions of greenhouse gases. Paris, 12 - 14 April 1989.

BOLIN, B., 1987. The role of carbon dioxide and of other greenhouse gases for climatic variations and associated impacts. Paper at the European Workshop on Interrelated Bioclimatic and Landuse Changes, Noordwijkerhout, Netherlands, 16 - 21 Oct. 1987. (see also KWADIJK, J. & H. DE BOOIS, 1989).

BOLIN, B., E.T. DEGENS, S. KEMPE & P. KETNER, 1979. The global carbon cycle. SCOPE 13; Wiley & Sons, Chichester.

BOLIN, B., B. R. DÖÖS, J. JÄGER & R. A. WARRICK, 1987. The greenhouse effect, climatic change and ecosystems. SCOPE Vol. 29. Wiley & Sons, New York.

BOLLE, H.J., W. SEILER, & B. BOLIN, 1987. Other greenhouse gases and aerosols. In: B. BOLIN, B.R. DÖÖS, J. JÄGER & R.A. WARRICK. The greenhouse effect, climatic change and ecosystems. SCOPE Vol. 29. Wiley & Sons, New York.

BOUWMAN, A.F., 1988. (Draft) background paper for the International Conference on Soils and the Greenhouse Effect. ISRIC report, Feb. 1988.

BROUNS, J.J.W.M., 1988. The impact of elevated carbon dioxide levels on marine and coastal ecosystems.— NIOZ report 1988-7 and RIN report 88-58.

CLARK, W.C. & R.E. MUNN, 1986. Sustainable development of the biosphere. Cambridge Univ. Press and IIASA, Laxenburg.

CRUTZEN, P.J. & T.E. GRAEDEL, 1986. The role of atmospheric chemistry in environment - development interactions. In: W.C. CLARK & R.E. MUNN, 1986. Sustainable development of the biosphere. Cambridge Univ. Press and IIASA, Laxenburg.

DICKINSON, R.E., 1986. Impact of human activities on climate - a framework. In: W.C. CLARK & R.E. MUNN, 1986. Sustainable development of the biosphere. Cambridge Univ. Press and IIASA, Laxenburg.

HEKSTRA, G.P., 1986. Will climatic change flood the Netherlands? Effects on agriculture, land use and well-being. Ambio **15** (6): 316-326.

——, 1988. Prospects of sea level rise and its policy consequences. Discussion paper for the workshop on 'Controlling and adapting to greenhouse warming', Resources for the Future, Washington, D.C., 14-15 June 1988.

IIASA, 1989. IIASA Climatic change study: development of a strategy for reducing greenhouse gas emissions and thereby delaying a climatic change. International Institute for Applied Systems Analysis undated working paper with plan of action, Spring 1989.

JÄGER, J., 1988. Developing policies for responding to climatic change. The Beijer Institute, Stockholm.

JELGERSMA, S., 1986. Coastal lowlands working group: starting note. Paper for the preparation of the European Workshop at Noordwijkerhout, Netherlands, 16 - 21 Oct. 1987. To be published in J. KWADIJK & H. DE BOOIS, 1989.

JONES, P.D., T.M.L. WIGLEY & P.B. WRIGHT, 1986. Global temperature variations between 1861 and 1984. Nature **322:** 430-434.

KRAMER, C.J.M., 1987. Effects of increased solar UV-B radiation on coastal and marine ecosystems: a literature survey. MT-TNO report R87/223.

KWADIJK, J. & H. DE BOOIS, 1989. Selected papers, conclusions and findings of the European Workshop on Interrelated Bioclimatic and Landuse Changes. RIVM, Bilthoven.

LEVINE, D.G., 1989. The potentially enhanced greenhouse effects: status, projections, concerns and need for constructive approaches. Paper for the IEA/OECD expert seminar on energy technologies for reducing emissions of greenhouse gases. Paris, 12 - 14 April 1989.

LOVINS, A.B., L.H. LOVINS, F. KRAUSE & W. BACH, 1981. Least-cost energy: solving the CO_2-problem. Andover.

OERLEMANS, J., 1987. Possible changes in the mass balance of Greenland and Antarctica ice sheets and their effects on sea level. IN: T.M.L. WIGLEY & R.A. WARRICK. Norwich Workshop, 1-4 Sept. 1987 (to be published).

RAMANATHAN, V.R., J. CICERONE, H.B. SINGH & J.T. KUHL, 1985. Trace gas trends and their potential in climate change.—J. Geophys. Res. **90:** 5547-5566.

RASMUSSEN, R.A. & M.A.K. KHALIL, 1984. Atmospheric methane in the recent and ancient atmospheres: Concentrations, trends and interhemispheric gradients.—J. Geophys. Res. **89** (D7): 11599-11605.

RONDE, J.G. DE, 1987. What will happen to the Netherlands if sea level accelerates? In: T.M.L. WIGLEY & R.A. WARRICK. Norwich Workshop, 1-4 Sept. 1987 (to be published).

SIEGENTHALER, U. & H. OESCHGER, 1987. Biospheric CO_2 emissions during the past 200 years reconstructed by deconvolution of ice core data.—Tellus **39B:** 140-154.

STIGLIANI, W.M., 1988. Changes in valued 'capacities' of soils and sediments as indicators of nonlinear and time-delayed environmental effects. IIASA WP-88-38, Laxenburg.

TERAMURA, A.H., 1986. Overview of our current state of knowledge of UV effects on plants. In: J.G. TITUS. Effects in changes of stratospheric ozone and global climate. Vol. 1: Overview. Proceedings of the International Conference on Health and Environmental Effects of Ozone Modification and Climate Change. UNEP and EPA (Washington D.C.) 1986.

TORRENS, I.M., 1989. Global greenhouse warming: role of the power generation sector and mitigation strategies. Paper for the IEA/OECD expert seminar on energy technologies for reducing emissions of greenhouse gases. Paris, 12 - 14 April 1989.

USEPA, 1989. Policy options for stabilizing global climate. Draft report to Congress. Executive Summary. United States Environmental Agency, Office of Planning and Evaluation, Feb.1989.

VELLINGA, P., 1987. Sea level rise, consequences and policies. Paper for Villach Workshop, 28 Sept - 2 Oct. 1987. Beyer Inst. Stockholm.

WIND, H.G., 1987. Impact of sea level rise on society. Balkema, Rotterdam: 191 pp.

WORREST, R.C. 1986. The effects of solar UV-B radiation on aquatic systems: an overview. In: J.G. TITUS. Effects in changes of stratospheric ozone and global climate. Vol. 1: Overview. Proceedings of the International Conference on Health and Environmental Effects of Ozone Modification and Climate Change. UNEP and EPA (Washington D.C.) 1986.

ECOLOGICAL INTERPRETATION OF CLIMATE PROJECTIONS

HANS DE BOOIS

National Institute for Public Health and Environmental Protection RIVM, P.O. Box 1, 3720 BA Bilthoven, The Netherlands

ABSTRACT

General Circulation Models offer indications of possible climate change. European-scale climate-change scenarios can be formulated. For assessment of ecological impacts two approaches are considered: analogue and analytical. The analogue assessment is limited by many biogeographical differences between regions, other than climate itself. For the analytical assessment many details are needed with regard to extremes of climate which operate the ecological selection. From available climate data many relevant details can be derived, which can be related to projected changes of mean values of climate. Attention should be paid to the time scales of climatic events and climate change in relation to time scales of ecological processes.

1. INTRODUCTION

Climate change due to the greenhouse effect is generally referred to as the increase of the global mean equilibrium temperature by doubled concentration of CO_2. It is obvious, that change of mean temperatures by several degrees centrigrades will have large impacts on ecosystems throughout the world. The spatial resolution and the specification of climate changes in projections at a global scale do not allow for ecological interpretation of the forthcoming climate events. Assessment of the ecological effects requires more specific information, both on climate changes and on ecology. In this paper options for specification of links between climate change and ecology are surveyed.

2. GLOBAL TEMPERATURE CHANGE

Increasing concentrations of greenhouse gases (GHGs) in the atmosphere can be translated in changes in the earth's radiation budget. The atmospheric blancket around the earth becomes more effective. General Circulation Models (GCMs) demonstrate that doubling the concentration of CO_2 (from 280 to 560 ppmv) may lead to 1.5–4.5°C increase of the global mean equilibrium temperature (WASHINGTON & MEEHL, 1984; BOLIN *et al.*, 1986; HANSEN *et al.*, 1988). Model projections starting from the situation in 1900 suggest that the total increase from all GHGs together is already half of this range today (Fig. 1). About 60% of this increase is caused

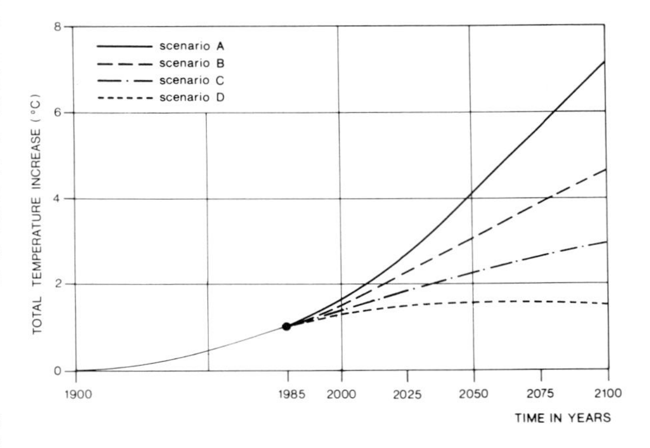

Fig. 1. Global mean equilibrium temperature increase according to ROTMANS *et al.*, 1988. Real temperature increase will be delayed by several decades. The scenarios apply to trends in emissions of greenhouse gases, varying from unrestricted increase (A) to forced reductions (D) at a global scale.

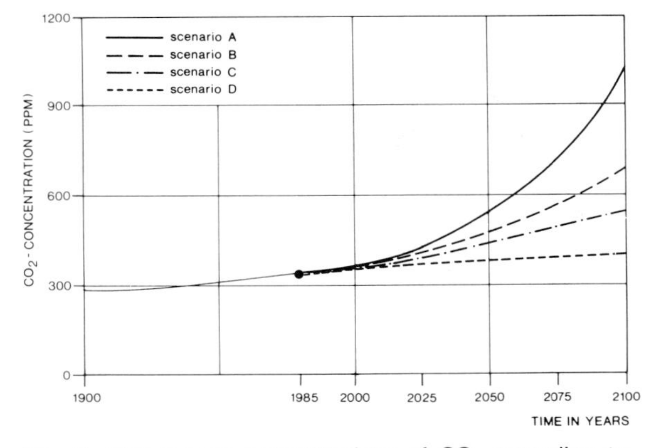

Fig. 2. Atmospheric concentrations of CO_2 according to ROTMANS *et al.*, 1988.

J.J. Beukema et al. (eds.), Expected Effects of Climatic Change on Marine Coastal Ecosystems, 17–22.

by increased concentrations of CO_2, and the remaining 40% by increased concentrations of CH_4, N_2O, CFCs and ozone. A change of the IR-radiation absorption capacity, equivalent to the effect of doubled concentration of CO_2, may be expected to occur around 2010, if emissions grow according to 'business as usual' (scenario A). The concentration of CO_2 itself may be doubled about 2060 (ROTMANS *et al.*, 1990) (Fig. 2).

There is a difference between this theoretical change in the climate forcing and the actual effects. In reality changes are delayed by thermal inertia of the oceans by several decades, depending from the scenario. Or, from a different point of view, the transient temperature rise can be indicated as about 70% of the equilibrium temperature rise (WIGLEY *et al.*, 1986).

3. REGIONAL TEMPERATURE CHANGE

It is quite important, that GCMs also indicate that global mean temperature change will be unevenly distributed over the globe and over the year. At higher latitudes and in wintertime the largest increases in temperature are expected, indicating high latitude averages of 2.4 times the global mean and regional effects being even larger.

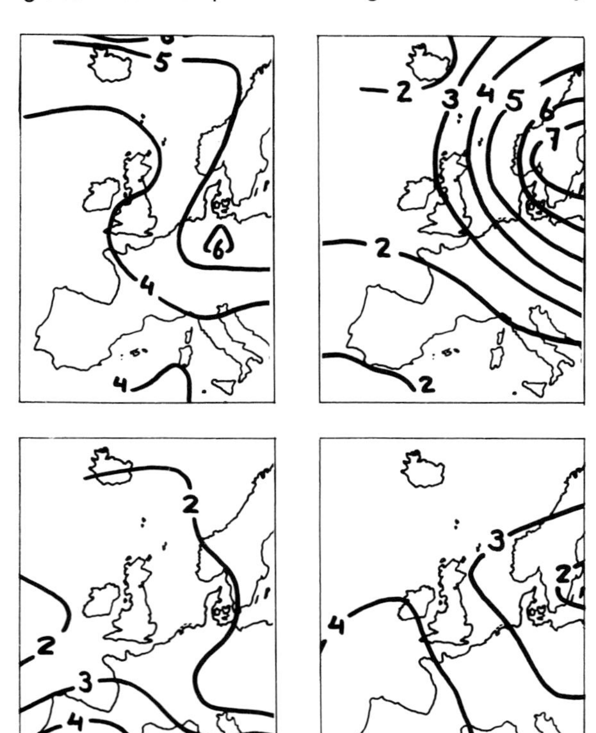

Fig. 3. Examples of GCM results for temperature change in Europe (2x CO_2 -1x CO_2). Left: BMO model; right: GISS model (1981). The upper figures apply to the winter temperature change, the lower figures to the summer temperature change.

Table 1 shows that the results of different GCMs vary in details, but agree on major points. Still one should keep in mind what the British climatologist T.M.L. Wigley once said about GCMs: 'They are all different and they are all wrong' (WIGLEY, 1987). The main argument for this statement is that the ability of GCMs to simulate the features that dominate today's climate is still unsatisfactory. For that reason Wigley prefers the use of instrumental data based scenarios for specifying changes at a regional scale: analogues of the climatological conditions in relatively warm historical years offer a more detailed image of possible changes of climates at a European scale. Also this approach has limited value however: the relations are statistical but are not based on causal relationships with changes in *e.g.* ocean currents or the location of areas of high and low barometric pressure. Nevertheless, the general pattern of an analogue climate change in Eurcpe (PALUTIKOF *et al.*, 1984) agrees with the general picture of CGMs for 2x CO_2 (Fig. 3), but the analogue scenario includes also a possible decline of temperatures between the latidutes 50°–60°N.

4. SCENARIOS FOR EUROPEAN COASTAL AREAS

To study the impact of climate change we must accept the results as produced by the best available methods, whether they have large uncertainties or not. Statements about the possible consequences global change may have for climates in coastal areas are needed to make progress in such studies. For that reason we present in Table 2 two scenarios for the equilibrium 2x CO_2-climate change in three types of Atlantic coastal areas in Europe, indicated as +/+ and +/–. In both scenarios the summer temperature increases. In scenario +/+ the winter temperature also increases, but in scenario +/– the winter temperature remains the same or decreases. Because of the uncertainties we should consider both scenarios equally plausible.

The figures presented in Table 2 for scenario +/+ fit within the ranges of the figures in Table 1. The alternative lower figures for the change of winter temperature reflect an extreme interpretation of the instrumental data based scenario, implying increase of the continentality of climate. The figures are meant as examples of possible changes which we can use for the assessment of the impact of climate changes. The time by when such changes may occur, varies

TABLE 1

Regional scenarios for climate change (source: Jaeger, 1988). Temperature changes as multiples of the global average.

Region	*Temperature change*		*Precipitation change*
	summer	*winter*	
High latitudes (60-90 deg)	0.5 / 0.7x	2.0 / 2.4x	enhanced in winter
Mid latitudes (30-60 deg)	0.8 / 1.0x	1.2 / 1.4x	possibly reduced in s.
Low latitudes (0-30 deg)	0.9 / 0.7x	0.9 / 0.7x	enhanced in places with heavy rainfall today

TABLE 2

Scenarios for possible climate change in European atlantic coastal areas for the equivalent of 2 x CO_2 (for 3°C global mean equilibrium temperature rise).

Area	*T summer warmer*	*T winter warmer*	*T winter colder*	*Precipitation P summer*	*P winter*
N: boreal	+ 2°C	+ 6°C	0°C	0%	+ 10%
C: central	+ 3°C	+ 4°C	- 2°C	0%	+ 10%
S: mediterranean	+ 4°C	+ 3°C	0°C	0%	0%

from 2030 till beyond 2100, depending on uncertainties about emission scenarios and uncertainties about the rate of delay of the impact of changes in the forcing of climate (ROTMANS *et al.*, 1990) (Fig. 1). Focusing at *e.g.* 2050 the changes may be far greater or smaller than is indicated by these figures. Despite these uncertainties, the figures can be used as a starting point for ecological assessment of the changes.

Two approaches to reason the consequences of climate changes are presented: an analogue and an analytical approach.

5. ANALOGUE ASSESSMENT

The most obvious way for ecological assessment of climate changes is to consider comparable land systems in areas where today's climate corresponds with the present-plus-changes climate in the assessed area. Ecological consequences of the climate change may than be derived by comparison of the two areas.

Simply stated we can look for the impact of scenario +/+ by looking about 500 km south (Fig. 4).

In scenario +/− we meet more problems. In fact this scenario includes a shift towards a more continental climate. It is difficult to find more continental coastal ecosystems that can be compared with the Atlantic coastal ecosystems along the North Sea and the French-Iberian Atlantic coast. Best opportunities for comparison exist in the Baltic by the west-east orientation. The northern parts of the atlantic coasts of Ireland and the U.K. may be compared with more continental coasts in SW-Sweden and S-Norway.

If we use this analogue approach, we should realize, that the ecological comparability of areas is limited, as DE VOOYS (1988) shows in his report while comparing the fauna of the Waddensea with the fauna of the Gironde. Apart from climate also other environmental conditions may be quite different: the physical and chemical characteristics of the environment, including tidal movements, river inflow and water quality. Besides, the biogeographical conditions of areas may restrict or delay realisation of potential ecological changes.

6. ANALYTICAL ASSESSMENT

The school books teach us that 'climate is the average of weather over a period of about 30 years'. However, from an ecological point of view the average values do not matter for selection but rather the variability and the extremes that limit the possiblities for species to occur. The data of Table 2

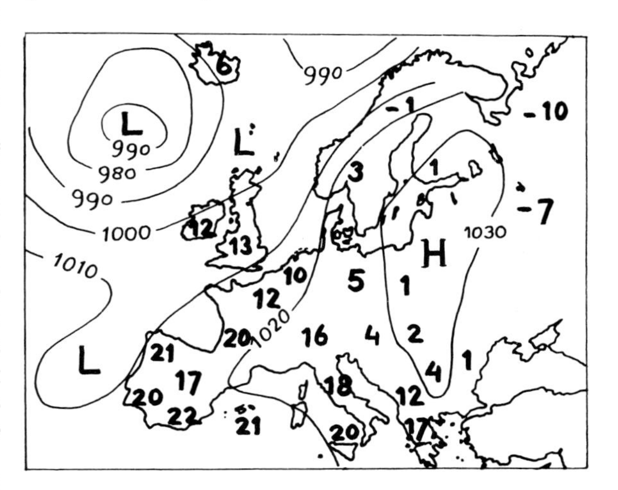

Fig. 4. A typical pattern of the pressure over Europe and the temperature differences (9-11-1988).

TABLE 3

Climate at De Bilt, NL, 1921-1950.

parameter	*month*											
	J	*F*	*M*	*A*	*M*	*J*	*J*	*A*	*S*	*O*	*N*	*D*
av. T ⁰C	1.9	2.3	4.9	8.4	12.4	15.2	17.2	16.7	14.2	9.8	5.5	2.4
av. daily Tmax ⁰C	4.6	5.7	9.8	13.4	18.1	20.9	22.6	22.0	19.4	14.1	8.6	4.9
av. daily Tmin ⁰C	-0.7	-0.6	1.1	4.2	7.5	10.4	12.7	12.4	9.8	6.5	2.9	0.0
highest T ever ⁰C	13	18	22	27	34	37	36	36	34	27	19	14
lowest T ever ⁰C	-25	-22	-12	-6	-4	1	3	4	-1	-8	-14	-21
nr.ice-days (max 0⁰C)	5	3	0	.	.	.	.	.	.	0	4	12
nr.frost days (min 0⁰C)	14	14	12	3	1	.	.	.	0	2	7	14
nr.days Tmax>25⁰C	.	.	.	0	3	5	8	6	2	0	.	.
nr.days Tmax>30⁰C	.	.	.	.	0	1	2	1	0	.	.	.
hours of sunshine	54	74	133	155	211	218	206	192	149	105	53	45
duration of sunshine %	21	27	36	37	43	43	41	42	39	32	20	18
nr.days without sunshine	13	9	5	3	2	1	1	1	2	5	11	14
av.rel.humidity	87	83	77	72	69	68	71	75	79	84	88	89
precipitation mm	62	49	41	50	55	58	74	82	75	74	76	65
hours prec.	62	50	39	38	35	31	34	35	36	47	60	56
nr.days with > 1 mm prec.	13	10	9	10	9	9	11	12	11	12	13	12
nr.days snowing	5	5	4	1	0	.	.	.	.	0	1	4
predominant wind direction	sw	sw	sw	w	no	nw	sw	sw	sw	sw	s	s
av.wind velocity m/sec	6	6	5	5	5	4	4	4	4	5	5	5

TABLE 4

Climatological selection mechanisms (+ = increase; 0 = no change; - = decrease; */ = summer; /* = winter).

climatic factor	*selective factor*	*scen. +/+*	*scen. +/-*
high temperatures	warm water in pools	+/0	+/0
	oxygen deficit in water	+/0	+/0
	(relative) draught	+/0	+/0
low temperatures	frozen tidal flats	0/-	0/+
	ice cover	0/-	0/+
draught	increase of salt concentrations	+/+	+/0
wind	frequency and duration of:		
	low water levels	+/0	+/+
	high water levels	-/?	-/-
	storms	?/?	?/?

must be considered inadequate for ecological assessments. Analysis of changes that affect specific relations between climate and organisms is more complicated and difficult, but perhaps also more reliable then the analogue approach. It enables the use of more scientific knowledge.

In this perspective we are more interested in weather and in particular in the extreme weather conditions rather than in climate.

In this context we will briefly discuss some of the details about climate, and how these may be used for the analysis of ecological impacts. Averages conceal interesting variations at a detailed level. Table 3 gives an impression of a more comprehensive description of climate. More details can already be recognized, which are important from an ecological point of view. The table enables comparison of temperatures: monthly means, daily means, daily maximum and minimum.

More details of other parameters are also available. With regard to figures of precipitation and evapotranspiration the sum of a season is of less interest than the risks of *e.g.* excessive run-off of rivers, the chance of long dry periods and the probability and severity of storms.

7. THE ECOLOGICAL MEANING OF CLIMATE CHANGE

Also the ecological approach should be quite different if we look analytically. We should not look for comparable suitable circumstances in which ecosystems can occur, but on the contrary we might look to changes in the mechanisms that select species (see *e.g.* Table 4). This requires a detailed knowledge of tolerance ranges of species on one hand and detailed projections of selective climatic parameters on the other hand.

In both scenarios we may expect more frequent hot periods. Perhaps also hotter and dry, which is of major importance for the survival of plant and animal species above the water level and in shallow waters: dry soils, increased salt concentrations and water temperatures higher than *e.g.* 30°C are ecologically relevant consequences.

In the +/+ winter season we may expect less extreme low temperatures. This may be seen as a release of environmental constraints for species. Competition and food web relations between populations will be affected by changed survival rates. It is a challenge for ecology to guess what are the possibilities and the consequences of immigration of new species in the present ecosystems.

In the +/– scenario the winters will be more severe, with more frequent and longer periods of ice cover, of low water episodes and of frozen soils. It may be of considerable importance if extreme events were introduced in areas where they do not occur yet. Equally it would be of importance if such events would occur in periods that were not struck before, since they have strong impacts on organisms and other environmental conditions.

8. QUANTITATIVE RELATIONS BETWEEN CLIMATE PARAMETERS

For the translation of the general statements about climate change (see Table 2) to ecologically important specific parameters we can look for quantitative correlations between (the variations in) both parameters in basically comparable climates. A prerequisite for such translations should be, that one should always be able to explain the relation by causality. Some examples can be presented as illustrations.

- One might wonder whether the relation between mean temperatures and extremes varies as mean temperatures change. HANSEN *et al.* (1988) show that the GISS GCM produces for the USA distributions of the deviation of daily mean temperatures from the monthly mean at 2x CO which are similar to those at 1x CO_2. This makes sense since both parameters are evidently related.
- Some more examples of translating mean temperatures into more specific temperatures can be produced by using present-day climate data. Fig. 5 shows tentative relations between monthly mean temperature and several ecologically relevant parameters: number of ice days, number of days with temperatures exceeding a certain level, average daily minimum and daily maximum temperature during a month. These relations could already be derived from Table 3. However, before fully quantifying the scales it should be wise to include data from other stations as well.

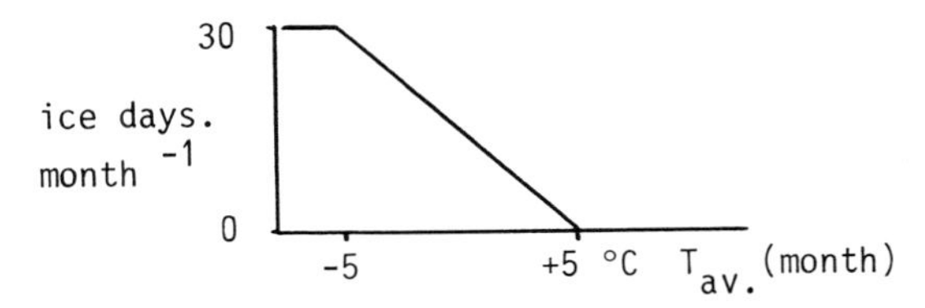

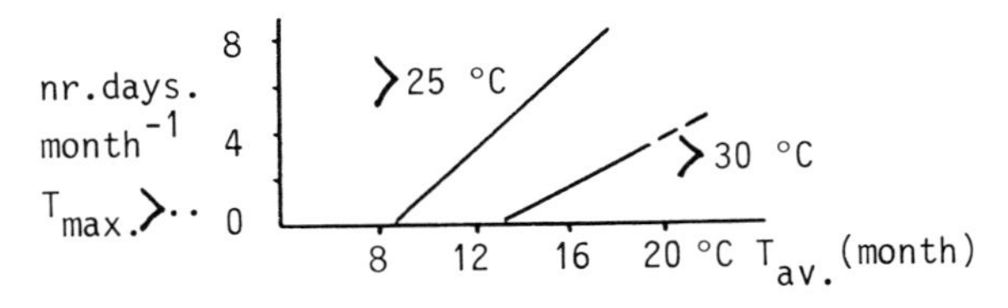

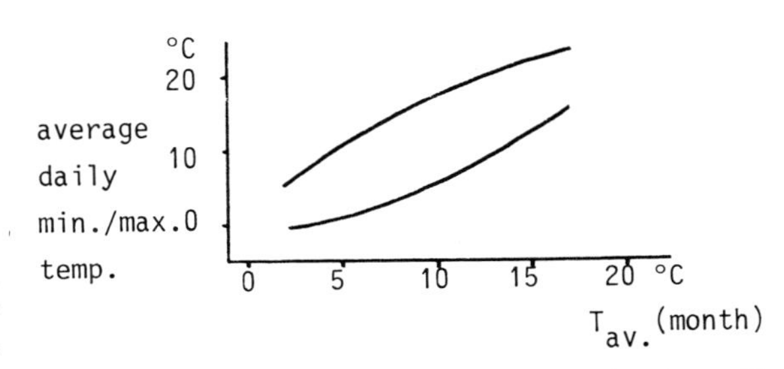

Fig. 5. Examples of relations between general and specific temperature parameters: monthly mean temperature - number of ice days/month; monthly mean temperature - days/month with high temperature; monthly mean temperature - mean daily minimum and maximum.

If causal relations between the parameters are obscure or complex it may be difficult to create a reliable trade-off function:

- Temperature and precipitation in summer are negatively correlated along the atlantic coast between 35°–60°N. This can also be changed into the statement that temperature decrease and precipitation increase are correlated in a positive sense. One may think that the relation exists because of the latitudinal differences. This may seem plausible, but still doesn't explain much. The influence of predominating wind directions may be more essential, as is the influence of surface relief. Therefore, the observed relation between temperature and precipitation cannot be used for indicating the impact of future temperature rise on precipitation.
- The relation between the predominating wind direction and temperature varies from place to place and varies throughout the year. A singular relation, to be

used for assessment of the influence of temperature rise on wind in coastal areas, cannot be derived from observed data on both parameters. The distribution of high and low barometric pressure throughout the continent and the ocean is paramount for the wind direction. Both GCMs and Instrumental data based scenarios provide information on changes in pressure and the resulting wind. From this it should be possible to derive scenarios for the role of wind in future coastal climate.

9. TIME SCALE PROBLEMS

Returning from the details to the long range climate we should be aware of two more problems which are related to the time scale of ecological processes:
- BEUKEMA (this issue) shows in his paper that the recovery of a mollusc population after a severe winter may take several years. It is a relevant question, whether the frequency of the occurrence of extreme events that controls populations in ecosystems keeps pace with the potential for recovery of populations and even entire ecosystems.
- Another question is whether the time scale for ecological adaptation of the present ecosystem and the time scale for development of a new ecosystem keep pace with the rate of climate change. Migration and colonisation of new species often take several decades or even centuries. Also soil development is a slow proces. The projected rate of climate change however, exceeds any precedent known from geological records.

10. FINAL REMARKS

There are optimists, who believe in the stability and self-regulating capacity of the earth's systems. That is not in dispute here. But it should be recalled, that this world houses hot deserts and cold polar regions; that ice-ages have occurred; that tornados, draught and river floods occur frequently, irregularly and unpredictably. We know very little of what causes such events. 18 000 years ago the global average temperature was only about 9°C lower then at present, and ice covered large areas of Europe and North-America. This would be difficult to imagine, if there were no experts, who tell us so on the basis of reliable data. We can not yet imagine what we may expect from +4°C global temperature rise. There are few experts yet, and no reliable data. In spite of great differences in the time scales of past and future climate changes we tend to think of regular changes. But we do so only because we cannot forcast surprises and our imagination is not compatible with modern science.

ACKNOWLEDGEMENTS

The author is indepted to Dr. H.A.M. de Kruijf (RIVM) and Dr. A.P.M. Baede (KNMI) for reviewing this paper.

11. REFERENCES

BOLIN, B., B.R. DÖÖS, J. JÄGER & R.A. WARWICK , 1986. The greenhouse effect, climatic change and ecosystems.—SCOPE 29. John Wiley and Sons, Chichester, UK.

HANSEN, J., I. FUNG, A. LACIS, D. RIND, S. LEBEDEFF, R. RUEDY, G. RUSSELL & P. STONE, 1988. Global climate changes as forecast by Goddard Institute for Space Studies three-dimensional model.—J. geophys. Res. **93** No. D8: 9341-9364.

JÄGER, J. (ed), 1988. Developing policies for responding to climatic change. World Climate Impact Studies WCIP-1, WMO/TP-No.225, WMO, UNEP.

PALUTIKOF, J.P., T.M.L. WIGLEY & J.M. LOUGH, 1984. Seasonal scenarios for Europe and North America in a high-CO_2, warmer world. U.S. Dept. of Energy, Carbon Dioxide Research Division, Technical report TR012.

ROTMANS, J., H. DE BOOIS & R.J. SWART, 1990. An integrated model for the assessment of the greenhouse effect: the Dutch approach.—Climate Change (in press).

VOOYS, C. DE, 1988. Expected biological effects of long-term changes in temperatures on marine ecosystems in coastal waters around the Netherlands.—Neth. Inst. Sea Res., rapport 1988-6.

WASHINGTON, W.M. & G.A. MEEHL, 1984. Seasonal cycle experiment on the climate sensitivity due to a doubling of CO_2 with an atmospheric general circulation model coupled to a simple mixed-layer ocean model.—J. geophys. Res. **89** No. D6: 9475-9503.

WIGLEY, T.M.L., P.D. JONES & P.M. KELLY, 1986. Emperical climate studies. Warm world scenarios and the detection of a Co_2-induced climatic change. In: BOLIN, B., B.R. DÖÖS, J. JÄGER & R.A. WARWICK. The greenhouse effect, climatic change and ecosystems. SCOPE 29. John Wiley and Sons, Chichester, U.K.

WIGLEY, T.M.L., 1987. Climate scenarios. Paper presented at the European Workshop on Interrelated Bioclimatic and Land Use Changes. Noordwijkerhout, The Netherlands. RIVM, Bilthoven, The Netherlands.

WESTERN EUROPEAN REGIONAL CLIMATE SCENARIOS IN A HIGH GREENHOUSE GAS WORLD AND AGRICULTURAL IMPACTS

C.M. GOODESS and J.P. PALUTIKOF

Climatic Research Unit, University of East Anglia, Norwich NR4 7TJ, U.K.

ABSTRACT

Three approaches to the construction of high greenhouse gas world scenarios are described: the General Circulation Model (GCM) approach, the Past Climate approach and the use of Arbitrary Scenarios. Seasonal grid point data for western Europe from CO_2-perturbed and control runs for five GCMs are presented. It is concluded that the regional climate scenarios currently available are not sufficiently reliable at the required spatial resolution for the purposes of coastal ecosystem impact studies. Modelling techniques developed for the study of climate/agricultural links are described. Techniques which are appropriate for the study of climate and coastal ecosystems are identified. The construction of regional scenarios and of a crop-climate model are illustrated by a Netherland's case-study.

1. INTRODUCTION

The investigation of the potential effects of climate change on marine coastal ecosystems requires an interdisciplinary approach. Researchers working in the areas of marine biology and salt-marsh ecology, for example, need from climatologists projections of future changes in temperature, carbon dioxide and UV-B; from oceanographers they want estimates of sea level changes. The spatial resolution of these input parameters for impact studies is of the order of tens of kilometres.

In this paper the ability of scientists to provide such detailed projections of the climate changes likely to be associated with the increasing concentration of greenhouse gases in the atmosphere is assessed. Three approaches can be used in the construction of high greenhouse gas world scenarios: the General Circulation Model approach, the Past Climate approach and the use of Arbitrary Scenarios. In Section 2 each of these approaches is considered in turn, with emphasis on the construction of regional scenarios for north west Europe and for the Netherlands in particular.

A search of the climatological literature has not revealed any existing European studies of climate change linked to coastal ecosystem impacts. There is, however, an extensive literature on climate change and agricultural impacts. It is considered that some of the modelling techniques developed for the study of climate/agriculture links may be appropriate for the study of climate and coastal ecosystems. A hierarchy of climate/agriculture models is outlined in Section 3 and the construction of a statistical crop-climate model for the Netherlands described.

Previous studies of agricultural impacts in north west Europe in a high greenhouse gas world are outlined in Section 4. Consideration is given to the possible use of the Netherlands climate scenario and the crop-climate model in impact assessment.

Interactions between the climate and direct carbon dioxide effects on agricultural crop species are considered in Section 5. Finally, in Section 6, a number of future research directions relevant to the study of the impacts of greenhouse gas-induced climate change on coastal ecosystems are proposed.

2. CONSTRUCTION OF HIGH GREENHOUSE GAS WORLD SCENARIOS

Three approaches to scenario development can be identified (WIGLEY *et al.*, 1986; LAMB, 1987). The simplest is the arbitrary approach where, for example, a temperature increase of 2°C and a precipitation increase of 10% might be assumed. This approach can sometimes be useful in climate sensitivity studies but is not considered in any more detail here. The two commonest approaches to scenario development, the General Circulation Model and Past Climate approaches, are now described.

2.1. THE GENERAL CIRCULATION MODEL APPROACH

General Circulation Models (GCMs) have been developed by climatologists from numerical

J.J. Beukema et al. (eds.), Expected Effects of Climatic Change on Marine Coastal Ecosystems, 23–32.

TABLE 1

Key characteristics of general circulation models.

	UKMO	*GISS*	*NCAR*	*GFDL*	*OSU*
lat x long	5 x 7.5	7.83 x 10	4.5 x 7.5	4.5 x 7.5	4 x 5
vertical layers	11	9	9	9	2
insolation	annual & diurnal	annual & diurnal	annual	annual	annual
ocean model	prescribed heat exchange	prescribed heat exchange varying mixed layer	mixed layer	mixed layer	6-layer ocean GCM
global warming 2x CO_2	5.2°	4.2°	3.5°	4.0°	2.8°
global precipitation change 2x CO_2	+15%	+11%	+7.1%	+8.7%	
for model description see:	Wilson & Mitchell, 1987	Hansen *et al.*, 1984	Washington & Meehl, 1984	Wetherald & Manabe, 1986	Schlesinger *et al.*, 1985

meteorological forecasting models. Since the publication of the earliest studies (MANABE & WETHERALD, 1975, 1980; MANABE & STOUFFER, 1980), they have been widely used to investigate the climatic effects of rising atmospheric carbon dioxide (CO_2) concentrations. The standard approach is to run the models with the present day atmospheric CO_2 concentration (the control run) and then to rerun the model with a doubling or quadrupling of CO_2 (the perturbed run). In both, the models are allowed to reach equilibrium before the results are recorded. Improvements in GCMs have been made over the years and recent reviews conclude that the best estimates of a global temperature rise associated with a doubling of CO_2 lie in the range of 1.5 to 4.5°C (MACCRACKEN & LUTHER, 1985; BOLIN *et al.*, 1986).

Five principal groups have developed GCMs. These are the UK Meteorological Office (UKMO), Goddard Institute of Space Studies (GISS), National Center for Atmospheric Research (NCAR), Geophysical Fluid Dynamical Laboratory (GFDL) and Oregon State University (OSU). There are many different versions of the models developed by each of these groups. However, in this paper we consider only one version of each model. The key characteristics of each model version are outlined in Table 1. All of these models have a realistic land/ocean distribution and topography. All have predicted sea ice and snow. Clouds are calculated in each atmospheric layer in all models. With a doubling of CO_2 the models predict an increase in global mean temperature in the range of 2.8 to 5.2°C and an increase in global precipitation of the order of 7 to 15%.

Despite the improvements that have been implemented over the years, there remain a number of outstanding problems which must be considered when using GCM output. This is particularly the case in impact studies, where individual grid point values may be used to give an estimate of regional climate changes.

Firstly, all models have low horizontal resolution. The grid scale of the 5 models in Table 1 ranges from 4° latitude x 5° longitude (OSU) to 7.83° latitude x 10° longitude (GISS). GCMs, therefore, have a spatial resolution of several hundred kilometres whereas a coastal ecologist, for example, ideally requires information with a resolution of only a few kilometres. Some statistical techniques are available for the estimation of sub-grid scale values (KIM *et al.*, 1984) and it is possible to nest a fine-mesh model within a coarse GCM grid. However, the physical representation in GCMs of sub-grid scale processes, such as convective precipitation and cloud formation, is poor, which must affect the accuracy of the predicted grid-point values. Present day GCMs are generally considered to have realistic geography, but, here again, local detail is lacking. The UKMO GCM, for example, is not unusual in having no relief over the UK.

The second major problem concerns the failure of models to reproduce the details of present day climate. Comparison of GCM control runs with observed features of the climate system suggests that many models perform particularly badly over Europe (GATES, 1985). Discrepancies in the predicted and observed strength and location of the Icelandic Low and Azores High are, for example, commonly observed (SANTER, 1988).

All models suffer from poor representation of feedback processes, particularly those involving clouds. There is still some debate as to whether cloud feedbacks will be positive or negative. Recent research suggests that low and middle level clouds will have a negative effect whilst the high-level cirrus type clouds will tend to have a positive feedback effect (HENDERSON-SELLERS & ROBINSON, 1986).

Whilst all models concur that, with doubling of CO_2, global temperature and precipitation will increase and that the increases will be greatest at high latitudes, there are significant regional differences between models. These discrepancies are particularly notable for parameters such as precipitation and soil moisture. The GFDL model, for example, predicts a 30 to 50% reduction in summer soil moisture over western Europe, but this summer drying is not apparent in the NCAR model simulations (MANABE & WETHERALD, 1987; MEEHL & WASHINGTON, 1988). In this case, differences in the perturbed runs of the two models can be directly linked to differences existing in the control runs.

Many of these outstanding problems are acknowledged by climatologists but there are, as yet, no widely available quantitative parameters for assessing model reliability. Statistical methods are, however, being developed (SANTER, 1988).

Finally it must be noted that GCMs are equilibrium response models only and do not fully take into account either the thermal inertia of the oceans or its role as a carbon dioxide sink. Transient response

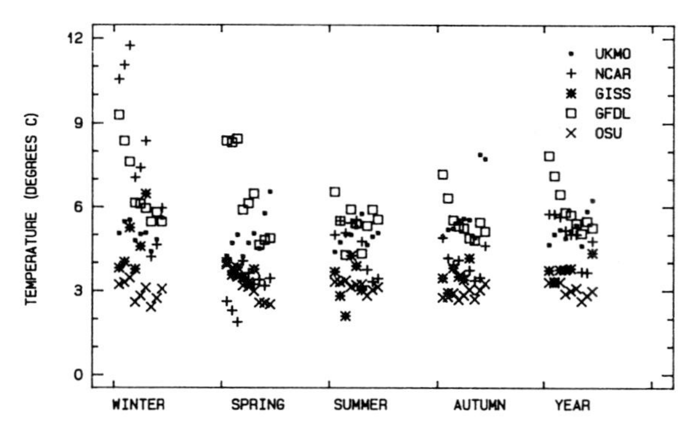

Fig. 1. Temperature increase predicted by 5 GCM studies for western European grid points 2 × CO_2.

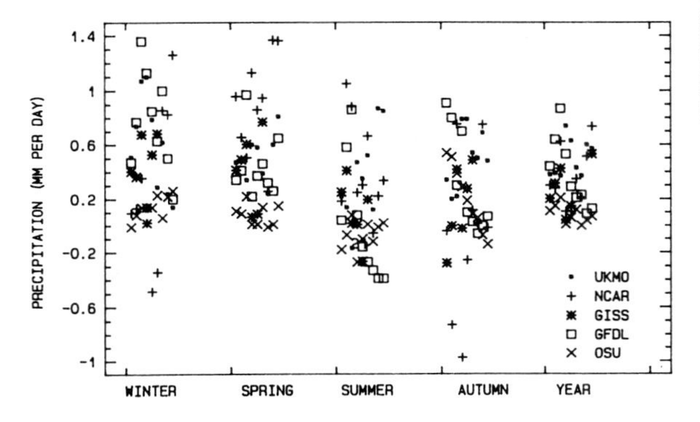

Fig. 2. Precipitation change predicted by 5 GCM studies for western European grid points 2 × CO_2.

models have been developed but, in the near future, are even less likely to provide reliable regional scenarios than the equilibrium models (WIGLEY, 1987a).

The uncertainties inherent in GCM results, particularly in their regional detail, are widely recognised by the modellers themselves. A modeller from the UKMO group, for example, writes:

'There are many uncertainties in the modelling of climate change which remain to be narrowed... The physical and biological processes involved in climate change are not yet modelled adequately... although the simulated changes in regional climate from current climate models are of considerable interest, they cannot yet be relied on for making quantitative estimates of economic, agricultural or social impacts' (MITCHELL, 1988).

GCM results may, however, be used (with an awareness of their limitations) in climate sensitivity studies and in the development of methodologies for impact studies.

Seasonal grid point data have been obtained for Western Europe from CO_2-perturbed and control runs for each of the 5 GCMs (Table 1). For this study, the definition of the Western European area is constrained by the GCM grid scales and is taken to be 50 to 67.5°N x 11.25°W to 10°E. The UKMO, GFDL, NCAR and OSU models have 9 grid points within this area, whilst the GISS model has only 6. The grid point data in general indicate a temperature rise in the range of 2 to 8°C in all seasons (Fig. 1). A notable exception is the NCAR GCM, however, which predicts winter temperature increases of up to 12°C. Most models predict increases of precipitation in the range of 0 to 1.5 mm per day (Fig. 2). All models, however, predict a small decrease in summer for at least one grid point. Precipitation decreases are also predicted in autumn.

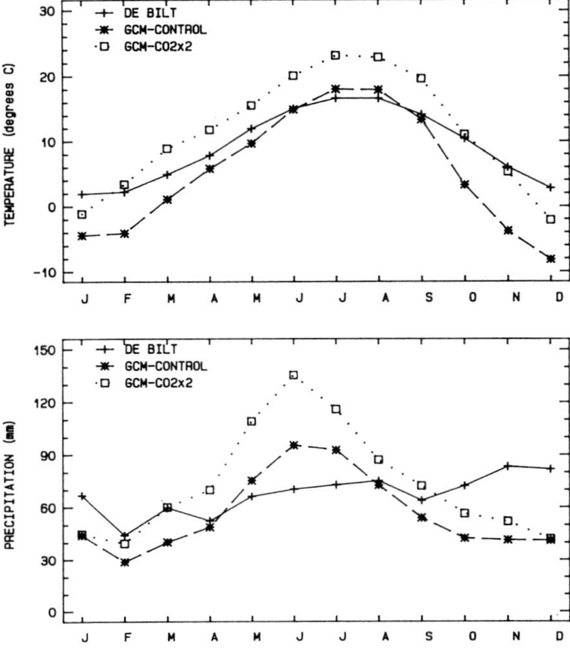

Fig. 3. Control and perturbed (2 × CO_2) run output for UKMO GCM grid point 9 (Netherlands) compared with mean monthly data for De Bilt (1961-1986). a) temperature, b) precipitation.

In many areas of the natural and human environment, changes in the seasonal cycle may have more significance than mean annual changes. In the next section we consider the ability of one GCM to simulate seasonal variations at one grid point.

2.1.1. NETHERLANDS CASE STUDY

As an illustration of the problems encountered in using GCM output at the high spatial and temporal resolutions required for impact studies we consider monthly data for one grid point from the UKMO GCM. The chosen grid point (GP9) is 52.5°N 3.75°E and lies just off the Dutch coast. Since, however, this par-

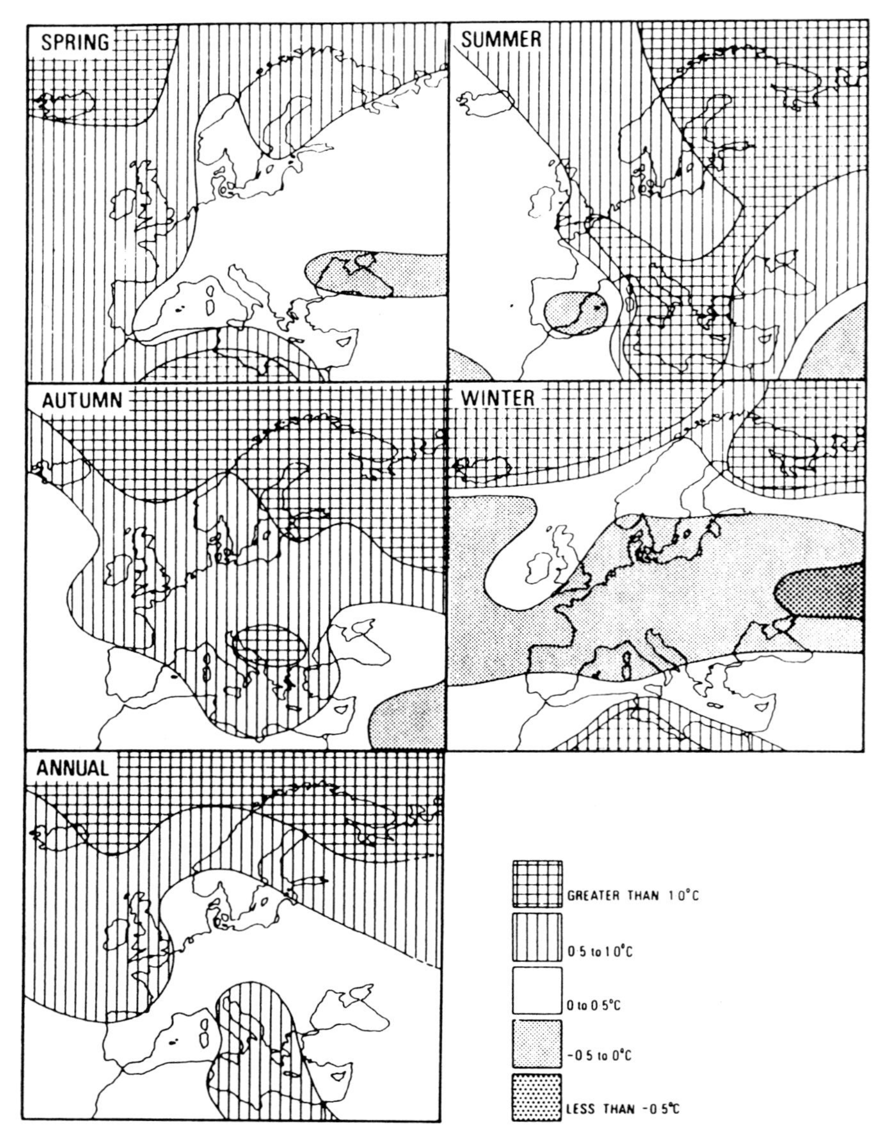

Fig. 4. Western Europe warm world instrumental scenario. Temperature differences, warm period minus cold period. (From LOUGH *et al.*, 1983).

ticular grid square is treated by the model as a land box, not as an ocean box, model output for GP9 can be compared with a land-based station. Comparison of the control run output for GP9 with data from De Bilt indicates that the UKMO GCM fails to reproduce the observed seasonal cycle (Fig. 3). In the case of temperature the model overemphasises the seasonal cycle whilst in the case of precipitation an unrealistic summer peak is produced.

Given the poor control run results, predictions of the magnitude of climatic variables by the perturbed run of the UKMO GCM cannot be used with confidence. Similar problems exist with the other 4 GCMs. Further analysis is necessary to determine whether or not GP9 data are representative of the general region, and to relate the control-run performance of the UKMO model to the other four GCMs. Only on the basis of such an analysis can the absolute and relative performance of the 5 GCMs be fully assessed.

The seasonal distribution of changes between the control and perturbed runs may, however, be more reliable. With a doubling of CO_2, UKMO model temperature increases at GP9 reach a maximum in autumn (+9.1°C) and late winter (+7.8°C). The smallest temperature increases occur in mid-winter

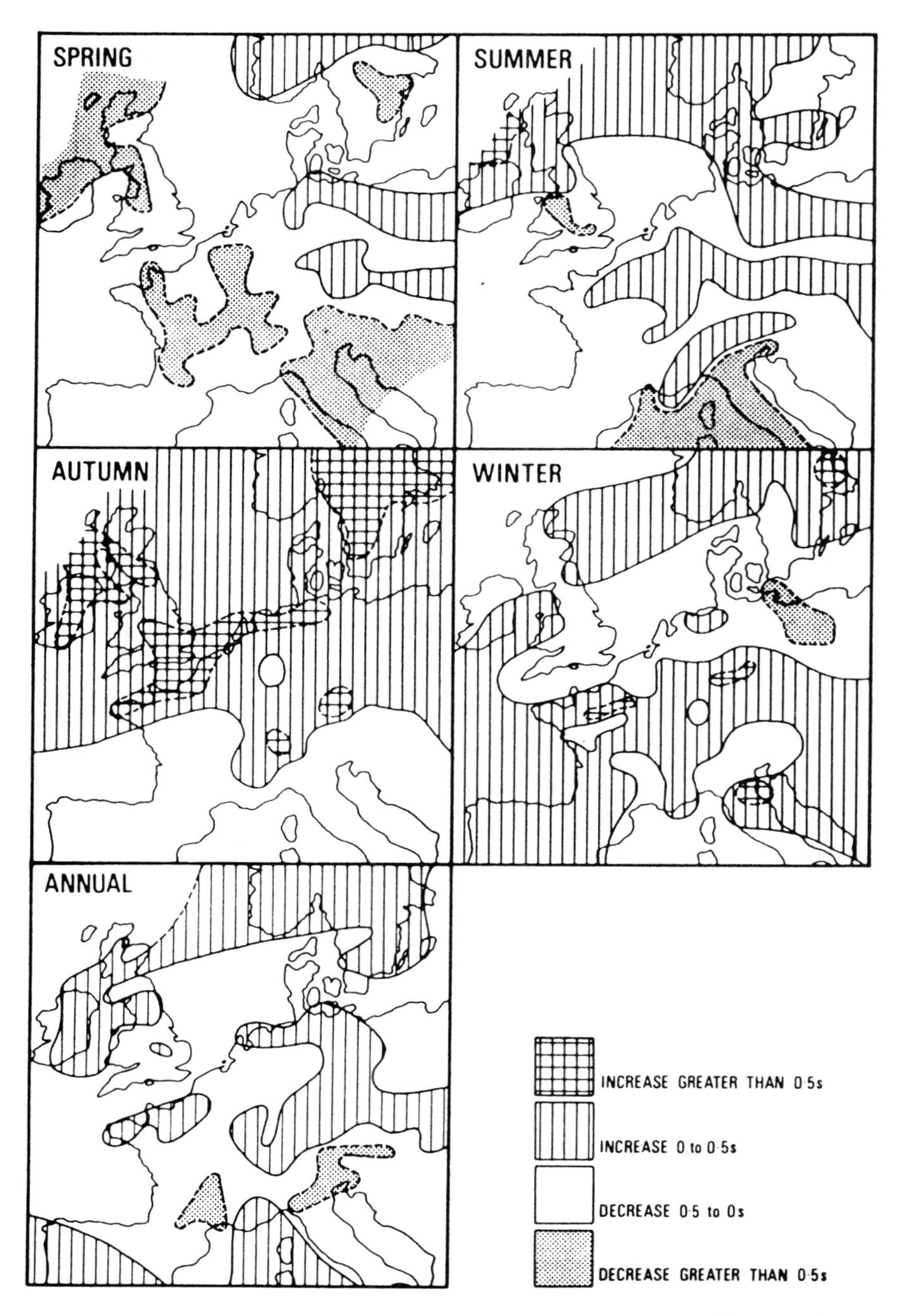

Fig. 5. Western Europe warm world instrumental scenario. Precipitation differences (warm period minus cold period) as multiples of the standard deviation. (From LOUGH *et al.*, 1983).

(+3.3°C) and summer (+4.9°C). Precipitation changes range from zero in winter to a summer peak at +1.3 mm per day.

2.2. THE PAST CLIMATE APPROACH

The problems associated with the development of model-based scenarios have prompted the use of a second approach to scenario construction. This alternative involves the use of past warm climate periods as analogues of a future high greenhouse gas world (WEBB & WIGLEY, 1985). The fundamental assumption of the analogue approach is that, given constant boundary conditions, the climate system responds in a similar way to different forcing factors, whether an increase in the solar constant or an increase in the atmospheric concentration of greenhouse gases. Warm world analogues can be drawn from the historical/geological record of palaeoclimate or from the instrumental record (WIGLEY *et al.*, 1986).

The Holocene thermal maximum or Altithermal, at 6000 Before Present, when the mean global temperature was 2°C warmer than today, has been suggested as a suitable warm period analogue. However, the presence of residual ice sheets means that boundary conditions may have been significantly different from those of the present day. Furthermore, lack of regional detail in the available proxy data limits the use of the Altithermal, and the previous interglacial (the Eemian), as appropriate analogues. The Medieval Warm Period has also been proposed as a suitable analogue period, but again there are data problems. The extent to which the Medieval warmth was a local or global event is also controversial.

The data problems described above can be avoided if the instrumental rather than the palaeoclimate record is used. There are a number of different ways in which warm periods can be selected from the instrumental record for scenario construction (WIGLEY *et al.*, 1986). In the case of the study described here, (LOUGH *et al.*, 1983; PALUTIKOF *et al.*, 1984) the warmest and coolest 20-year periods in the record of mean Northern Hemisphere surface temperatures (JONES *et al.*, 1982) were identified.

The JONES *et al.* temperature series is derived from gridded temperature data, with a grid resolution of 5° latitude by 10° longitude. The differences between the two periods, 1901–1920 (cold) minus 1934–1953 (warm), were calculated for the grid points relevant to the continent of Europe.

The resulting warm world annual temperature scenario indicates warming over Europe, reaching a maximum of over 1°C in the north of the region. Differences are, however, evident in the subregional and seasonal pattern of change (Fig. 4). Spring, summer and autumn are all generally warmer but winters tend to be cooler and to show an increase in interannual variability. The pattern of precipitation change is more complex, but there is a tendency for spring and summer to be drier and for autumn and winter to be wetter (Fig. 5).

The use of the instrumental scenario described here is limited by the relatively small difference in mean Northern Hemisphere temperature, only 0.4°C, between the two periods. In contrast, the GCM studies predict a global warming of 1.5 to 4.5°C. The instrumental scenarios of LOUGH *et al.* (1983) can, therefore, only be used as a guide to conditions during the early stages of a greenhouse gas-induced warming.

TABLE 2

Critique of crop-climate models (based on Warrick *et al.*, 1986).

	Model type		
	Statistical	*Mechanistic*	*Simulation*
Too many disposable constants.	+	+	+
Too many disparate sources.	-	-	+
Too few critical validations.	+	+	+
Too site/species specific.	+	-	-
Too many physiological forcing functions.	-	+	+
Too comprehensive to understand.	-	-	+
Sinks rather than sources of understanding.	+	-	+

\+ = applicable
\- = not applicable

3. UNDERSTANDING PRESENT DAY CLIMATE/AGRICULTURAL RELATIONSHIPS

3.1. MODEL HIERARCHY

A prerequisite of any agricultural impact study is the development of a model which describes the present day relationships between climate and agriculture. A hierarchy of models can be identified. A number of excellent reviews describe this hierarchy and provide a critique of the models and existing model-based studies (see, for example, WMO, 1985; WARRICK *et al.*, 1986). A brief summary of the models available is presented here.

Firstly, we can identify those models which describe first order effects: crop-climate models. Three types of crop-climate model are available and their particular merits and disadvantages are summarised in Table 2. The most commonly used are the statistical, generally multiple regression, models. These are 'black box' models which tend to be site specific and therefore cannot be applied outside the region or data range from which they were originally developed. We describe the development of such a model for the Netherlands in the next section. They are relatively easy to construct, in contrast to the complex computer simulation models. Simulation models attempt to describe the physical, chemical and physiological mechanisms important for plant growth, together with their interactions. Their sheer complexity tends to limit their value as a tool for developing an understanding of climate relationships (WMO, 1985). Finally, we can identify mechanistic schemes, which are intermediate in complexity between the statistical and simulation approaches.

Crop-climate models are concerned only with the primary effects of environmental conditions on crop yields. Changes in yields at the crop margins and spatial shifts in cropping patterns are incorporated into marginal spatial analysis models (WARRICK *et al.*, 1986). In these studies the spatial boundaries of particular crop types are defined in terms of climate parameters. In central Europe both temperature and precipitation act as limiting factors whilst to the north, temperature is the dominant limiting factor and to the south, precipitation is dominant (FLOHN & FANTECHI, 1984). The spatial analysis technique is best confined to coherent and spatially-simple ecological regions. For this reason most studies have concentrated on the North American Corn Belt. Two such studies, for example, are based on the assumption that the northern limit of the Corn Belt is defined by the number of accumulated growing degree days per growing season, whereas the western limit is defined by inadequate soil moisture content (NEWMAN, 1980; BLASING & SOLOMON, 1983). It is estimated that warming would displace the Corn Belt north-by-northeastwards by up to 175 kilometres per degree C.

Neither crop-climate nor marginal spatial analysis models consider the secondary effects of climate change on the economic, technological and social aspects of the agricultural system. Changes in crop varieties, irrigation and fertilization practices and in the availability of financial subsidies may all occur in response to changes in crop yields. Such indirect and feedback effects are incorporated into agricultural sector analysis models (WARRICK *et al.*, 1986). The most comprehensive agricultural systems impact studies to be completed are those undertaken for IIASA/UNEP in marginal cool temperate and cold regions and in semi-arid regions (PARRY *et al.*, 1988). Three scenarios are considered in the temperate and cold region studies: 1) an extreme decade from the instrumental record, 2) a single extreme year from the instrumental record, and 3) a GCM CO_2x2 simulation. The scenarios selected for semi-arid regions include a 1-in-10 year drought event.

The final approach to climate impact assessment involves the use of historical case studies. Data concerning the quality of grass growth and hay yields for the period 1601–1780 have, for example, been extracted from Icelandic documentary sources and a statistical link found between climate parameters and the crop data (OGILVIE, 1984). Interpretation of documented agricultural changes in terms of climatic change is, however, complicated by the influence of social, political and economic factors.

It is considered that marginal spatial analysis and crop-climate models are appropriate for use in coastal ecosystem impact studies. Both statistical and mechanistic crop-climate models have been used in the studies of European agricultural impacts in a high greenhouse gas world described in Section 4. The stages involved in the construction of a statistical crop-climate model are described in the next section.

3.2. A STATISTICAL CROP-CLIMATE MODEL FOR THE NETHERLANDS

In order to demonstrate a modelling technique widely used by climatologists, the authors have developed a multiple regression model for Dutch winter wheat yields.

Annual wheat yields for the period 1967–1987 were obtained from the EEC Agricultural Yearbooks. The observed strong trend in yields per hectare over this period is considered to be due to 'technological' factors and is, therefore, removed by fitting a quadratic trend. The resulting residual series is used as input to a stepwise multiple regression analysis, together with monthly temperature and precipitation data for De Bilt as the independent variables. No trend is evident in the climate data. Typically, over 96% of the Dutch wheat crop is winter wheat and so it is reasonable to assume that the residual yield series represents the year-to-year variability in winter wheat yields alone. A number of different models were tested and the preferred model is given by Equation-1:

$$Y_{res} = -54.5 - 0.09\ R_{Oct} - 0.23\ R_{Apr} + 4.31\ T_{June} - 0.06\ R_{July} + 0.77\ T_{July}$$

where Y_{res} = annual residual yield
R_{Oct} = October rainfall
R_{Apr} = April rainfall
T_{June} = June temperature
R_{July} = July rainfall
T_{July} = July temperature

Equation-1 explains 77% of the observed variance and has 10 degrees of freedom. The model can be used to identify growing conditions conducive to high yields. During the sowing period a dry October is beneficial. Dry conditions in April, together with a warm June (the flowering period), are important. Finally, the ripening process is aided by a warm and dry July.

4. AGRICULTURAL IMPACTS IN A HIGH GREENHOUSE GAS WORLD

4.1. PREVIOUS NORTH WEST EUROPEAN STUDIES

Three previous studies have been identified where estimates of changes in crop yields in a warmer, high greenhouse gas, world have been made for western Europe.

In the first study, a regression model was developed for wheat yields in EEC countries (SANTER, 1984, 1985). The climate scenario for a doubling of CO_2 was based on output from the GISS GCM. Over the study area, temperature was predicted to rise by 4°C and precipitation to increase by 20%. No change in Dutch wheat yields was predicted but decreases were expected to occur elsewhere. A maximum reduction in yield of 17.5% was predicted for Ireland.

In the second study, a combination of mechanistic and regression models was used to investigate crop yield changes for a number of arbitrary scenarios

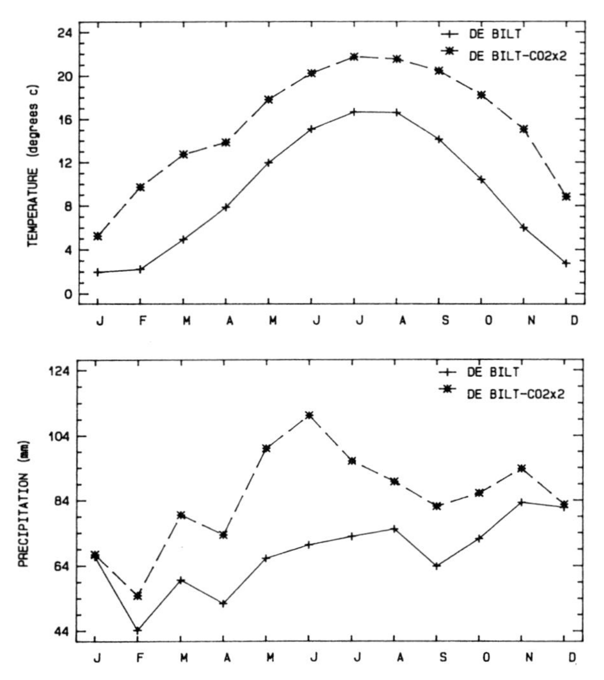

Fig. 6. De Bilt mean monthly data (1961-1986) and De Bilt 2 × CO_2 scenario (see text for details). a) temperature, b) precipitation.

(MONTEITH, 1981). Cereal and potato yields were found to decrease by 5% for each 5° temperature rise.

In the third study, the results from the instrumental climate scenario described in Section 2.2 were applied to regression models developed for western Europe (LOUGH *et al.*, 1983). No significant changes were found in wheat yields but for other crops a moderate to large decrease was predicted.

To summarise, previous western European agricultural impact studies predict no change or a decrease in crop yields. These results are consistent with 'best estimates' for the main mid-latitude North American and European cereal regions based on a temperature rise of 2°C (WARRICK *et al.*, 1986). In these major agricultural regions this would lead to a reduction in cereal yields of 3 to 17%. Any decline in yields would be accentuated by a decrease in precipitation but offset by an increase in precipitation.

4.2. REGIONAL AGRICULTURAL IMPACT STUDIES: NETHERLANDS CASE STUDY

In Section 2.1.1 we described the development of one possible warm world scenario for the Netherlands and in Section 3.2 a statistical crop-climate model applicable to Dutch winter wheat yields was developed. Is it possible to apply the results from these two analyses to an assessment of Dutch wheat yield changes in a high greenhouse gas world?

It was concluded that the magnitude and seasonal cycle of the UKMO GCM control run output was unrealistic and that, therefore, the predictions of the model for climate in a high greenhouse gas world at the Dutch grid point (GP9) cannot be regarded as reliable (Fig. 3). In this respect, the UKMO model is no different from any other GCM. The magnitude and seasonal distribution of the differences between the control and perturbed runs may, however, be more reliable. For each month, the predicted change in temperature and precipitation can be added to a 'real' instrumental data series. This approach to scenario construction can be described as a 'change in mean' approach and results for De Bilt temperature and precipitation are given in Fig. 6.

It is apparent that the predicted increase in temperature and precipitation is large enough to bring the adjusted, warm world, De Bilt mean series outside the range of year-to-year variability observed in the original series. The latter series was used to construct the wheat regression model and for this reason it would be statistically invalid to calculate warm world yields from Equation 1 using the De Bilt scenario (Fig. 6).

It is also notable that the projected De Bilt temperature rise indicates that, in a high greenhouse gas world, mean monthly temperatures would only fall below 6°C, the base line temperature widely used in crop growing degree day calculations, during January (MONTEITH, 1981). How might the agricultural impacts of climate changes of this magnitude be assessed?

First, it is considered that the development of mechanistic models (MONTEITH, 1981) which are sensitive to the phenological stages of crop growth (SANTER, 1983) will provide a useful tool for impact assessment. Second, it is possible to produce climate scenario data in a more appropriate form for impact assessment than simply changes in monthly mean temperature and precipitation. Daily data time series can be generated from monthly GCM output using Markov Chain modelling techniques (WILKS, 1988). Relevant variables such as soil moisture, sunshine, maximum and minimum temperature or accumulated temperature, can be extracted or estimated from GCM output (GATES, 1985). Finally, changes in the year-to-year variability can be considered. This is sometimes referred to as the 'shift in risk' approach, as opposed to the 'change in mean' approach (WARRICK *et al.*, 1986). This will allow analysis of long-term variability in yields and production in a high greenhouse gas world.

5. CLIMATE AND DIRECT CO_2 EFFECTS

The models and studies discussed in the previous

sections are all concerned with the response of agricultural crops to the climate changes likely to be associated with a greenhouse gas-induced warming. In one recent assessment it is estimated that, by 2030, close to 50% of the combined greenhouse gas radiative forcing will be attributable to methane, nitrous oxides and chlorofluorocarbons, with the remainder attributable to carbon dioxide (WIGLEY, 1987b). A number of, mainly laboratory, studies have investigated the direct effects of CO_2 on agricultural plant species.

Laboratory experiments indicate that, for C_3 crops, the best estimates of the increase in yield with a doubling of CO_2 lie in the range of +10% to 50% (BOLIN *et al.*, 1986). Estimates are also available for individual crops (CURE, 1985; CURE & ACOCK, 1986). Wheat yields are found to increase by 35 $\pm$ 14% and barley by 70% (CURE & ACOCK, 1986). The beneficial effects of enhanced CO_2 levels have been quantified in a modelling study: for every 1 ppm rise in CO_2 it is estimated that crop yields will increase by 0.1% (MONTEITH, 1981).

Both modelling and laboratory studies confirm the CO_2 'fertilization' effect upon plant growth. However, a number of questions remain. How do plants respond under field, rather than laboratory, conditions? It has, for example, been suggested that direct CO_2 effects will be reduced outdoors and on the larger scale (MORISON, 1987). How do the climate and direct CO_2 effects interact, particularly with respect to the hydrological cycle? A reduction in precipitation may, for example, be offset by the increased water use efficiency of plants in a high CO_2 world (WIGLEY *et al.*, 1984; PALUTIKOF, 1987).

6. FUTURE RESEARCH DIRECTIONS

The main aim of this paper has been to describe and to assess the research techniques which are available to climatologists for the construction of regional climate scenarios of a high greenhouse gas world specifically oriented towards plant growth impact studies. Specific methodologies have been illustrated, using the Netherlands as a case study. The regional climate scenarios currently available are not considered to be sufficiently reliable at the required spatial resolution for the purposes of coastal ecosystem impact studies. Four areas worthy of future research effort are listed below:

-(i) the development of more reliable regional and sub-grid scale scenarios through GCM studies,

-(ii) the construction of scenarios of more appropriate climate and climate-related parameters,

-(iii) the development of mechanistic and physiological plant growth/climate models, and

-(iv) the development of techniques for investigating the interaction of direct CO_2 and climate effects.

ACKNOWLEDGEMENTS

Dr M. Hulme of the Climatic Research Unit and Dr J. Mitchell of the UK Meteorological Office are thanked for making available the GCM grid point data presented in this paper.

7. REFERENCES

BLASING, T.J. & A.M. SOLOMON, 1983. Response of North American corn belt to climatic warming. DoE/NBB-004 U.S. Dept. of Energy, Office of Energy Research, Carbon Dioxide Research Division, Washington, D.C.

BOLIN, B., B.R. DÖÖS, J. JÄGER & R.A. WARRICK, 1986. The greenhouse effect, climatic change and ecosystems. Wiley, Chichester: 1-542.

CURE, J.D., 1985. Carbon dioxide doubling responses: A crop survey. In: B.R. STRAIN & J.D. CURE. Direct effects of increasing carbon dioxide on vegetation. DoE/ER-0238, U.S. Dept. of Energy, Office of Energy Research, Carbon Dioxide Research Division, Washington, D.C.: 99-116.

CURE, J.D. & J. ACOCK, 1986. Crop responses to carbon dioxide doubling: A literature survey.—Agric. Forest Met. **38:** 127-145.

E.E.C. EUROSTATS, 1967-1987. Agriculture, forestry and fisheries year books.

FLOHN, H. & R. FANTECHI, 1981. The climate of Europe: Past, present and future. D. Reidel, Dordrecht: 1-356.

GATES, W.L., 1985. The use of General Circulation Models in the analysis of ecosystem impacts of climatic change.—Clim. Change **7:** 267-284.

HANSEN, J.E., A. LACIS, D. RIND, G. RUSSELL, P. STORE, I. FUNG, R. RUEDY & J. LERNER, 1984. Climate sensitivity: Analysis of feedback mechanisms. In: J.E. HANSEN & T. TAKAHASHI. Climate processes and climate sensitivity. Geophysical Monograph 29, Maurice Ewing Vol. 5. American Geophysical Union, Washington, D.C.

HENDERSON-SELLERS, A. & P.J. ROBINSON, 1986. Contempory climatology. Longman Scientific and Technical.

JONES, P.D., T.M.L. WIGLEY & P.M. KELLY, 1982. Variations in surface air temperature: Part 1. Northern Hemisphere, 1881-1980.—Mon. Wea. Rev. **110:** 59-70.

KIM, J.W., J.T. CHANG, N.L. BAKER, D.S. WILKS & W.L. GATES, 1984. The statistical problem of climate inversion: Determination of the relationship between local and large-scale climate.—Mon. Wea. Rev. **112:** 2069-2077.

LAMB, P.J., 1987. On the development of regional climatic scenarios for policy-oriented climatic-impact assessment.—Bull. Am. Met. Soc. **68:** 1116-1123.

LOUGH, J.M., T.M.L. WIGLEY & J.P. PALUTIKOF, 1983. Climate impact scenarios for Europe in a warmer world.—J. Clim. App. Met. **22:** 1673-1684.

MACCRACKEN, M.C. & F.M. LUTHER, 1985. Projecting the climatic effects of increasing carbon dioxide. DoE/ER-0237, U.S. Dept. of Energy, Office of Energy Research, Carbon Dioxide Research Division, Washington, D.C.

MANABE, S. & R.J. STOUFFER, 1980. Sensitivity of a Global

Climate Model to an increase of CO_2 concentration in the atmosphere.—J. Geophys. Res. **85:** 5529-5554.
MANABE, S. & R.T. WETHERALD, 1975. The effects of doubling the CO_2 concentration on the climate of a General Circulation Model.—J. Atmos. Sci. **32:** 3-15.
——, 1980. On the distribution of climate change resulting from an increase in the CO_2 content of the atmosphere.—J. Atmos. Sci. **37:** 99-118.
——, 1987. Large-scale changes of soil wetness induced by an increase in atmospheric carbon dioxide.—J. Atmos. Sci. **44:** 1211-1235.
MEEHL, G.A. & W.M. WASHINGTON, 1988. A comparison of soil-moisture sensitivity in two Global Climate Models.—J. Atmos. Sci. **45:** 1476-1492.
MITCHELL, J.F.B., 1988. Local effects of greenhouse gases.—Nature **332:** 399-400.
MONTEITH, J.C., 1981. Climatic variation and the growth of crops.—Quart. J. R. Met. Soc. **107:** 749-774.
MORISON, J.I.L., 1987. Plant growth and CO_2 history.—Nature **327:** 560.
NEWMAN, J.E., 1980. Climate change impacts on the growing season of the North American corn belt.—Biomet. **7:** 728-742.
OGILVIE, A.E.J., 1984. The impact of climate on grass growth and hay yield in Iceland: A.D. 1601 to 1780. In: J.A. MÖRNER & W. KARLEN. Climatic changes on a yearly to millennial basis. D. Reidel, Dordrecht: 343-352.
PALUTIKOF, J.P., 1987. Some possible impacts of greenhouse gas induced climatic change on water resources in England and Wales. In: The influence of climate change and climatic variability on the hydrologic regime and water resources. IAHS Publ. No. **68:** 585-596.
PALUTIKOF, J.P., T.M.L. WIGLEY & J.M. LOUGH, 1984. Seasonal climate scenarios for Europe and North America in a high-CO_2, warmer world. DoE/EV/10098-5, TR012, U.S. Dept. of Energy, Office of Energy Research, Carbon Dioxide Research Division, Washington, D.C.
PARRY, M.L., T.R. CARTER & N.T. KONIJN, 1988. The impact of climatic variations on agriculture. Vol. 1: Assessments in cool temperate and cold regions: 876 pp. Volume 2: Assessments in Semi-arid regions: 764 pp. Kluwer Academic Publishers.
SANTER, B., 1983. The socio-economic impacts of climatic change: An analysis of the impacts of a CO_2-induced climatic change on the agricultural sector of the EEC 9. Dornier Systems, prepared for the Commission of the European Communities: 1-141.
——, 1984. The impacts of a CO_2-induced climatic change on the agricultural sector of the European Communities. In: H. MEINL & W. BACH. Socioeconomic impacts of climatic changes due to a doubling of atmospheric CO_2 content. Commission of the European Communities Contract No.CL1-063-D: 456-642.
——, 1985. The use of General Circulation Models in climate impact analysis. - A preliminary study of the impacts of a CO_2-induced climatic change on West European agriculture.—Clim. Change **7:** 71-94.
——, 1988. Regional validation of General Circulation Models. Climatic Research Unit Research Publication No. 9, CRURP9: 1-375.
SCHLESINGER, M.E., W.L. GATES & Y.J. HAN, 1985. The role of the ocean in CO_2-induced climatic change. In: J. NIHOUL. Coupled Ocean-Atmosphere Models. Elsevier: 447-479.
WARRICK, R.A., R.M. GIFFORD & M.L. PARRY, 1986. CO_2, climatic change and agriculture. In: B. BOLIN, B.R. DÖÖS, J. JÄGER & R.A. WARRICK. The greenhouse effect, climatic change and ecosystems. Wiley, Chichester: 393-473.
WASHINGTON, W.M. & G.A. MEEHL, 1984. Seasonal cycle experiments on the climate sensitivity due to a doubling of CO_2 with an Atmospheric General Circulation Model coupled to a Simple Mixed Layer Ocean Model.—J. Geophys. Res. **89:** 9475-9503.
WEBB, T. & T.M.L. WIGLEY, 1985. What past climates can indicate about a warmer world. In: M.C. MACCRACKEN & F.M. LUTHER. Projecting the climatic effects of increasing carbon dioxide. DoE/ER-0237, U.S. Dept. of Energy, Office of Energy Research, Carbon Dioxide Research Division, Washington, D.C.: 237-257.
WETHERALD, R.T. & S. MANABE, 1986. An investigation of cloud cover change in response to thermal forcing.—Clim. Change **8:** 5-24.
WIGLEY, T.M.L., 1987a. The effect of model structure on projections of greenhouse-gas-induced climatic change.—Geophys. Res. Let. **14:** 1135-1138.
——, 1987b. Relative contributions of different trace gases to the greenhouse effect.—Climate Monitor **16:** 14-28.
WIGLEY, T.M.L., K.R. BRIFFA & P.D. JONES, 1984. Predicting plant productivity.—Nature **312:** 102-103.
WIGLEY, T.M.L., P.D. JONES & P.M. KELLY, 1986. Empirical climate studies: Warm world scenarios and the detection of climatic change induced by radiatively active gases. In: B. BOLIN, B.R. DÖÖS, J. JÄGER & R.A. WARRICK. The greenhouse effect, climatic change and ecosystems. Wiley, Chichester: 271-322.
WILKS, D.S., 1988. Estimating the consequences of CO_2-induced climatic change on North American grain agriculture using General Circulation Model information.—Clim. Change **13:** 19- 42.
WILSON, C.A. & J.F.B. MITCHELL, 1987. A doubled CO_2 sensitivity experiment with a Global Climate Model including a simple ocean.—J. geophys. Res. **92:** 13315-13343.
WMO (World Meteorological Organization), 1985. Report of the WMO/UNEP/ICSU-SCOPE expert meeting on the reliability of Crop-Climate Models for assessing the impacts of climatic change and variability. WCP-90, WMO/TD-no.39. WMO, Geneva.

THE PRIMARY PRODUCTIVITY OF *PUCCINELLIA MARITIMA* AND *SPARTINA ANGLICA*: A SIMPLE PREDICTIVE MODEL OF RESPONSE TO CLIMATIC CHANGE

S.P. LONG

Department of Biology, University of Essex, Colchester, CO4 3SQ, U.K.

ABSTRACT

A simple analytic model in which primary production is predicted from solar radiation interception and conversion efficiencies was developed for two salt-marsh grasses which utilise different photosynthetic pathways: *Spartina anglica* a C_4 species and *Puccinellia maritima* a C_3 species. Interception efficiency was predicted as a function of the leaf area index and the canopy extinction coefficient. Leaf area index in turn was predicted as a function of thermal time, development stage and leaf longevity. Conversion efficiency was considered a function of photosynthetic pathway, atmospheric CO_2 concentration and temperature pre-history. The model was validated by predicting production from climatic data for eastern England in 1978. These predictions agreed closely with the measured primary production of both species in that year. Assuming a doubling of atmospheric CO_2 concentrations and a 3°C increase in air temperatures throughout the year the model was then used to predict, and analyse the causes of change in, production for the year 2050. The model suggests that although the productivity of the C_3 species will increase, due primarily to increased conversion efficiency through reduced photorespiratory losses, productivity of the C_4 species may increase similarly through the effect of increased air temperature decreasing the incidence of photoinhibitory damage and accelerating leaf canopy development. The potential advantages of analytic models in predicting effects of climate change on primary production are discussed.

1. INTRODUCTION

Puccinellia maritima and *Spartina anglica* form large monospecific stands on the coastal salt marshes of N.W. Europe from N. France northwards to S. Norway. They are also major components of mixed salt marsh vegetation (LONG & MASON, 1983). *S. anglica* is unusual within the flora of N.W. Europe in being the only C_4 plant of significance, in terms of the area that it occupies (LONG *et al.*, 1975; THOMAS & LONG, 1978). Under conditions of adequate water supply photosynthesis in C_4 plants is CO_2 saturated, thus elevation of atmospheric CO_2 is not expected to increase productivity through increased photosynthesis. By contrast, elevation of CO_2 will increase photosynthesis in C_3 plants both under light limiting and light saturating conditions leading to increased productivity. C_4 photosynthesis is believed to have evolved in the semi-arid tropics or sub-tropics, and C_4 plants in general are poorly adapted to the present climatic conditions of N.W. Europe (LONG, 1983). *S. anglica*, has evolved in, and is limited in distribution to, cool temperate climates. However, the responses of photosynthesis and growth to temperature suggest it to be poorly adapted to a cool temperate climate by comparison to native C_3 grasses (LONG, 1983; DUNN *et al.*, 1987). In E. England, development of the shoots of *S. anglica* in the spring lags 1–2 months behind that of the C_3 grass, *P. maritima* (DUNN *et al.*, 1981). In addition, leaves which develop during the low temperatures of spring show impaired photosynthetic capacity (LONG & INCOLL, 1979; DUNN, 1981). Thus, whilst the productivity of *S. anglica* might not be enhanced by increased atmospheric CO_2 levels, the increased temperatures associated with elevated atmospheric concentrations of greenhouse gases may be expected to benefit this species.

The biomass dynamics, above- and below-ground net primary productivity, leaf demography and leaf area indices have been reported for *S. anglica* in the Stour Estuary (DUNN *et al.*, 1981; DUNN, 1981; JACKSON *et al.*, 1986; Long *et al.*, 1988) and for *P. maritima* at Colne Point (HUSSEY, 1980; HUSSEY & LONG, 1982). Both sites are in eastern England on the southern North Sea coast. Simultaneously with these productivity measurements, the temperature and light responses of CO_2 assimilation of *S. anglica* were determined at monthly intervals (DUNN *et al.*, 1981; DUNN, 1981).

Controlled environments were used to assess the responses of plant dry matter production and leaf

J.J. Beukema et al. (eds.), Expected Effects of Climatic Change on Marine Coastal Ecosystems, 33–39.

area development to temperature in *S. anglica* (DUNN *et al.*, 1987), salinity in *S. anglica* (YAAKUB, 1980), and nitrogen and phosphorous in both species (OTHMAN, 1980). In addition, the responses of leaf photosynthesis to light, temperature, carbon dioxide and water vapour pressure deficits have already been described for *S. anglica* grown in controlled environments (LONG & WOOLHOUSE, 1978a, b). Thus much is already known about the growth and photosynthesis of these two species under controlled environment conditions. From this, the response of productivity to climatic change could be predicted. The productivity of the same genetic material in the field provides data for model validation.

Two mathematical approaches to predicting the response of production to climatic change could be used. 1) A dynamic model incorporating the major physiological determinants of productivity and their responses to environmental variables could be constructed (*e.g.* LONG & WOOLHOUSE, 1979; REYNOLDS *et al.*, 1980). Such an approach is useful to an understanding of the key mechanisms underlying productivity and in identifying deficiencies in knowledge of the system, however the approach cannot yield predictions of response to future change with any quantitative precision (CHARLES-EDWARDS, 1982). Given that the responses of individual physiological processes to environmental parameters are unlikely to be established with a precision better than ±10%, the multiplication or addition of these in the iterative procedures of dynamic model simulation will lead to uncertainty in the final prediction which will be almost certain to exceed the prediction in magnitude. 2) An alternative approach is to use an analytic model. MONTEITH (1977) proposed that the dry matter productivity of a wide range of agricultural crops could be described as a simple function of light interception and conversion efficiency. This is summarised by Eqn. 1:

$$P_n = S.Q.\phi/J \qquad (1)$$

Where:
P_n=The net primary production, in term of dry matter ($kg.m^{-2}.d^{-1}$)
S=Total solar radiation receipt ($MJ.m^{-2}.d^{-1}$)
Q=Interception efficiency
ϕ=Conversion efficiency
J=The energy content of the plant biomass ($MJ.kg^{-1}$)

Using this basic equation, this study develops a simple model for estimating P_n for *S. anglica* and *P. maritima*, examines its validation against field measurements of P_n, and examines predictions of P_n for the temperature and carbon dioxide increases that might have occurred by the year 2050.

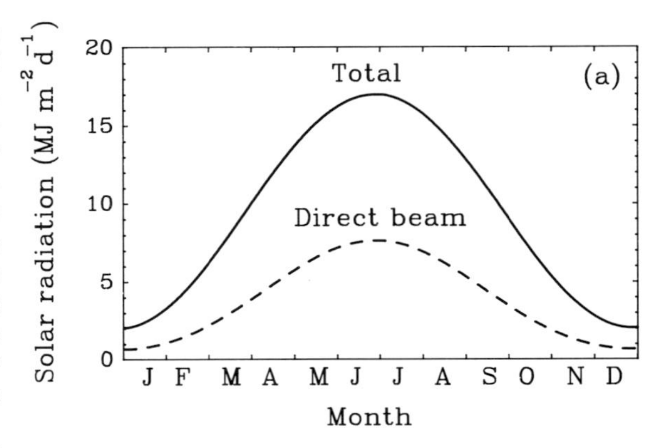

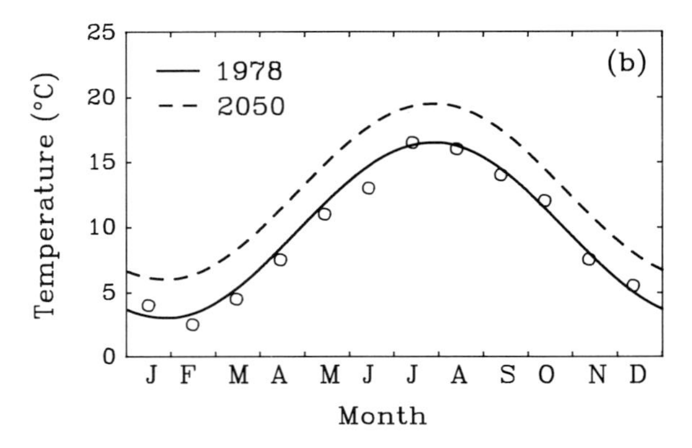

Fig. 1. (a) Total and direct beam radiation for N.E. Essex, calculated from the minimum and maximum monthly averages in 1978 (METEOROLOGICAL OFFICE, 1978) using Eqn. 2. (b) Calculated mean air temperatures for N.E. Essex calculated from minimum and maximum monthly values (Table 1) using Eqn. 3. Circles indicate the measured monthly means for 1978 (DUNN, 1981).

2. THE MODEL

The model was constructed by considering the 3 basic components of equation 1.

2.1. SOLAR RADIATION (S)

Solar radiation was considered to follow a sine function, with a minimum average daily value (S_{min}) of 2 $MJ.d^{-1}$ on the winter solstice and a maximum (S_{max}) of 18 $MJ.d^{-1}$ at the summer solstice:

$$S = S_{min} + [S_{max} - S_{min}] \cdot [\sin(2\pi\{d+11\}365 - 1.5) + 1]/2 \qquad (2)$$

Where d is number of days from January 1.

Predicted solar radiation receipts agreed closely with recorded values for the Essex coast. Solar radiation receipts for 2050 are assumed to be unchanged from those of 1978 (Fig. 1a).

Temperature (T) was similarly predicted by a sine function of time, assuming a 30 day delay behind the predicted peaks and troughs in solar radiation. The minimum of the average monthly temperatures (T_{min}) was 3°C for January and the maximum (T_{max}) was 16°C for July based on the 30 year averages for Clacton in Essex (Fig. 1b).

$$T=T_{min}+[T_{max}-T_{min}]\cdot[\sin(2\pi\{d+19\}365-1.5)+1]/2 \quad (3)$$

This equation provided a close fit to the recorded monthly values for 1978 (Fig. 1b).

Prediction of daily mean temperatures for 2050 was achieved simply by changing T_{min} and T_{max} in Eqn. 3.

2.2. INTERCEPTION EFFICIENCY (Q)

The proportion of incident solar radiation intercepted by a plant canopy may be related to its leaf area index (L) and exponential extinction coefficient (k) by analogy to Beer's low (MONSI & SEAKI, 1953; CHARLES-EDWARDS, 1982; NOBEL & LONG, 1985).

$$Q=1-e^{-kL} \quad (4)$$

Values of k were estimated at 0.6 for *P. maritima* and 0.5 for *S. anglica* from measurements of radiation absorption with tube solarimeters (TSL and TSLM, Delta T devices) for canopies of known leaf area index (NOBEL & LONG, 1985).

Leaf area index (L) was predicted using the concept of thermal time, *i.e.* leaf growth is linearly dependent on the degree days. Where degree days are calculated by summing, for each day, the number of degrees above the threshold for leaf growth (T_{th}). Field observations of *P. maritima* and *S. anglica* suggested thresholds for leaf extension growth (T_{th}) of 5°C and 8°C, respectively. Field observations suggested that the number of degree days required for the formation of one unit leaf area index (t_L) were 250°Cd for *P. maritima* and 300°Cd for *S. anglica* (calculated from HUSSEY, 1980; DUNN, 1981). The daily increment of degree days is given by:

if $[T>T_{th}]$ then:

$$°Cd=°C-T_{th} \quad (5)$$

if $[T<T_{th}]$ then:

$$°Cd=0 \quad (6)$$

Whilst leaf growth is strongly dependent on temperature, leaf longevity (l) appeared to vary little through the year, mean values based on those observed in the field are chosen (Table 1). Leaf area loss on a given day (d) may therefore be predicted by substracting the leaf area produced l days earlier from the total.

The daily increment of leaf area index (δL) is given by

$$\delta L=t_L/°Cd - t_L/°Cd_{(d-l)} \quad (7)$$

Where $t_L/°Cd_{(d-l)}$ is used to calcutate the leaf area that would therefore be lost through death at this point. L is computed by adding the daily increment (δL) to the existing leaf area index, starting from an initial value L_0 on Jan. 1. Values of L_0 chosen are those that were observed in the field in 1978. Since autumn growth of leaves would be enhanced by the increased degree days forecast for 2050, L_0 is increased proportionately (Table 1).

2.3. CONVERSION EFFICIENCY (ϕ)

Conversion efficiency has been shown to be a constant for a wide range of C_3 crops grown under temperate conditions, at 0.014. The conversion efficiency of C_4 crops has been shown to be 0.020, this higher value being attributed to the lack of photorespiration in C_4 plants (MONTEITH, 1978; BEADLE *et al.*, 1985). Elevation of CO_2 to 600 ppm will substantially reduce photorespiration and thus the of the C_3 plants under these conditions will approach or exceed the 0.020 of C_4 plants.

The major determinant of conversion efficiency is considered to be the light limited rate of CO_2 assimilation, which is commonly expressed as the maximum quantum yield. The quatum yield of leaves of C_4 leaves developed at low temperatures has been shown to be depressed (FARAGE & LONG, 1986; BAKER *et al.*, 1989). This has been correlated with a reduced light conversion efficiency (FARAGE & LONG, 1986; BAKER *et al.*, 1988, 1989). Although the quantum yield of photosynthesis in C_4 leaves is theoretically constant at 0.065 moles CO_2 per mole of photon (PEARCY & EHLERINGER, 1984), maize leaves developed during the early summer were found to have quantum yields of <0.040 (FARAGE & LONG, 1986), whilst DUNN (1981) found that the quantum yield of *S. anglica* plants from the field ranged from 0.020 in the late winter only rising to the theoretical 0.065 in July, when most leaves present developed at mean temperatures of >12°C. Assuming that the correlation between quantum yield and conversion efficiency applies in *S. anglica*, then conversion efficiency is assumed to be decreased for leaves developed at temperatures below 12°C, declining to zero for leaves developed at the growth threshold of 8°C.

if $[T_{(d-30)}>12°C]$ then:

TABLE 1

Parameter values for model. S_{max} and S_{min} are maximum and minimum daily radiation receipts; T_{max} and T_{min} are maximum and minimum monthly temperatures; L_0 is the leaf area index at the winter solstice; k is the canopy solar radiation extinction coefficient; t_L is the number of degree days required for the development of one unit of leaf area index; l is the longevity of a unit of leaf area in days; T_{th} is the threshold temperature for leaf extension growth; and ø is the efficiency of conversion of intercepted radiation into biomass.

	S. anglica	*P. maritima*
S_{max}	18 $MJ.m^{-2}.d^{-1}$	18 $MJ.m^{-2}.d^{-1}$
S_{min}	3 $MJ.m^{-2}.d^{-1}$	3 $MJ.m^{-2}.d^{-1}$
T_{min} (1978)	3°C	3°C
T_{max} (1978)	16°C	16°C
T_{min} (2050)	6°C	6°C
T_{max} (2050)	19°C	19°C
L_0 (1978)	0.3	0.6
L_0 (2050)	0.6	1.0
K	0.5	0.6
t_L	300°Cd	250°Cd
l	60d	50d
T_{th}	8°C	5°C
ø (340ppm CO_2, > 12°C)	0.020	0.014
ø (340ppm CO_2, < 8°C)	0	0.014
ø (600ppm CO_2, < 12°C)	0.020	0.020
ø (600ppm CO_2, < 8°C)	0	0.020

$$\phi=0.020 \tag{8a}$$

if [$T_{(d-30)}<12°C$ and $>8°C$] then:

$$\phi=0.020\ \{1 - (12-T_{(d-30)})/4\} \tag{8b}$$

if [$T_{(d-30)}<8°C$] then:

$$\phi=0 \tag{8c}$$

Where $T_{(d-30)}$ is the mean temperature one month earlier, as an approximate estimate of the mean temperature at the time when most of the existing canopy at time d is developed.

The estimate that conversion efficiency (ϕ) would decline to zero for leaves developed at 8°C is supported by the evidence from controlled environment studies of net assimilation rates of *S. anglica*. Plants transferred to a controlled environment with a mean temperature of 9°C showed a decline in net assimilation rate from an initially high value to zero by 60 days, when most of the leaves present would have developed at 9°C temperature (Dunn *et al.*, 1987). The observation also emphasises the importance of temperature pre-history over instantaneous temperature in determining net assimilation rate and therefore conversion efficiency.

3. MODEL VALIDATION

The predictions of the model were obtained by making daily estimates of net primary production for 365 days from Jan. 1. For each day calculations were conducted in the following sequence:

1) Mean temperature (T) for the day was calculated (Eqn. 3).
2) The daily increment of day degrees (°Cd) was calculated (Eqns. 5 or 6), substituting T from step 1 and using the appropriate threshold temperature (T_{th}) of Table 1.
3) The daily change in leaf area index (δL) was calculated (Eqn. 7) using the appropriate values of °Cd calculated in step 2 and leaf longevity (l) from Table 1. Leaf area index was then updated as the sum of δL and L for the previous day, where L for day 0 is given in Table 1.
4) The efficiency of solar radiation interception (Q) was calculated (Eqn. 4) using L determined in step 3 and the appropriate value of k from Table 1.
5) Conversion efficiency (ϕ) was considered a constant (Table 1), except for *S. anglica* when $T<12°C$ in the preceding 30 days, in which case the value is then adjusted according to Eqn. 8.
6) The daily total receipt of solar radiation (S) was then calculated (Eqn. 2).
7) Net production of dry matter (P_n) was calculated (Eqn. 1) using Q, and S determined in steps 4, 5 and 6, respectively.
8) Cumulative net production was obtained by summation of the daily values. The procedure was then repeated from step 1 for the subsequent day, terminating with the 365th day of the year.

The model was run for both *P. maritima* and *S. anglica*, first with the climatic variables of 1978 and secondly with the estimated climatic variables for 2050 (Table 1). Predictions for 1978 were then compared with measured values of Pn.

Fig. 2a illustrates leaf area index predicted for *S. anglica*, from Eqn. 7 and that observed in the field in 1978. Whilst Eqn. 7 provided a close fit to observed data for *S. anglica*, in *P. maritima* the equation overestimated leaf area from late June onwards. In the field it was observed that recruitment of new leaves was markedly reduced in late June and did not recover until about 40 days later (Hussey, 1980).

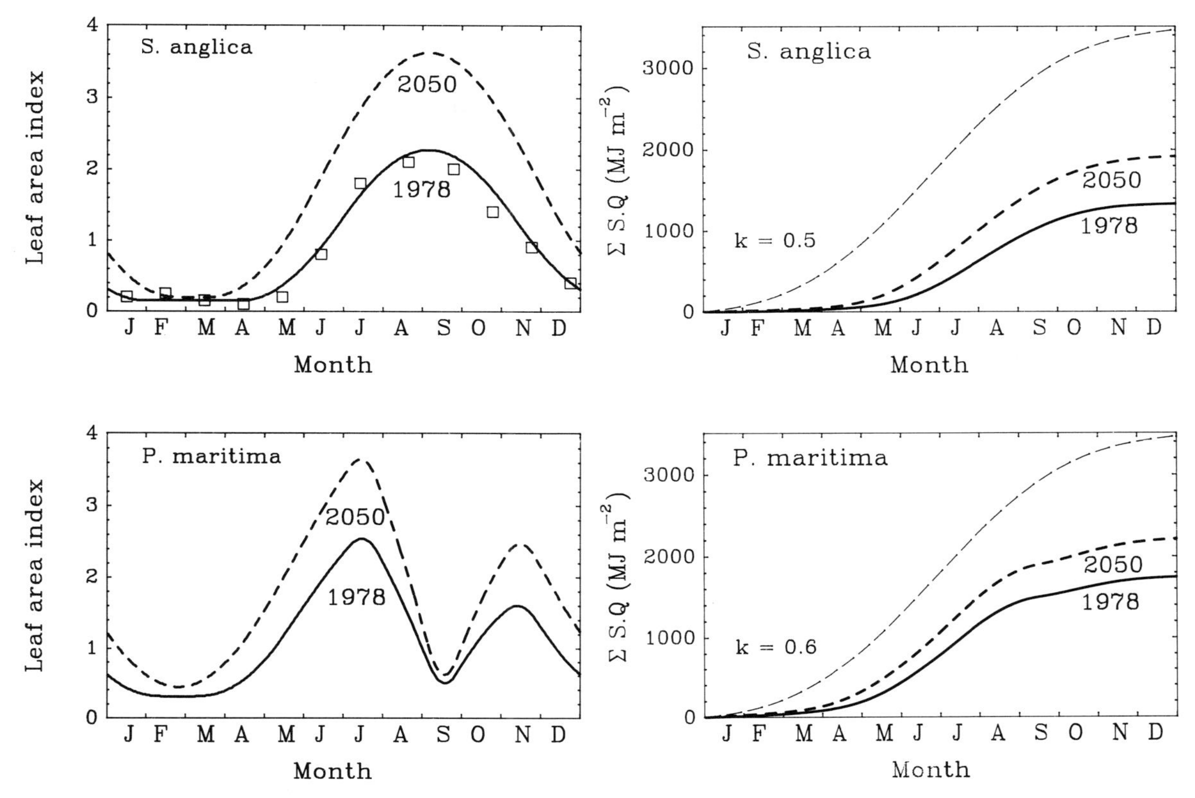

Fig. 2. Leaf area index (L) for *S. anglica* and *P. maritima* calculated by integrating Eqn. 4 with respect to time, using the parameters of Table 1 and temperatures as illustrated in Fig. 1b. In *P. maritima* leaf area development is assumed to be zero between days 180 and 220, as explained in the section on model validation. Open squares indicate measured values of L for *S. anglica* in N.E. Essex in 1978 (DUNN, 1981).

Fig. 3. The calculated cumulative quantity of solar radiation intercepted (ΣS. Q.) by *S. anglica* and *P. maritima*, using the parameters of Table 1 and summing the daily products of interception efficiency (Eqn. 4) and solar radiation (Eqn. 2). The solid line and broken line indicate interception for 1978 and 2050, respectively. The dotted line indicates the cumulative incident solar radiation, *i.e.* the total quantity of solar radiation (S) available for interception, for comparison.

This was incorporated into the model by assuming that leaf growth ceased at day 180 and recommenced on day 220 (Fig. 2b). By this adjustment, leaf areas agreed closely with those calculated from HUSSEY (1980). Extrapolating from leaf area index to light interception using Eqn. 4 it is apparent that *P. maritima* would intercept about 50% of the incoming solar radiation and *S. anglica* only 34% (Fig. 3). Net production (Eqn. 9), shows good agreement between estimates and measured values for both *P. maritima* and *S. anglica* (Fig. 4). However, a good fit was only obtained for *S. anglica* when ϕ was assumed to decline at low temperatures (<12°C). If ϕ was assumed constant throughout the year then P_n of *S. anglica* was overestimated, particularly during the spring period (Fig. 4).

The major difference between the two species is that the leaf canopy develops earlier in the year in *P. maritima*, than in *S. anglica* (Fig. 2), such that whilst L is 1 by the summer solstice in *P. maritima* it is only 0.5 in *S. anglica*, leading to a greatly decreased efficiency of light interception at the time of peak radiation values (Fig. 3).

4. PREDICTED CHANGES IN PRODUCTIVITY WITH CLIMATE CHANGE

To illustrate the potential of the model, a single climate change scenario has been used. The principles used could be simply applied for other scenarios. The Oregon State University General Circulation Model (OSU-GCM) adopted by the Netherlands Landscape Ecological Impact of Climatic Change program (LICC) predicts, for a doubling of greenhouse gases in the atmosphere, that temperatures through the year in maritime W. Europe will increase by *ca* 3°C in all months. This scenario has therefore been used to estimate what

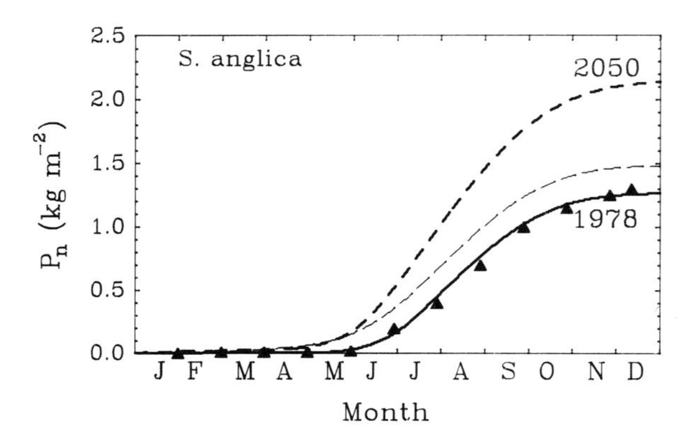

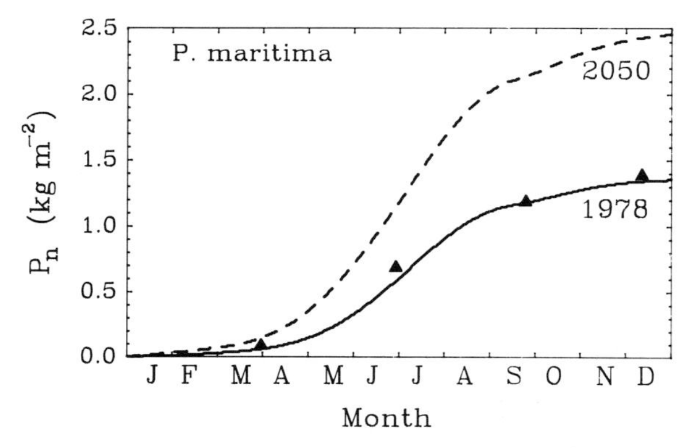

Fig. 4. Cumulative net primary production (P_n) of *S. anglica* and *P. maritima* over 12 months calculated from Eqn. 9 using calculated daily values of solar radiation interception and conversion efficiency calculated from the parameters of Table 1. For *S. anglica* the thin long-dashed line indicates Pn in 1978 assuming a constant ϕ according to Eqn. 8. Solid triangles indicate measured NPP for both species in N.E. Essex in 1978 (calculated from: HUSSEY, 1980; DUNN, 1981).

changes in productivity may be expected in 2050 assuming that the greenhouse gas concentrations have doubled by this time in relation to ability to absorb long-wave infra-red radiation (Table 1).

The results suggest that leaf area development in both species will be accelerated by increased temperatures. In *P. maritima* the point where LAI will be sufficient to intercept 30% of the incoming radiation will be attained 35 days earlier and 50 days earlier in *S. anglica*. This is of particular significance in *S. anglica* since whereas now its late leaf development and impaired photosynthetic capacity prevent it from capturinng and utilising the high inputs of solar radiation during May and June, a 3°C elevation of temperature would markedly alter both the size and photosynthetic capacity of the canopy. Thus, whilst *S. anglica* may derive no direct benefit from the increase in CO_2, the increase in temperature will substantially increase both its interception of radiation and its efficiency of conversion of the trapped energy into biomass, with a predicted increase in production over the year from 1.3 kg.m^{-2} in 1978 to 2.1 kg.m^{-2} in 2050 (Fig. 4a). *P. maritima* will gain some benefit from increased leaf growth in the Spring and Autumn, increasing the proportion of radiation intercepted (Fig. 3b). However, the major predicted gain results from the higher conversion efficiency in a high CO_2 environment raising annual production from 1.4 kg.m^{-2} in 1978 to 2.5 kg.m^{-2} in 2050.

In producing a simplified model of response to climatic change several factors which may affect either Q and/or have necessarily been ignored. In particular the effects of salinity and water level change, both of which may change with sea-level rises and precipitation changes that are expected to accompany global warming. Ability to incorporate these possible effects into the model will depend on the availability of data on the responses of Q and to these environmental variables. The model also assumes that the relationship of leaf area growth with degree days and leaf longevity remain constant, however decreased water use in a high CO_2 environment might increase both parameters. Again more information on the responses of leaf growth and longevity in high CO_2 and constant temperatures will be necessary. Finally, the model assumes that leaf growth is not nitrogen limited, whilst this appears not to affect the predictions for 1978, increased productivity might be expected to increase the amount of nitrogen sequestered by the vegetation leading to nitrogen limitation at least in the latter part of the growing season. Again this will require further information on the interaction of N and Q under conditions of elevated temperature and CO_2.

5. CONCLUSION

This study illustrates that a simple analytic model developed from light interception and conversion efficiencies, can provide a close prediction of net primary production for two salt marsh species. Although the approach greatly over-simplifies the underlying physiological processes determining production, this appears to have little affect on its predictive capacity. Its major advantage over more complex dynamic models lies in its simplicity. This is illustrated in the present study, by the ease with which deviations in predicted production from measured production, could be traced back to mid-season depressions of leaf growth in *P. maritima* and early season depressions of photosynthetic capacity in *S. anglica*. It thus provides a simple framework in which subsequent findings of the effects of other key factors, which may be affected by climate change, in particular nutrient availability and period of inundation may be easily integrated and their effects predicted.

6. REFERENCES

BAKER, N.R., M. BRADBURY, P.K. FARAGE, C.R. IRELAND & S.P. LONG, 1989. Measurements of quantum yield of carbon assimilation and chlorophyll fluorescence for assessment of photosynthetic performance of crops in the field.—Phil. Trans. R. Soc. London: **B 323:** 295-308.

BAKER, N.R., S.P. LONG & D.R. ORT, 1988. The effects of temperature on photosynthesis. In: S.P. LONG & F.I. WOODWARD. Plants and Temperature. Cambridge Univ. Press, Cambridge: 347-375.

BEADLE, C.L., S.P. LONG, S.K. IMBAMBA, R.J. OLEMBO & D.O. HALL, 1985. Photosynthesis in relation to plant production in terrestrial ecosystems. United Nations Environmental Programme, Tycooly International, Oxford: 1-156.

CHARLES-EDWARDS, D.A., 1982. Physiological determinants of crop growth. Academic Press, Sydney: 1-161.

DUNN, R., 1981. The effects of temperature on the photosynthesis, growth and productivity of *Spartina townsendii* (*sensu lato*) in controlled and natural environments. Ph.D. Thesis, Univ. Essex, Colchester, U.K.

DUNN, R., S.P. LONG & S.M. THOMAS, 1981. The effects of temperature on the growth and photosynthesis of the temperate C_4 grass *Spartina townsendii*. In: J. GRACE, E.D. FORD & P.G. JARVIS. Plants and their atmospheric environment. Blackwell, Oxford: 303–311.

DUNN, R., S.M. THOMAS, A.J. KEYS & S.P. LONG, 1987. A comparison of the growth of the C_4 grass *Spartina anglica* with the C_3 grass *Lolium perenne* at different temperatures.—J. Exp. Bot. **38:** 433–441.

FARAGE, P.K. & S.P.LONG, 1986. Damage to maize photosynthesis in the field during periods when chilling is combined with high photon fluxes.—In: W.J. Biggins. Proceedings VIIth International Congress on Photosynthesis Research Vol. IV. Martinus Nijhoff, Dordrecht: 139-143.

HUSSEY, A., 1980. The net primary production of an Essex salt marsh, with particular reference to *Puccinellia maritima*. Ph.D. Thesis, Univ. Essex, Colchester, U.K.

HUSSEY, A. & S.P. LONG, 1982. The net primary production of emergent salt-marsh at Colne Point , Essex. I. Seasonal changes in plant mass.—J. Ecol. **70:** 757–772.

JACKSON, D., S.P. LONG & C.F. MASON, 1986. Net primary production, decomposition and export of *Spartina anglica* on a Suffolk salt-marsh.—J. Ecol. **74:** 647–662.

LONG, S.P., 1983. C_4 photosynthesis at low temperatures.—Plant, Cell and Environment **6:** 345–363.

LONG, S.P. & N.R. BAKER, 1986. Photosynthesis in saline environments. In: N.R. BAKER & S.P. LONG. Photosynthesis in contrasting environments. Elsevier, Amsterdam: 63–102.

LONG, S.P. & L.D. INCOLL, 1979. The prediction and measurement of photosynthetic rate of *Spartina townsendii* in the field.—J. Appl. Ecol. **16:** 879–891.

LONG, S.P. & C.F. MASON, 1983. Ecology of salt marshes. Tertiary Level Biology Series. Blackie, Glasgow: 1-180.

LONG, S.P. & H.W. WOOLHOUSE, 1978a. The responses of net photosynthesis to vapour pressure deficit and CO_2 concentration in *Spartina townsendii*, a C_4 species from a cool temperate climate.—J. Exp. Bot. **29:** 567–577.

——, 1978b. The responses of net photosynthesis to light and temperature in *Spartina townsendii*, a C_4 species from a cool temperate climate.—J. Exp. Bot. **29:** 803-814.

——, 1979. Primary production in *Spartina* marshes. In: R.L. JEFFERIES & A.J. DAVY. Ecological processes in coastal environments. Blackwell, Oxford: 333-352.

LONG, S.P., R. DUNN, D. JACKSON, S.B. OTHMAN & M.H. YAAKUB, 1988. The primary productivity of *Spartina anglica* in an East Anglian estuary. In: A.J. GRAY. *Spartina*-Review of current research. Inst. of Terrestrial Ecology, H.M.S.O.

LONG, S.P., T.M. EAST & N.R. BAKER, 1983. Chilling damage to photosynthesis in young *Zea mays* I. Effects of light and temperature variation on photosynthetic CO_2 assimilation.—J. Exp. Bot. **34:** 177-188.

LONG, S.P., L.D. INCOLL & H.W. WOOLHOUSE, 1975. C_4 photosynthesis in plants from cool temperate regions, with particular reference to *Spartina townsendii*.—Nature **257:** 622-624.

METEOROLOGICAL OFFICE, 1978. Monthly weather reports. Meteorological office, Bracknell.

MONSI, M. & T. SAEKI, 1953. Über der Lichtfactor in den Pflanzengesellschaft und seine Bedeutung für die Stoffproducktion.—Jap. J. Bot. **14:** 22-52.

MONTEITH, J.L., 1977. Climate and efficiency of crop production in Britain.—Phil. Trans. R. Soc. London. Ser. B. **281:** 277-294.

——, 1978. Reassessment of maximum growth rates for C_3 and C_4 crops.—Exp. Agric. **14:** 1-5.

NOBEL, P.S. & S.P. LONG, 1985. Canopy structure and light interception. In: J. COOMBS, D.O. HALL, S.P. LONG & J.M.O. SCURLOCK. Techniques in bioproductivity and photosynthesis, 2nd Edn. Pergamon, Oxford: 229-250.

OTHMAN, S.B., 1980. Distribution of salt marsh plants and its relation to adaptive factors with particular reference to *Puccinellia maritima* and *Spartina townsendii*. Ph. D. Thesis, Univ. Essex, Colchester.

PEARCY, R.W. & J. EHLERINGER, 1984. Comparative ecophysiology of C_3 and C_4 plants.—Plant Cell Environment **7:** 1-13.

REYNOLDS, J.F., B.R. STRAIN, G.L. CUNNINGHAM & K.R. KNOERR, 1980. Predicting primary productivity for forest and desert ecosystem models. In: J.D. HESKETH & J.W. JONES. Predicuting photosynthesis for ecosystem models. Vol. II. CRC Press, Boca Raton: 169-207.

YAAKUB, M.H., 1980. Growth, photosynthesis and mineral nutrition of the halophytes *Aster tripolium* and *Spartina townsendii* in response to salinity. Ph. D. Thesis, Univ. Essex, Colchester.

THOMAS, S.M. & S.P. LONG, 1978. C_4 photosynthesis in *Spartina townsendii* at low and high temperatures.—Planta **142:** 171-174.

DIRECT EFFECTS OF ELEVATED CO_2 CONCENTRATION LEVELS ON GRASS AND CLOVER IN 'MODEL-ECOSYSTEMS'

D. OVERDIECK

Universität Osnabrück, Fachbereich Biologie/Chemie, Arbeitsgruppe Ökologie, Barbarastr. 11, D–4500 Osnabrück, BRD

ABSTRACT

In long-term experiments (up to 2.5 vegetation periods) grass/clover-mixtures (1 : 1) were exposed to 4 CO_2 concentration levels (340, 450, 600 and 800 $mm^3 \cdot dm^{-3}$) in acrylic-miniglasshouses which were climatized according to the microclimate outside. At 600 $mm^3 \cdot dm^{-3}$, plant growth and production were enhanced by 20-40% compared to cultures at 340 $mm^3 \cdot dm^{-3}$. Only the seed weight of the clover species increased by max. 28% with elevated CO_2 concentration levels. Without clippings, the clover species tended to be more enhanced by additional CO_2. With clippings, the grass was more successful in competition. The C/N-, C/P-, C/Ca- and C/K-relationships were higher at elevated CO_2 concentration levels. The CO_2 net fixation of the whole canopy increased by 40%, when the CO_2 concentration was raised from 340 to 600 $mm^3 \cdot dm^{-3}$. This enhancement decreased until the end of the third vegetation period to about 10%. The ecological consequences of these findings are discussed.

1. INTRODUCTION

The upward trend of annual, global tropospheric CO_2 concentration values is strongly evident since KEELING (1986) started continuous measurements in 1958 at Mauna Loa station, Hawaii. At this station the average was 342 mm^3 CO_2 per dm^3 in 1984 (SOLOMON *et al.*, 1985). Based on the measurements of air entrapped in bubbles of dated Greenland and Arctic ice, the preindustrial value was between 275 and 285 mm^3 CO_2 per dm^3 (NEFTEL *et al.*, 1985). Thus there has been an increase of about 62 mm^3 CO_2 per dm^3 in the global CO_2 concentration within the last century almost parallel with the fossil fuel consumption and the global forest devastation. Locally, the trend to a still higher increase in CO_2 concentration is most probable as we were able to measure during the last few years on a small hill in the surroundings of the town of Osnabrück in Northwest-Germany (Fig. 1).

As the availability of CO_2 is increasing everywhere, perhaps even reaching two or three times the concentration of the preindustrial time during the next century, plants and all other organisms dependent on them for food will be affected.

For this reason, the objects of our study were the direct effects of carbon dioxide on single plants and on small vegetation units under the CO_2 concentrations which might occur in the near future. Taking simplified mixtures of grass and clover for study we chose a vegetation type which dominates the agriculturally used land in some parts of the Fed. Rep. of Germany (Fig. 2).

2. MATERIAL AND METHODS

The main experiment was conducted from 1984 until 1988 in 4 acrylic mini-glasshouses in the field containing homogenized garden soil and as vegetation units mixtures (1:1, 40 seeds per dm^2) of *Trifolium repens* L. (var: Milka Paybjerg), *Lolium perenne* L. (var: Printo) and of *Trifolium pratense* L. (var: Lero), *Festuca pratensis* Huds. (var: Cosmos 2) (1:1, 20 seeds per dm^2). The soil volume was 0.38 m^3 and the air volume 0.51 m^3. The climate inside was adjusted to the outside one by means of automatical temperature and air humidity regulation (temperature: ± 0.5°C, rel. air humidity: ± 15%). Dependent on the irradiation angle the light intensity behind the glass was decreased between 10 and 20%. Until 1986 the interior air velocity was constant at 0.5 $m.s^{-1}$. During 1987 and 1988 also the air movement was approximated automatically to the exterior one. The systems were watered with the precipitation amounts measured at the station. In order to control cuvette effects open plots with the same soil and plant mixtures were situated in front of the mini-glasshouses.

First the systems developed under unchanged ambient CO_2 concentrations and at 600 mm^3 CO_2 per dm^3 and during 1987 and 1988 also at 450 and 800 mm^3 CO_2 per dm^3. In preliminary experiments dur-

J.J. Beukema et al. (eds.), Expected Effects of Climatic Change on Marine Coastal Ecosystems, 41–47.

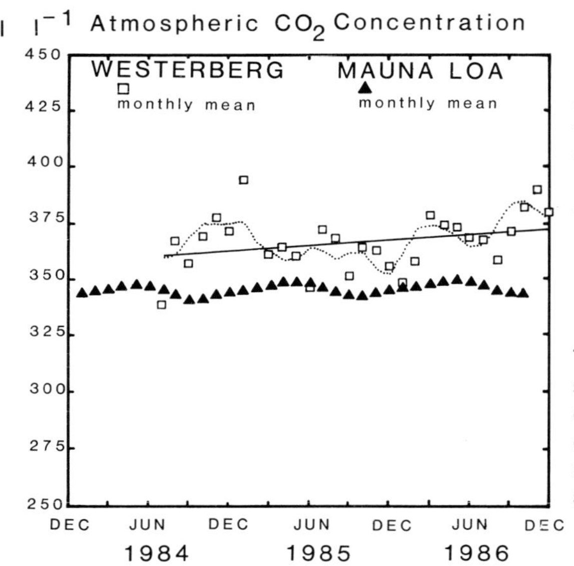

Fig. 1. Concentration of atmospheric CO_2 at Mauna Loa Observatory (Hawaii) after KEELING (1986) and at Westerberg Station (Osnabrück, North-West Germany, after BERLEKAMP & OVERDIECK, unpublished).

ing 1981 and 1982 the same experimental set-up was used with only two CO_2 concentration levels: unchanged ambient air and 620 mm^3 CO_2 per dm^3. Some of the results from those first experiments are also presented in the following text.

The construction and the working mode of the experimental equipment is described by OVERDIECK & BOSSEMEYER (1985).

The experimental set-up allowed to obtain two kinds of information: - 1st about development, growth, production and performance of the single species in competition, and - 2nd information about the long-term CO_2 gas exchange of entire small vegetation units with the upper soil included under varying microclimatic conditions.

3. RESULTS

3.1. DRY MATTER PRODUCTION

Selected, as one example from a couple of long-term productivity measurements, Fig. 3 shows the total dry matter amounts of white clover/perennial ryegrass mixtures at regulary applied clippings. After 100 days the difference between the two cultures became rather constant, stabilizing itself around 68% of the ambient air control (OVERDIECK & F. REINING, 1986). Comparable amounts of additional CO_2 (+ 320 mm^3 CO_2 per dm^3) caused enhancements of 20 to 40% in unclipped systems (unchanged systems) only.

3.2. SEED PRODUCTION

Seed weight of red-clover and meadow fescue was enhanced by the slight elevation of the CO_2 concentration level from 340 to 450 mm^3 CO_2 per dm^3. Further CO_2 increase had no effect (Table 1). An increase of 28% in seed weight was also found for white clover, but not for perennial ryegrass (OVERDIECK, 1986).

3.3. PERFORMANCE IN COMPETITION

The enhancement of net primary production of each species depended on the clipping (Fig. 4). Whitout clippings white clover was more enhanced by the additional CO_2 and overgrew the ryegrass. After regular clippings, however, the grass became stronger in competition.

3.4. C/N-RELATIONSHIP

The C/N-relationship was selected as one aspect from the results of the chemical determinations of mineral contents of the experimental plants as an example for a general tendency (OVERDIECK *et al.*, 1988). The CO_2 enrichment led to more carbon accumulation which was not followed by a greater nitrogen uptake (Fig. 5). In principle this was found for the relation of carbon to phosphorus, calcium and potassium, too (OVERDIECK & E. REINING, 1986).

3.5. CO_2 GAS EXCHANGE OF THE WHOLE MODEL-ECOSYSTEM

Fig. 6 shows a typical daily course of the measured parameters in relation to the CO_2 gas exchange rates from the control mini-glasshouses and the glasshouse with the elevated CO_2 concentration level ($\sim$600 mm^3 CO_2 per dm^3) as half-hour means. In this case the cultures were not cut. The greatest differences between the two experimental set-ups coincided with light peaks, and the CO_2 losses during the night were greater in the cabinets with additional CO_2 as also during the day with low light intensities. In young cultures (and shortly after clipping) the differences between the CO_2-enriched and the control plot were extremely great (Fig. 7).

In order to answer the question if our model-ecosystems were a source or a sink for the additional atmospheric CO_2 in the long-run the daily uptake rates of CO_2 were balanced against the daily losses over 3 years (Fig. 8). The difference between the two experimental set-ups remained positive until the end of the first year. At the end of the second year 600 g CO_2 per m^2 were fixed additionally under high CO_2. However, this CO_2 was completely given back to the

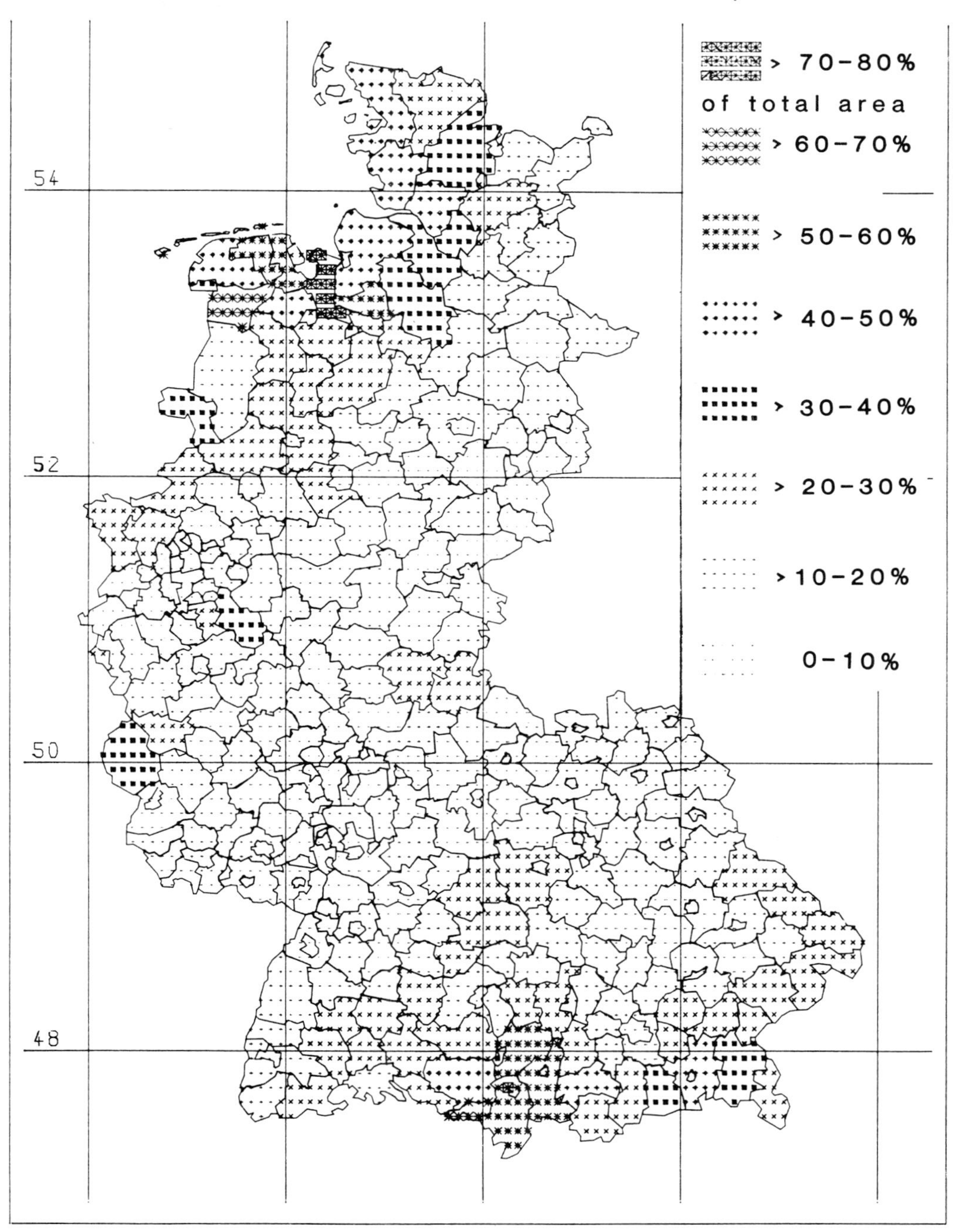

Fig. 2. Present distribution of managed grassland of the Federal Republic of Germany on the basis of departments (= Kreise) (after OSTENDORF & OVERDIECK, unpublished).

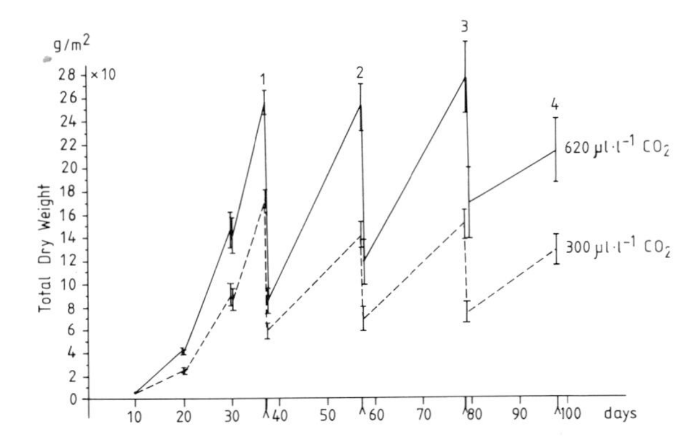

Fig. 3. Total dry weight per m² ground area of mixtures (1:1) of *Lolium perenne* and *Trifolium repens* grown at ~ 300 and at ~ 620 mm³ CO_2 per dm³ under field-like conditions before and after clippings: 1–4 (OVERDIECK & F. REINING, 1986).

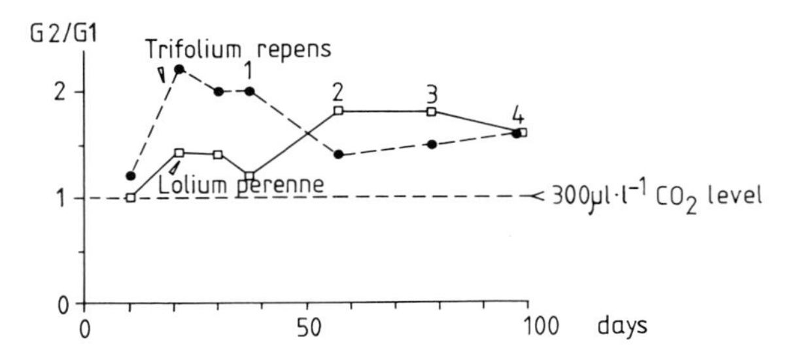

Fig. 4. Enhancement of net primary production by ~ 620 mm³ CO_2 per dm³ (G2) versus ~ 300 mm³ CO_2 per dm³ (G1) for *Lolium perenne* and *Trifolium repens* mixtures (1:1); 1–4: clipping numbers (OVERDIECK & F. REINING, 1986).

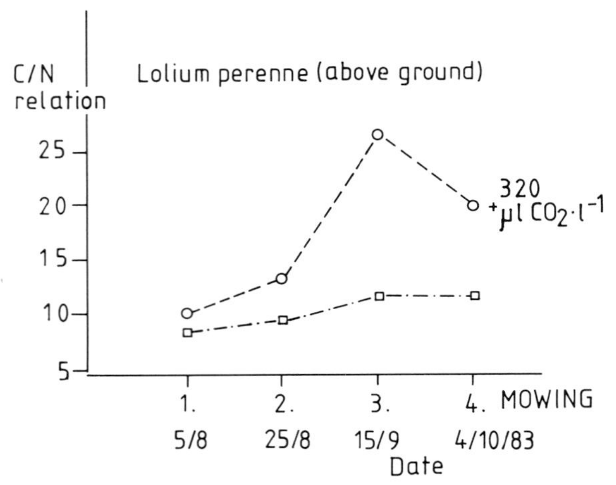

Fig. 5. Carbon/nitrogen ratio in dry matter cut from mixtures of *Lolium perenne* and *Trifolium repens* (1:1) grown at ~ 300 mm³ CO_2 per dm³ (squares) and at ~ 620 mm³ CO_2 per dm³ (circles) under field-like conditions (19–21 days interval between the clippings; OVERDIECK & E. REINING, 1986).

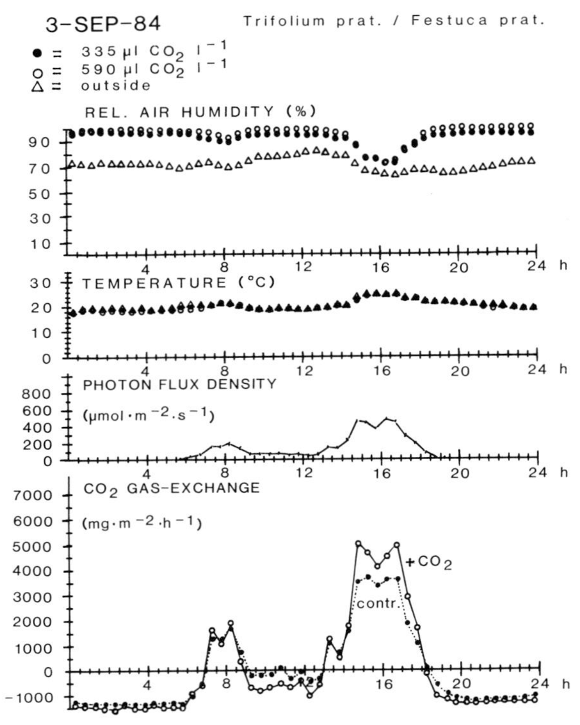

Fig. 6. 24-h course of the CO_2 gas exchange rates of two 'model- ecosystems' with the same mixtures (1:1) of *Trifolium pratense* and *Festuca pratensis* grown at unchanged ambient air (Contr.: ~ 345 mm³ CO_2 per dm³ and at ~ 600 mm³ CO_2 per dm³ (+ CO_2). Inside and outside of the gas exchange cabinets the rel. air humidity, the temperature and (only inside) the photon flux density of the photosynthetically active radiation were measured parallely (all values are half-hour means; OVERDIECK & FORSTREUTER, 1987).

TABLE 1

Weight of seeds and caryopses after long-term exposure to different CO_2 concentration levels (n=10·10 seeds resp. caryopses; after Forstreuter, unpubl.).

	mean weight of thousand seeds [g/1000]	
CO_2 concentration level µl·l⁻¹	Trifolium pratense	Festuca pratensis
340	1.98 ± 0.11	1.05 ± 0.21
450	2.14 ± 0.13	1.62 ± 0.27
600	2.25 ± 0.09	1.40 ± 0.33
800	2.26 ± 0.10	1.63 ± 0.30

atmosphere during winter, and after the third vegetation period only 1/3 of the level of the year before was reached. These findings show that the enhancement of CO_2-uptake decreases in the long-run. During this long-term experiment the systems were only clipped once a year.

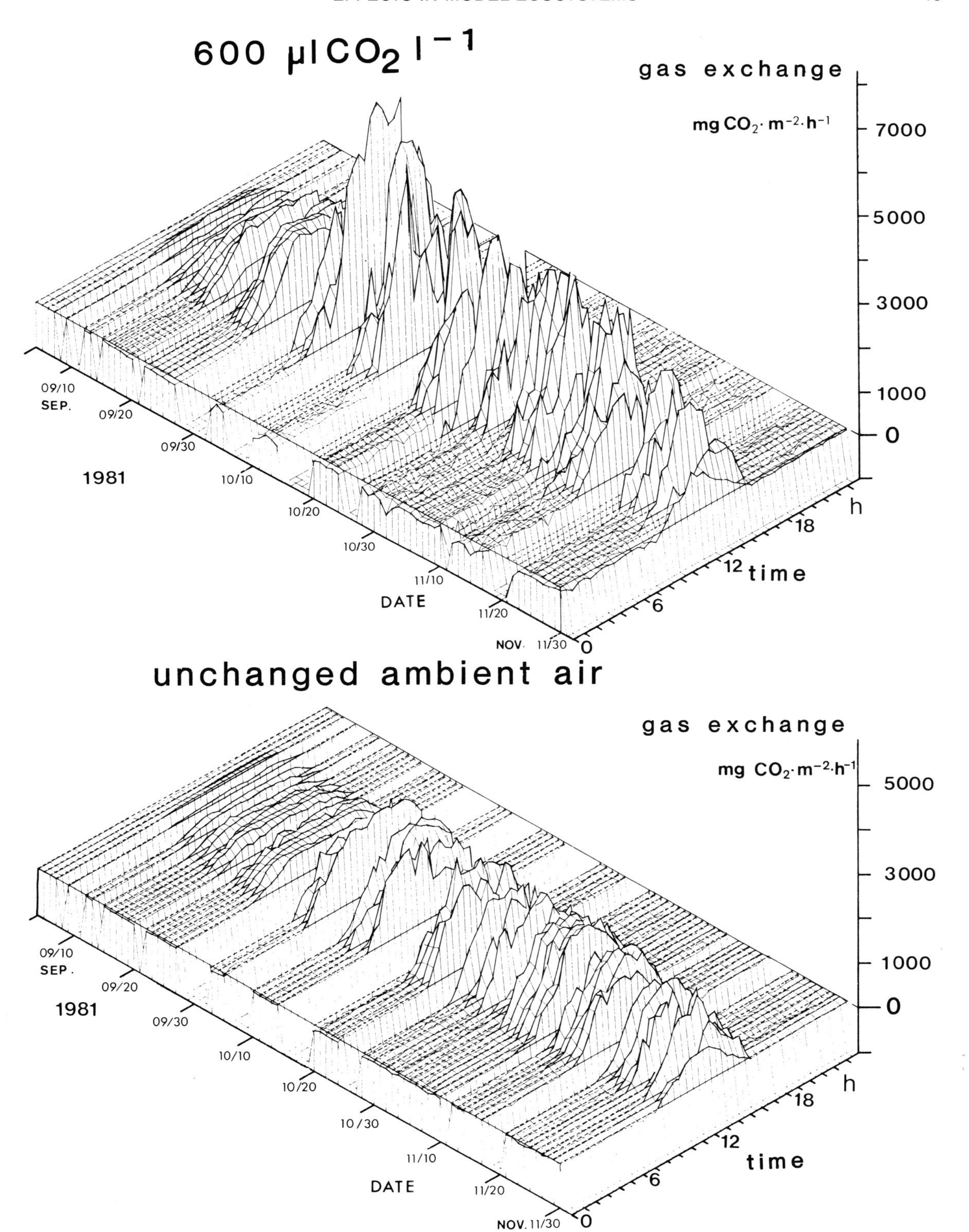

Fig. 7. 24-h courses of the CO_2 gas exchange from two 'model- ecosystems' (Sept.-Nov.) with the same mixtures (1:1) of *Trifolium repens* and *Lolium perenne* at unchanged ambient air and at ~ 600 mm^3 CO_2 per dm^3 (all values are half-hour means; OVERDIECK & BOSSEMEYER, unpublished).

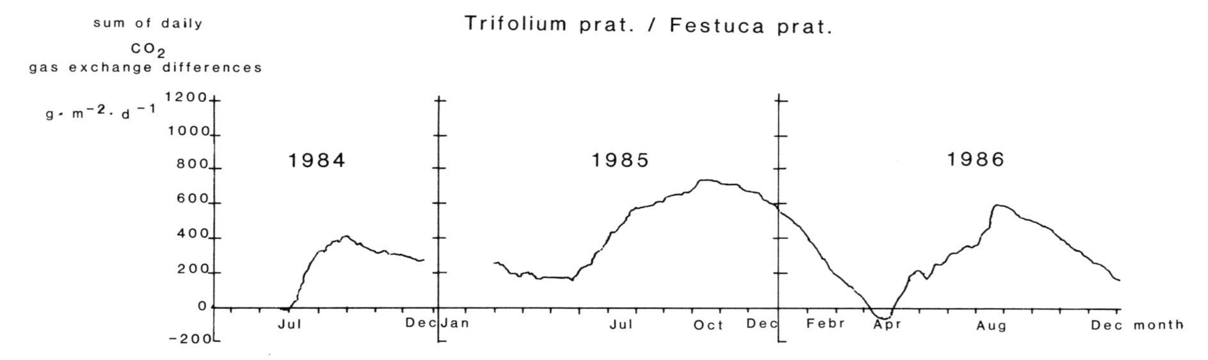

Fig. 8. Daily differences summed up over 3 vegetation periods between the CO_2 gas exchange rates of two 'model-ecosystems' with the same mixture (1:1) of *Trifolium pratense* and *Festuca pratensis* grown at unchanged ambient air (~ 345 mm^3 CO_2 per dm^3) and at ~ 600 mm^3 CO_2 per dm^3 (OVERDIECK & BERLEKAMP, unpublished).

TABLE 2

Summary of the results and their ecological consequences.

Results	Probable Consequences
1. plant growth and production is enhanced	biomass accumulation (denser and thicker vegetation cover)
2. grassland is a sink for additional atmospheric CO_2 in its juvenile phase	often mowed grassland will lower the CO_2 concentration level to a certain degree (systems in steady state not)
3. the performance of plant species in competition is affected	spontaneous and managed vegetation will change in species composition
4. seeds grow bigger	long-term advantages for more enhanced species in coming generations
5. mineral contents are changed	reduced quality for herbivores, and reduced decomposition rates

Table 2 summarizes some results of our long-term experiments.

4. RECOMMENDATION

In further research, long-term CO_2 enrichment studies should be combined with pollution studies which focus mainly on toxic atmospheric trace gases.

Another important scarcely investigated complex is the water budget of model-ecosystems under atmospheric CO_2 enrichment.

There are many hints in literature that plants reduce their transpiration rates at elevated CO_2 levels. This should influence the atmospheric water cycle as a whole.

5. REFERENCES

KEELING, C.D., 1986. Atmospheric CO_2 concentrations - Mauna Loa Observatory, Hawaii 1958-1986. Prepared by T.A. Boden. U.S. Department of Energy. (Oak Ridge (Oak Ridge National Laboratory, Environmental Science Division), TN). Publication No. 2798.

NEFTEL, A., E. MOOR, H. OESCHGER & B. STAUFFER, 1985. Evidence from polar ice cores for the increase in atmospheric CO_2 in the past two centuries.—Nature **315:** 45-47.

OVERDIECK, D., 1986. Long-term effects of an increased CO_2 concentration on terrestrial plants in model-ecosystems. Morphology and reproduction of *Trifolium repens* L. and *Lolium perenne* L.—Int. J. Biometeor. **30:** 323-332.

OVERDIECK, D. & D. BOSSEMEYER, 1985. Langzeit-Effekte eines erhöhten CO_2-Angebotes auf den CO_2-Gaswechsel eines Modell-Ökosystems.—Angew. Bot. **59:** 179-198.

OVERDIECK, D. & M. FORSTREUTER, 1987. Langzeit-Effekte eines erhöhten CO_2-Angebotes bei Rotklee-Wiesenschwingelgemeinschaften.—Verhandl. Ges. f. Ökologie (Giessen, 1986), **XVI:** 197-206.

OVERDIECK, D. & E. REINING, 1986. Effect of atmospheric CO_2 enrichment on perennial ryegrass (*Lolium perenne* L.) and white clover (*Trifolium repens* L.) competing in managed model- ecosystems. II. - Nutrient uptake.—Oecol. Plant. **7:** 367-378.

——, 1986. Effect of atmospheric CO_2 enrichment on perennial ryegrass (*Lolium perenne* L.) and white clover (*Trifolium repens* L.) competing in managed model- ecosystems. II. - Nutrient uptake.—Oecol. Plant. **7:** 357-366.

OVERDIECK, D., CH. REID & B.R. STRAIN, 1988. The effects of preindustrial and future CO_2 concentrations on growth, dry matter production and the C/N relationship in plants at low nutrient supply: *Vigna unguiculata* (cowpea), *Abelmoschus esculentus* (okra) and *Raphanus sativus* (radish).—Angew. Bot. **62:** 119-134.

SOLOMON, A.M., J.R. TRABALKA, D.E. REICHLE & L.D. VOORHEES, 1985. 1. The global cycle of carbon. In: J.R. Trabalka. Atmospheric carbon dioxide and the global carbon cycle. United States Department of Energy (Oak Ridge (Oak Ridge National Laboratory)) DOE/ER–0239: 5, TN.

EFFECTS OF ATMOSPHERIC CARBON DIOXIDE ENRICHMENT ON SALT-MARSH PLANTS

J. ROZEMA[1], G.M. LENSSEN[1], R.A. BROEKMAN[1] and W.P. ARP[2]

[1] *Department of Ecology and Ecotoxicology, Free University of Amsterdam, De Boelelaan 1087, 1081 HV Amsterdam, The Netherlands*

[2] *Smithsonian Environmental Research Center, P.O. Box 28, Edgewater, MD 21037, USA*

ABSTRACT

***Aster tripolium* and *Spergularia maritima* were cultivated at 340 ppm CO_2 (Ambient) and 580 ppm CO_2 (Elevated); salinity of the culture medium was varied at 10 mM NaCl and 250 mM NaCl. Culture solutions were flushed either with oxygen or nitrogen gas. In both species the mean relative growth rate was increased at elevated CO_2, but in the present paper there was no significant interaction with the salinity treatment.**

Flushing of the nutrient solution with nitrogen reduced the mean relative growth rate of both species under all conditions tested. Increased salinity reduced the mean relative growth rate of both species under all conditions tested. The rate of photosynthesis was increased with enriched CO_2 in *Spergularia maritima* and to a lesser extent in *Aster tripolium*.

Transpiration rates of both species decreased with CO_2 enrichment. The total water potential of the shoot (Ψ_T) was less negative at elevated CO_2. As a result of an increased photosynthetical rate and decreased stomatal conductance the water use efficiency was significantly increased in *Spergularia maritima* and less pronounced so in *Aster tripolium*.

1. INTRODUCTION

The atmospheric concentration of carbon dioxiode is increasing and is expected to double during the next century (BAES *et al.*, 1977). The combustion of fossil fuels is a major source of the atmospheric carbon dioxide increase (STRAIN & CURE, 1985). Since the atmospheric concentration of carbon dioxide (about 340 vpm measured at the campus of the Vrije Universiteit, Amsterdam, autumn 1988) is limiting to plant growth and photosynthetic rate, both processes are expected to increase with elevated atmospheric carbon dioxide. Short-term responses of the photosynthetic rate to elevated CO_2 are reported by GAASTRA (1959) and it has been indicated that C_3-plants will respond stronger to increased atmospheric CO_2 than C_4-plant species (GATES *et al.*, 1983). In the horticultural practice of greenhouse farming, carbon dioxide enrichment is a common procedure to increase the yield of crops in addition to the application of fertilizers (ENOCH & KIMBALL, 1985). There is only scanty knowledge of effects of atmospheric CO_2 enrichment in interaction with other environmental stress factors limiting plant growth, such as reduced availibility of water (drought, salinity) (PAEZ *et al.*, 1983; SCHWARZ & GALE, 1984 BROUNS, 1988; VAN DIGGELEN, 1988; ROZEMA *et al.*, 1989) and mineral nutrient deficiency (WONG, 1979; GOUDRIAAN & DE RUITER, 1983). Neither there is descriptive nor experimental knowledge about the combined effects of increased atmospheric carbon dioxide, air pollution and the increase of UV-B irradiation in relation to destruction of the ozone layer (VAN DE STAAIJ *et al.*, this volume). In a cooperative research programme of the Free University, The Netherlands and the Smithsonian Environmental Research Center, Edgewater, Maryland, USA, both long-term field studies are made of growth and photosynthesis of American and European salt marsh species to elevated CO_2 in open top chambers (CURTIS *et al.*, 1989; CURTIS *et al.*, 1989) and more short-term experimental field, greenhouse and laboratory analyses of the interactive effects of increased CO_2, salinity and hypoxic conditions (ROZEMA *et al.*, 1989). Plant growth in the salt marsh environment is primarily limited by increased salinity and anaerobic, waterlogging conditions.

The aim of the present paper was to study the effects of enhanced atmospheric carbon dioxide, and possible interactions with salinity and anaerobic conditions, on growth, photosynthesis and water relations of two dicotyledonous salt marsh plant species *viz.*, *Aster tripolium* L. and *Spergularia maritima* (All.) Chior.

J.J. Beukema et al. (eds.), Expected Effects of Climatic Change on Marine Coastal Ecosystems, 49–54.

2. MATERIALS AND METHODS

2.1. SALT-MARSH PLANT SPECIES

Plants of *Aster tripolium* L. and *Spergularia maritima* (All.) Chior. (nomenclature follows HEUKELS & VAN DER MEYDEN, 1983) were grown from achenes and seed collected from the Stroodorpe salt marsh, The Netherlands, October 1987. Seedlings were precultured in the greenhouse 20°C day, 15°C night, 70% Relative Humidity, with a 12 h light (300 μEinstein.m^{-2}.s^{-1} PAR) and 12 h dark scheme on a 1:1 (v/v) sand:garden soil (calceolaria) mixture for three weeks.

2.2. CO_2 GROWTH CHAMBERS

Elevated (580 ppm CO_2) concentrations of carbon dioxide were reached and controlled by mixing pure CO_2 from an AGA-gas cylinder (200 atm, AGA, Amsterdam) with ambient air (340 ppm, collected from the University Campus) using Brooks mass flow-controllers. Growth chambers (1.10 m height, 0.7 m diameter, content 0.425 m^3) made from stainless steel, were coated white and illuminated with 400 Watt Philips HPIT lamps with 315 μEinstein.m^{-2}.s^{-1} at the plant level (PAR, measured with a Licor-Quantum sensor Li-185B), 60% R.H., 21°C during the light (12 h) and 15°C in the dark period.

Flow rate of the airstream in the CO_2-growth chambers was 100 dm^3.min^{-1}. Plants in the CO_2-growth chambers were cultivated in pots with 0.5 strength Hoagland's solution (3 plants per pot, culture solution refreshed every week). 250 mM NaCl (Saline:S) in the nutrient solution was achieved by a weekly increase of 40 mM NaCl in the solution. Salinity level in the control treatment (Fresh: F) was 10 mM NaCl. Hydrocultures were continuously flushed with oxygen (aerated: O) or nitrogen (anaerobic: N) supplied from gas cylinders. Nine replications per plant species (3 pots) were used in the growth chambers.

The mean relative growth rate (RGR) was calculated based on the total plant dry weight at the end (W_2) and start (W_1) of the growth period (6 weeks of growth in the growth chambers).

2.3. MEASUREMENT OF PHOTOSYNTHESIS AND TRANSPIRATION

Photosynthesis and transpiration were measured on plant leaves in the CO_2-growth chambers at 315 μEinstein.m^{-2}.s^{-1} using a portable ADC/LCA instrument (Analytical Development Co, Hoddesdon, Herts, U.K.) equipped with an Epson datalogger and Epson HX20 microprocessor. The flow through the ADC Assimilation unit was 325 cm^3.min^{-1}.

The total waterpotential Ψ_T, was estimated at harvest using a Scholander's Pressure bomb and leaf area was measured with a Licor-3000 leaf area meter.

Statistical analysis was carried out according to procedures given in SOKAL & ROHLF (1981).

3. RESULTS

There is a significant increase of the mean relative growth rate of *Spergularia maritima* under conditions of CO_2 enrichment both at low and increased salinity. Under saline (250 mM NaCl), anaerobic conditions there is no increase of the relative growth rate of *Spergularia* at elevated CO_2 (Table 1). The increase of the mean relative growth rate of *Spergularia* is somewhat less at 250 mM NaCl compared to low salinity (10 mM NaCl) of the nutrient solution. There is no significant interaction of the effect of CO_2 enrichment with the salinity treatment. Anaerobic conditions in the culture solution inhibit the mean relative growth rate of *Spergularia* and even more pronounced in *Aster tripolium*. Also, the mean relative growth rate of *Aster tripolium* is markedly reduced with increased salinity. There is an increase of the mean relative growth rate of *Aster* with elevated CO_2, particularly under conditions of low salinity (10 mM NaCl). Photosynthesis and transpiration measurements conducted with increased salinity in *Aster tripolium* (Table 3) and increased leaf temperature and lower calculated Ci-values (internal CO_2-concentrations) compared to *Spergularia* (Table 4). In *Spergularia*, the photosynthetic rate is increased with elevated CO_2 and transpiration rate reduced. Increased salinity does not reduce photosynthesis and transpiration rate such as in

TABLE 1

Relative growth rate (in mg.g.$^{-1}$.day^{-1}, dry weight basis) of *Spergularia maritima* and *Aster tripolium* with varied salinity and aeration at ambient (340 vpm) and elevated (580 vpm) carbon dioxide. Average values of 4 replicates with SD.

Treatment			Plant species	
CO_2	salinity mM	aeration	*Spergularia maritima*	*Aster tripolium*
340	10	aerated	84.2 ± 8.1	71.9 ± 8.2
		anaerobic	71.9 ± 23.1	61.1 ± 15.2
	250	aerated	74.1 ± 19.7	60.2 ± 18.4
		anaerobic	84.7 ± 3.1	46.2 ± 5.9
580	10	aerated	96.7 ± 11.3	87.5 ± 10.5
		anaerobic	91.1 ± 26.3	66.7 ± 10.9
	250	aerated	90.7 ± 15.2	57.1 ± 8.2
		anaerobic	81.8 ± 9.7	54.9 ± 6.7

TABLE 2

Percentage increase or decrease (-) of the relative growth rate due to CO_2 enrichment, based on the ratio of RGR elevated and RGR ambient for the salinity and aeration treatments (a), RGR 250 mM NaCl/RGR 10mM NaCl (b) or RGR O_2/RGR N_2 (c).

Treatment		Plant species	
salinity	aeration	*Spergularia maritima*	*Aster tripolium*
a. Effect CO_2 enrichment 580/340 ppm CO_2			
10	aerated	14.9	21.7
	anaerobic	26.7	9.2
250	aerated	22.4	-5.2
	anaerobic	-3.4	18.8
b. Effect increased salinity 250/10 mM NaCl			
CO_2	aeration		
340	aerated	-12.0	-6.3
	anaerobic	17.8	-24.4
580	aerated	-6.3	-35.7
	anaerobic	-10.7	-17.7
c. Effect aeration O_2/N_2			
CO_2	salinity		
340	10	17.1	17.5
	250	-12.6	30.3
580	10	6.1	31.2
	250	10.9	4.0

Aster. The water use efficiency (calculated as the ratio of the rate of photosynthesis and transpiration) increased with elevated CO_2 in *Spergularia* (Table 3). In *Aster tripolium*, the water use efficiency is enhanced with the raise of salinity. Elevated CO_2 increases the water use efficiency of *Aster tripolium* only under conditions of low salinity. Reduced transpiration and stomatal conductance in *Aster* due to increased salinity is associated with increased leaf temperature, in contrast to *Spergularia maritima*.

The total water potential (Ψ_T) of the shoot of *Spergularia maritima* and *Aster tripolium* becomes more negative with raised salinity, with no significant effect of aerated versus anaerobic conditions. With enriched CO_2, the total water potential tends to become less negative both under low (10 mM) and increased (250 mM NaCl) salinity (Table 5).

4. DISCUSSION

There is a continuous rise of the content of CO_2 in the atmosphere of the earth, starting from 270-280 ppm in the pre-industrial times to 345 ppm now. The most precise and accurate data base is referring to Keeling is measurements at the Mauna Loa observatory at Hawai (KEELING *et al.*, 1982; STRAIN & CURE, 1985). The atmospheric CO_2 rise has been confirmed to continue to at least until November 1988 (B.R. Strain, Department of Energy, Washington, USA, pers. comm.).

The increasing content of atmospheric CO_2 generally increases the photosynthetic and growth rate of plants. In a survey of many crop plants tested,

TABLE 3

Photosynthetic rate (in μmol CO_2 $m^{-2}.s^{-1}$), transpiration rate (in mmol $H_2O{\cdot}m^{-2}.s^{-1}$) and water use efficiency (photosynthesis/transpiration) of *Aster tripolium* and *Spergularia maritima* at ambient and elevated CO_2 conditions, low and increased salinity, aerated and anaerobic conditions. Average values of 3 replications with standard error of the mean.

Treatment			*Spergularia maritima*			*Aster tripolium*		
CO2 ppm	salinity mM NaCl	aeration	photosynthetic rate	transpiration rate	water use efficiency	photosynthetic rate	transpiration rate	water use efficiency
340	10	aerated	2.2 ± 0.5	8.9 ± 3.0	0.25	6.3 ± 0.6	6.2 ± 0.1	1.05
		anaerobic	2.7 ± 2.2	4.5 ± 1.4	0.60	4.8 ± 1.5	6.3 ± 0.8	0.76
340	250	aerated	2.7 ± 0.7	5.1 ± 1.2	0.53	5.0 ± 0.8	3.3 ± 0.5	1.52
		anaerobic	1.9 ± 1.5	5.0 ± 0.5	0.38	3.3 ± 0.5	2.1 ± 0.4	1.57
580	10	aerated	3.0 ± 0.7	4.6 ± 1.5	0.65	6.3 ± 0.5	5.3 ± 0.4	1.19
		anaerobic	3.7 ± 0.7	4.4 ± 0.4	0.84	3.4 ± 0.1	5.4 ± 0.7	0.63
580	250	aerated	4.1 ± 0.9	4.6 ± 0.7	0.89	4.2 ± 0.1	3.2 ± 0.1	1.31
		anaerobic	4.1 ± 0.7	4.2 ± 0.4	0.98	4.8 ± 1.4	1.5 ± 0.4	3.20

TABLE 4

Stomatal conductance for water vapour (in mol.m^{-2}.s^{-1}), leaf temperature (in °C), and internal CO_2 concentration (in ppm) of *Aster tripolium* and *Spergularia maritima* grown at varied levels of CO_2, salinity and aeration. Average values with standard errors are given.

			Spergularia maritima			*Aster tripolium*		
CO_2 ppm	salinity	aeration	stomatal conductance	leaf temperature	internal CO_2 (ppm)	stomatal conductance	leaf temperature	internal CO_2 (ppm)
340	10	aerated	0.61 ± 0.29	18.9 ± 1.6	324 ± 5.7	0.95 ± 0.04	20.8 ± 0.1	306 ± 3.5
		anaerobic	0.24 ± 0.70	19.7 ± 0.1	326 ± 27.6	0.77 ± 0.19	20.9 ± 0.3	309 ± 12.0
340	250	aerated	0.24 ± 0.08	21.6 ± 1.2	316 ± 2.8	0.16 ± 0.02	22.3 ± 0.1	274 ± 1.0
		anaerobic	0.26 ± 0.03	20.1 ± 0.5	327 ± 8.5	0.09 ± 0.01	23.3 ± 0.1	263 ± 5.6
580	10	aerated	0.25 ± 0.10	19.9 ± 0.8	538 ± 8.6	0.52 ± 0.11	21.9 ± 0.1	527 ± 2.8
		anaerobic	0.23 ± 0.03	20.4 ± 1.0	535 ± 9.2	0.74 ± 0.26	20.1 ± 0.1	549 ± 2.7
580	250	aerated	0.27 ± 0.09	19.2 ± 0.3	535 ± 19.1	0.14 ± 0.01	23.7 ± 0.0	503 ± 1.0
		anaerobic	0.24 ± 0.02	18.0 ± 0.5	537 ± 8.5	0.07 ± 0.02	23.3 ± 0.1	435 ± 9.9

doubling the present atmosphere CO_2 concentration, appeared to increase the yield by about 30% on average (KIMBALL, 1983). Plants with the C_4 pathway of CO_2 fixation generally respond significantly less to CO_2 enrichment than do C_3 plants (CURTIS *et al.*, 1989). Many studies on effects of CO_2 enrichment refer to agricultural species and it is difficult to extrapolate from crop species to species of natural ecosystems (BAZZAZ *et al.*, 1985). Longterm studies on the effects of CO_2 enrichment are rare (OVERDIECK & FORSTREUTER, 1987). Also, the interference with environmental stress (nutrient deficiency, drought and salinity stress) makes it uncertain to predict the response of plant species from natural vegetation to increased atmospheric CO_2. The present report demonstrates that the response of two succulent halophytic plant species *Aster tripolium* and *Spergularia maritima* to CO_2 enrichment interacts with two important environmental factors of the salt marsh ecosystem *viz.* salinity and waterlogging (anaerobiosis) conditions. The relative growth rate of *Aster tripolium* and *Spergularia maritima* decreased with increased salinity, also the growth rate increased with CO_2 enrichment, but this was not dependent on the salinity level used in the present study. Anaerobiotic conditions simulated by flushing the nutrient solution with nitrogen, tended to reduce the growth rate of both species significantly, compared to the aerobic conditions. In another paper the growth reduction due to increased salinity was reduced to some extent by CO_2 enrichment for the halophytic C_3-grass species *Scirpus maritima* and *Puccinellia maritima* (ROZEMA *et al.*, 1989). Also, SCHWARZ & GALE (1984) found increased salt tolerance with increased carbon dioxide treatments. In the present study *Aster* and *Spergularia* appeared to be less sensitive to increased salinity than *Scirpus* and *Puccinellia maritima*. Increased carbon dioxide

TABLE 5

Water potential of the shoot Ψ_T (MPa) of *Spergularia maritima* and *Aster tripolium* under varied conditions of CO_2, salinity and aeration. Average value of 3 replicates with standard error.

Treatment			Plant species	
CO_2 ppm	Salinity mM NaCl	aeration	*Spergularia maritima*	*Aster tripolium*
340	10	aerated	-1.78 ± 0.03	-0.97 ± 0.05
		anaerobic	-1.65 ± 0.04	-0.79 ± 0.02
340	250	aerated	-2.37 ± 0.16	-2.39 ± 0.03
		anaerobic	-2.39 ± 0.03	-2.20 ± 0.03
580	10	aerated	-1.05 ± 0.03	-0.57 ± 0.04
		anaerobic	-0.99 ± 0.03	-0.76 ± 0.02
580	250	aerated	-2.03 ± 0.13	-2.04 ± 0.01
		anaerobic	-2.32 ± 0.08	-1.91 ± 0.04

led to a reduction of stomatal transpiration in both species, and water use efficiency of both species increased markedly as a result of this and of an increased photosynthetic rate in *Spergularia maritima*. The total water potential of the shoot tended to become less negative at increased salinity with CO_2 enrichment for both species. For *Scirpus maritima* improved water relations under CO_2 enriched conditions appeared to relate to increased leaf elongation and this might provide an explanation for the increase of the growth rate at elevated CO_2 of this grass species. Experiments on the interactive effect of CO_2 with salinity on plant growth are scarce and the knowledge of a physiological mechanism is incomplete. However, it is likely to expect that under conditions where waterstress is limiting growth of crop plant species or species from natural ecosystems, CO_2 enrichment may increase drought and salt tolerance (ROGERS *et al.*, 1984). Therefore, it is to be predicted that shifts may occur in the competitive balance between plant species in vegetation of dry and saline areas.

In a field study in a Chesapeake Bay salt marsh, using open top chambers (CURTIS *et al.*, 1989) a C_3-grass species *Scirpus olneyii,* gained significantly more above ground and below ground biomass in response to CO_2 enrichment. There was no such an effect on the C_4-grass species *Spartina patens* and *Distichlis spicata.* Based on these results, a shift in the competitive balance between C_3 and C_4 plant species in favour of C_3- species is to be expected.

In addition to drought salinity and anaerobiosis, other environmental factors may interact significantly with the response of plants to carbon dioxide enrichment. At low levels of the photosynthetically active radiation (PAR), the response of plants to elevated CO_2 may be absent. Since radiation levels in greenhouse may be lower than levels of PAR obtained in the open, the responsiveness of plants to CO_2 may well relate to the radiation level applied (SIONIT *et al.*, 1982). Also the response of plants to CO_2 enrichment is temperature dependent. Below a mean air temperature of 18.5°C there is a negative interaction with the plant response to carbon dioxide. Above 18.5°C the interaction effect is positive (IDSO *et al.*, 1987).

The combined effect of altering environmental conditions, indicated as global change or climatic change is difficult to predict. Atmospheric CO_2 enrichment will favour C_4 plant species, those that are salt and/or drought tolerant in particular. C_4 plants will respond better to global warming (EHLERINGER & BJÖRKMAN, 1977). Little is known about the response of wild plant species to increased UV-B radiation as a result of stratospheric ozone depletion. Based on a limited set of experiments and a number of references VAN DE STAAIJ *et al.* (1989) indicated that there is evidence that monocots are less sensitive to increased UV-B than dicot plant species. In further experiments of ROZEMA *et al.* (1990) there seems to be no interaction between the effect of CO_2 enrichment and enhanced UV-B radiation. This is not remarkable since UV-B radiation affects Photosystem-II, while CO_2 enrichment influences stomatal opening and closure. To our opinion there is a need for rapid extension of experiments relating to the global change (CO_2, UV-B, temperature rise) to be carried out, such as with open top chambers (CURTIS *et al.*, 1989) or with an free air CO_2 enrichment (FACE) approach, if this set up proves technically and financially feasible. If so, the FACE set up for outdoor CO_2 enrichment studies allows the analysis of the changes in the water and energy balance in a enriched vegetation as well as the measurement of biosphere-atmosphere exchange with micrometeorological methods (BALDOCCHI *et al.*, 1988).

ACKNOWLEDGEMENTS

The authors are indebted to D. Hoonhout for typing the manuscript. The research participation of G.M. Lenssen is financed by the Netherlands Organization for the Advancement of Pure Research (NWO).

5. REFERENCES

BAES, C.F., H.E. GOELLER, J.S. OLSON & R.M. ROTTY, 1977. Carbon dioxide and climate: the uncontrolled experiment.—American Scientist **65:** 310-320.

BALDOCCHI, D.D., B.B. HICKS & T.D. MEYERS, 1988. Measuring biosphere-atmosphere exchanges of biologically related gases with micrometeorological methods.—Ecology **69:** 1331-1340.

BAZZAZ, F.A., K. GARBUTT & W.E. WILLIAMS, 1985. Effect of increased atmospheric carbon dioxide concentration on plant communities. In: B.R. STRAIN & J.D. CURE. Direct effects of increasing carbon dioxide on vegetation. DOE/ER-238, US Dept. Energy, Washington: 155-170.

BROUNS, J.J.W.M., 1988. The impact of elevated carbon dioxide levels on marine and coastal ecosystems.—NIOZ - RIN-Rapport 1988-7 / 88-58: 1-101.

CURTIS, P.S., B.G. DRAKE & D.F. WHIGHAM, 1989. Nitrogen and carbon dynamics in C_3 and C_4 estuarine plants grown under elevated CO_2 *in situ*.—Oecologia (in press).

CURTIS, P.S., B.G. DRAKE, P.W. LEADLY, W.J. ARP & D.F. WHIGHAM, 1989. Growth and senescence in plant communities exposed to elevated CO_2 concentrations on an estuarine marsh.—Oecologia (in press).

DIGGELEN, J. VAN, 1988. A comparative study on the ecophysiology of salt marsh halophytes. Ph. Thesis. Vrije Universiteit Amsterdam: 1-208.

EHLERINGER, T. & O. BJÖRKMAN, 1977. Quantum yields for CO_2 uptake in C_3 and C_4 plants: dependence on temperature, CO_2 and O_2 concentration.—Plant Physiol. **59:** 86-90.

ENOCH, H.Z. & B.A. KIMBALL, 1985. CO_2 enrichment of greenhouse crops. CRC Press, Boca Raton: 1-230.

GAASTRA, P., 1959. Photosynthesis of crop plants as influenced by light, carbon dioxide, temperature and stomatal diffusion resistance.—Meded. Landbouwhogesch. Wageningen **59:** 1-68.

GATES, D.M., B.R. STRAIN & J.A. WEBER, 1983. Ecophysiological effects of changing atmospheric carbon dioxide concentrations. In: O.L. LANGE, P.S. NOBEL, C.B. OSMOND & H. ZIEGLER. Physiological Plant Ecology IV. Springer, New York: 503-526.

GOUDRIAAN, J. & H.E. DE RUITER, 1983. Plant growth in response to CO_2 enrichment at two levels of nitrogen and phosphorus supply. I. Dry matter, Leaf area and development.—Neth. J. Agricult. Sci. **31:** 157-169.

HEUKELS, H. & R. VAN DER MEYDEN, 1983. Flora Van Nederland. Wolters Noordhoff, Groningen.

IDSO, S.B., B.A. KIMBALL, M.G. ANDERSON & J.R. MANNEY, 1987. Effects of atmospheric CO_2 enrichment on plant growth: the interactive role of air temperature.—Agriculture, Ecosystems and Environment **20:** 1-10.

KEELING, C.D., R.B. BACASTOW & T.P. WORF, 1982. Measurements of the concentration of carbon dioxide at Mauna Loa Observatory, Hawai. In: W.C. CLARK. Carbon dioxide review. Oxford University Press, New York: 377-385.

KIMBALL, B.A., 1983. Carbon dioxide and agricultural yield: an analysis of 430 prior observations.—Agron. Journal **75:** 779-788.

OVERDIECK, D. & M. FORSTREUTER, 1987. Langzeit-Effekte eines erhöhten CO_2 Angebotes bei Rotklee-Wiesen Schwingel Gemeinschaften.—Ges. f. Oekol. **16:** 197-206.

PAEZ, A., H. HELMERS & B.R. STRAIN, 1983. CO_2 enrichment, drought stress and growth of Alaska pea plants *(Pisum sativum)*.—Physiol. Plant **58:** 161-165.

ROGERS, H.H., N. SIONIT, J.D. CURE, J.M. SMITH & G.E. BINGHAM, 1984. Influence of elevated carbon dioxide on water relations of soybeans.—Plant Physiol. **74:** 233-238.

ROZEMA, J., F. DOREL, R. JANISSEN, G.M. LENSSEN, R.A. BROEKMAN, W.J. ARP & B.G. DRAKE, 1989. Effect of elevated atmospheric CO_2 on growth, photosynthesis and water relations of salt marsh grass species.—Aq. Bot. (in press).

ROZEMA, J, G.M. LENSSEN, W.J. ARP & J.W.M. VAN DE STAAIJ, 1990. Global change, the impact of increased UV-B radiation and the greenhouse effect on terrestrial plants. In: J. ROZEMA & J.A.C VERKLEIJ. Ecological responses to environmental stresses. Junk Publishers, Dordrecht (in press).

SCHWARZ, M. & J. GALE, 1984. Growth response to salinity at high levels of carbon dioxide.—J. Exp. Bot. **35:** 193-196.

SIONIT, N., H. HELMERS & B.R. STRAIN, 1982. Interaction of atmospheric CO_2 enrichment and irradiance on plant growth.—Agronomy Journal **74:** 721-725.

SOKAL, R.R. & F.J. ROHLF, 1981. Biometry. The principles and practice of statistics in biological research 2nd edition. W.H. Freeman and Company, San Francisco: 1-776.

STRAIN, B.R. & J. CURE, 1985. Direct effects of increasing carbon dioxide on vegetation. US Dept. Energy, DOE: 1-286.

WONG, S.C., 1979. Elevated atmospheric partial pressure of CO_2 and plant growth. I. Interactions of nitrogen and photosynthetic capacity in C_3- and C_4-plants.—Oecologia **44:** 68-74.

THE GEOGRAPHIC DISTRIBUTION OF SEAWEED SPECIES IN RELATION TO TEMPERATURE: PRESENT AND PAST

C. VAN DEN HOEK, A.M. BREEMAN and W.T. STAM

Department of Marine Biology, Biological Centre, P.O. Box 14, 9750 AA, Haren, The Netherlands

ABSTRACT

Dumontia contorta **exemplifies a number of temperate species occurring in the N Pacific Ocean and the N Atlantic Ocean and which were probably exchanged after the opening of the Bering Seaway (3.5 MA ago); their temperature responses indicate that their passage required a 6-8°C higher summer temperature than the present one in the Arctic Ocean. Some species in the above group, for instance** ***Desmarestia viridis*****, have also an amphipolar distribution. Their passage across the tropics would have required a ca 5°C lowering of the equatorial sea surface temperature. This same temperature lowering can explain the extinction, along the east coasts of the Atlantic and Pacific Oceans, of tropical seaweeds such as** ***Dictyosphaeria cavernosa*** **with a W Atlantic-Indo W Pacific distribution. This lowering probably did not occur (sub)recently (for instance during Pleistocene glaciations), but earlier in the Cenozoic. We approached the question of the timing of the above temperature lowerings by estimating genetic divergence of disjunct populations of the same species or species complexes using the DNA-DNA hybridization method. Disjunct Atlantic and Pacific populations of several tropical and warm temperate species appeared to have genetically diverged in a high degree thus indicating a highly conservative morphology; this accords with their hypothetical divergence by the mid-Miocene (or earlier) closure of the Tethys Ocean. The divergence of amphipolar populations of the same species was probably more recent. Temperature changes of the sea surface predicted for the next century have a similar extent as those which have caused, in a more or less distant geologic past, profound alterations in the composition of the world's seaweed floras.**

1. SEAWEED SPECIES: THEIR GEOGRAPHIC DISTRIBUTION IN RELATION TO CLIMATIC FACTORS

The differentiation of seaweed floras on a global scale is, as one would expect, primarily related to the latitudinal climatic gradients (VAN DEN HOEK, 1984; JOOSTEN & VAN DEN HOEK, 1986). However, climatic factors alone cannot explain the global distribution of seaweed floras. If they could, one would expect the same instead of the actually different seaweed floras in the corresponding climatic zones along the Atlantic and Pacific shores of both hemispheres, and one would expect seaweeds to be capable of unimpeded dissemination (which they are not) (VAN DEN HOEK, 1987). Speculative historic scenarios, involving the early Cenozoic distribution of continents and oceans and the then much more even climate, can be invoked to explain the evolution of the present day seaweed floras (VAN DEN HOEK, 1984; JOOSTEN & VAN DEN HOEK, 1986; PRUD'HOMME VAN REINE, 1988; PRUD'HOMME VAN REINE & VAN DEN HOEK, 1988, 1989).

We have tried to translate the overall idea that the differentiation of seaweed floras is related to the latitudinal climatic gradients, into a set of experimentally testable hypotheses explaining the climatic boundaries of each of a large number of seaweed species belonging to diverse geographic distribution types. In other words, we asked each species why it does not spread beyond its presumptive climatic boundaries. We submitted each species to combinations of temperature and photoperiod imitating natural conditions throughout and beyond its geographic range.

Over 60 species have now been tested in this way in our and other laboratories (see BREEMAN, 1988, for a review). Almost all boundaries could be explained on the basis of the species' temperature responses, some of these in combination with photoperiod responses, and generalized explanations for the diverse distribution types begin to emerge (VAN DEN HOEK, 1982a, b; BREEMAN, 1988, this issue).

We shall here present 4 examples with world-wide disjunct distributions and which are representative for larger groups of species. These 4 species have been selected because their geographic distributions offer the possibility to formulate hypotheses about sea temperatures in the geologic past, a topic which is particularly relevant in the context of the present

J.J. Beukema et al. (eds.), Expected Effects of Climatic Change on Marine Coastal Ecosystems, 55–67.

conference on 'Expected effects of climatic changes on marine coastal ecosystems'.

Before proceeding to the examples, it is necessary to point out that there are two fundamentally different types of temperature boundaries, namely 1) *lethal boundaries* and 2) *growth and/or reproduction boundaries*. Lethal boundaries are set by the species' capacity to survive during the adverse season. Growth and/or reproduction boundaries are set by the species' ability to grow and reproduce during the favourable season.

In the following account, a species' geographic boundary will be positioned on either a winter isotherm of the sea surface (corresponding with the mean temperature of the the coldest month), or on a summer isotherm (corresponding with the mean temperature of the warmest month). It should be kept in mind that these isotherms are only indicative of the real temperature requirements of the species, which will deviate to a greater or lesser degree from the isotherm values depending on the local nature of the inshore sea.

1.1. *DUMONTIA CONTORTA* RUPRECHT

The red alga *Dumontia contorta* (order Cryptonemiales) has an alternation between isomorphic gametophyte and tetrasporophyte generations, which both consist of a perennial crustose phase bearing an annual erect branched tubular phase. The initiation of the erect plants from the crusts is only possible at short photoperiods (≤13 h light per day) and low temperatures (< 14-16°C) (RIETEMA, 1982; RIETEMA & BREEMAN, 1982; RIETEMA & VAN DEN HOEK, 1984). In the field, initiation of erect plants mainly takes place in autumn, and their growth and maturation mainly in the next spring (KLEIN, 1987; POT *et al.*, 1989). Both in the N Atlantic and the N Pacific *Dumontia contorta* is widely distributed in the cold temperate regions;in the N Atlantic it extends into the warm temperate region along the Iberian NW coast (Fig. 1).

In the North Atlantic Ocean, *Dumontia contorta* has a composite southern boundary, comparable to the composite southern boundaries of *Desmarestia*

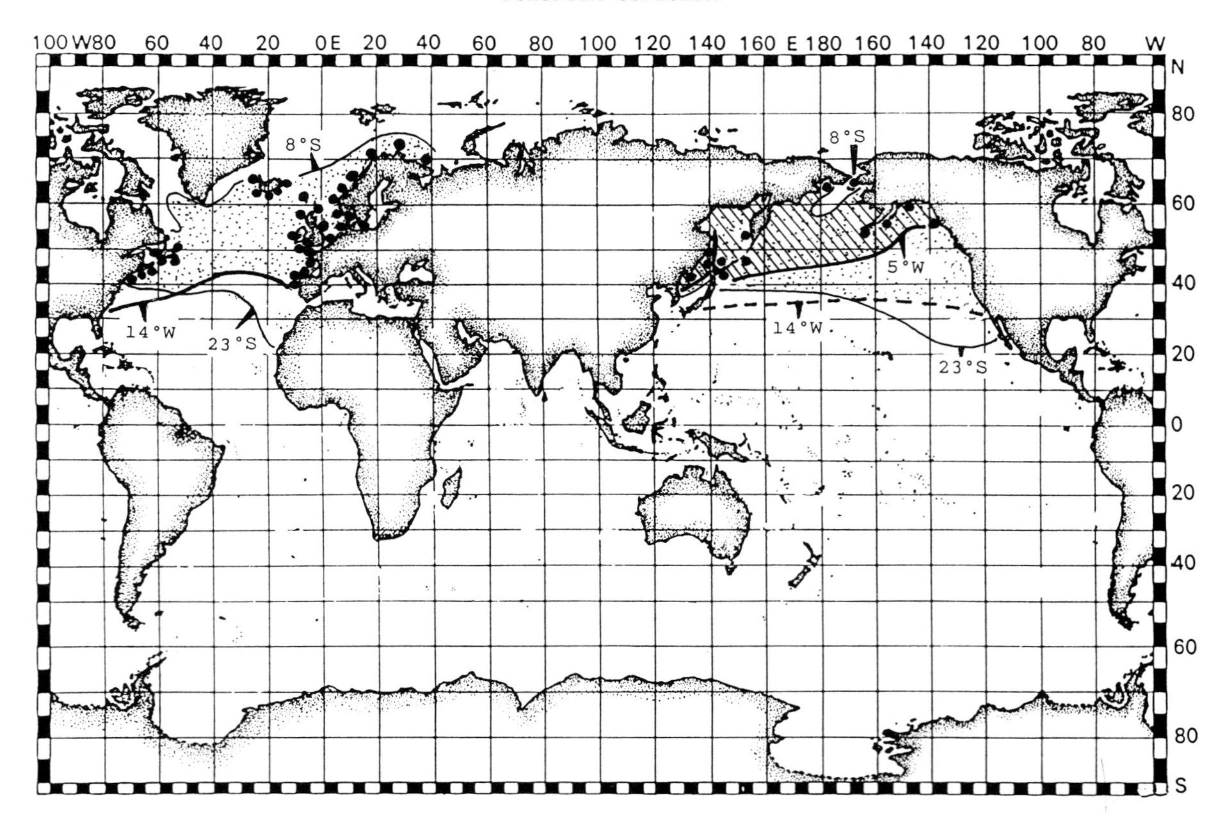

Fig. 1. Distribution of *Dumontia contorta* in relation to limiting temperatures. Thick line: limiting winter isotherm. Thin line: limiting summer isotherm. Stippled area: potential thermal area of N Atlantic material. Hatched area: (presumptive) potential thermal area of N Pacific material. Distribution as in VAN DEN HOEK (1982a) and LÜNING(1985) with additional data from LINDSTROM (1977) and SCAGEL *et al.* (1986).

viridis and other temperate N Atlantic species (compare Fig. 1 with Fig. 2). On the eastern coast of the N Atlantic Ocean, the southern boundary is situated close to the 14°C winter isotherm. This corresponds with a winter temperature in the 14-16°C interval which is just low enough to allow initiation of erect plants from crusts. As only erect plants are capable of reproduction, the southern boundary in the NE Atlantic is a growth/reproduction boundary.

On the western coast of the N Atlantic Ocean, the southern boundary is situated close to the 23°C summer isotherm, which corresponds with a lethal temperature in the 24-26°C interval. On both the east and west coast of the N Atlantic Ocean, the northern boundary of *Dumontia contorta* approximates the 8°C summer isotherm. This corresponds with a temperature, in the 4-8°C interval, which is just high enough during a sufficiently long period in early autumn (as soon as photoperiods are short enough) to allow the initiation of erect fronds from crusts. The northern boundary is consequently a growth/reproduction boundary (RIETEMA & VAN DEN HOEK, 1984).

Dumontia contorta also occurs in the N Pacific Ocean (Fig. 1, dots). In Fig. 1, the potential N Pacific distribution area within its Atlantic temperature boundaries (stippled area) does not fit the actual N Pacific distribution area, especially along the NE Pacific coast where one would expect the species to extend much farther in southward direction than it does (Fig. 1). Experimental data on the nature of the limiting temperatures for N Pacific *Dumontia contorta* are lacking, but its distribution suggests a southern boundary near the 5°C winter isotherm (instead of the 14°C winter isotherm). The N Pacific distribution does not disagree with the remaining two N Atlantic boundaries: a southern lethal boundary near the 23°C summer isotherm, and a northern growth/reproduction boundary near the 8°C summer isotherm. The thermal distribution area for the N Pacific *Dumontia contorta* is indicated in Fig. 1 by hatching.

DESMARESTIA VIRIDIS AND THE RELATED DESMARESTIA WILLEI

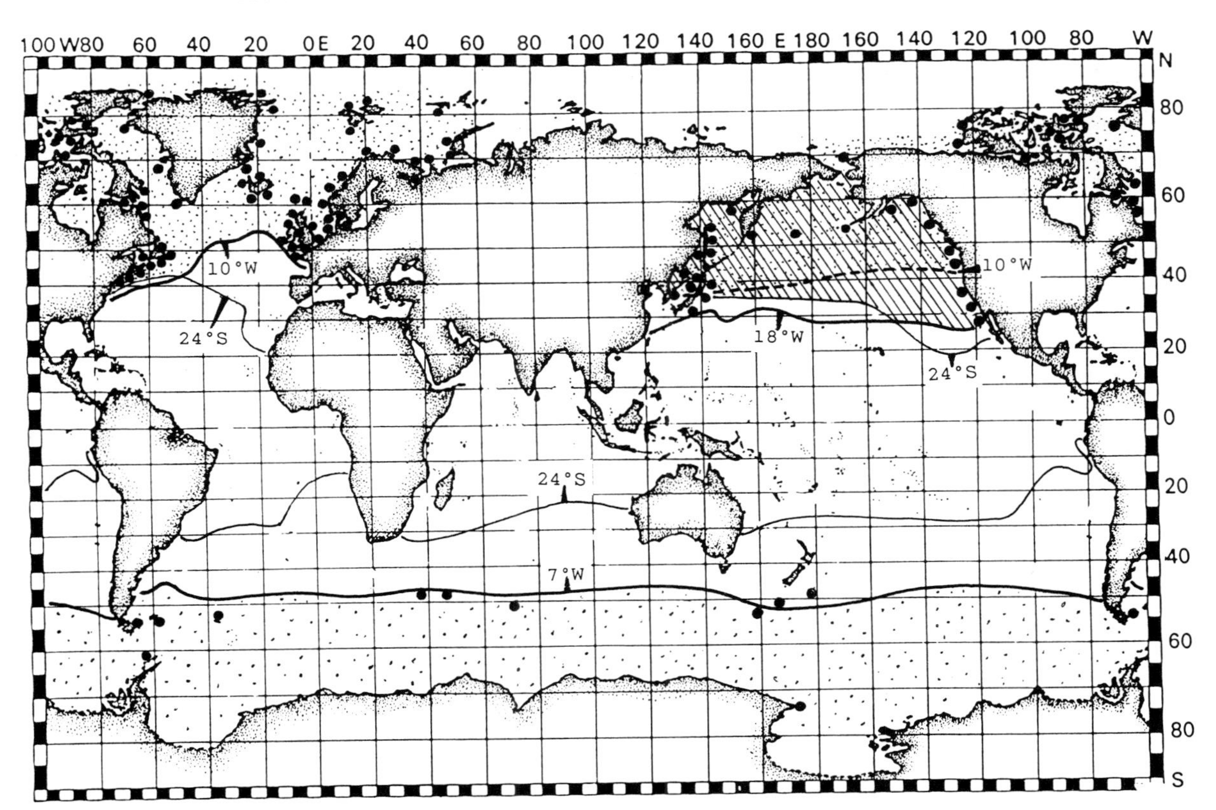

Fig. 2. Distribution of *Desmarestia viridis* and the related *D. willei* with respect to limiting temperatures. Stippled areas in the Northern Hemisphere: potential thermal area of N Atlantic material of *D. viridis*. Hatched area in the N Pacific: the thermal area of N. Pacific material of *D. viridis.* Stippled area in the Southern Hemisphere: (presumptive) thermal area of *D. willei*. Distribution according to HAY *et al.* (1985); VAN DEN HOEK (1982b); LÜNING (1985); RICKER (1987); SCAGEL *et al.* (1986); ZANEVELD (1966).

The considerable difference in latitudinal extent between the NE Atlantic and NE Pacific populations of the species suggest considerable genetic differences between both populations, and consequently a long isolation of the N Atlantic and N Pacific populations. During the period when *Dumontia contorta* migrated from the N Pacific to the N Atlantic or vice-versa (probably directly after the opening of the Bering Seaway, 3.5 MA ago, see MCKENNA, 1983), the temperature along the high Arctic coasts must have had the same values as presently at the species'northern growth/reproduction boundary, namely 8°C (average value) in summer (see above, RIETEMA & VAN DEN HOEK, 1984). This is 6°C higher than the present-day summer temperatures along large stretches of the Siberian Arctic coast and 8°C higher than present-day temperatures in the Canadian Arctic. In a similar way, Arctic summer temperatures in the range of 8-10°C would have been required for the passage of the N Pacific-N Atlantic temperate species *Gloiosiphonia capillaris* (MOROHOSHI & MASUDA, 1980) and *Nemalion helminthoides* (the latter species also occurs in the Southern Hemisphere; VAN DEN HOEK, 1982b). It should be pointed out here that the number of temperate *species* and even that of temperate-arctic species shared between the N Pacific and the N Atlantic are limited (LINDSTROM, 1987). The number of cold temperate to Arctic genera shared, however, is considerable (VAN DEN HOEK, 1984; LINDSTROM, 1987).

1.2. *DESMARESTIA VIRIDIS* (O. F. MÜLL.) LAMOUR.

The brown alga *Desmarestia viridis* (order Desmarestiales) is a common species in the 'Arctic-cold temperate North Atlantic flora' (JOOSTEN & VAN DEN HOEK, 1986); however, it is also widely distributed along the cold- and warm-temperate coasts on both sides of the N Pacific Ocean. In the N Atlantic its distribution is shared by many other species (VAN DEN HOEK, 1982a, b; BREEMAN, 1988; this issue). In its strongly heteromorphic life history, a macroscopic sporophyte alternates with a microscopic, monoecious oogamous gametophyte.

In the N Atlantic Ocean, *Desmarestia viridis* has a composite southern boundary (Fig. 2), comparable to the composite southern boundary of *Dumontia contorta* (see above, and Fig. 1). On the east coast of the Atlantic Ocean, the boundary is situated near the 10°C winter isotherm, which corresponds with a temperature in the 5-10°C interval which is just low enough to allow fertility of the gametophyte. On the west coast of the Atlantic Ocean, the southern boundary is situated close to the 24°C summer isotherm, which corresponds with a lethal temperature in the 23-25°C interval LÜNING, 1980, 1984; BREEMAN, 1988).

The potential distribution in the N Pacific between the Atlantic temperature boundaries (Fig. 2, stippled area in the N Pacific) does not agree with the real distribution there. This is particularly the case along the NE Pacific coast where one would expect *Desmarestia viridis* to extend much less far to the south than it actually does. Recent experimental evidence with Japanese *Desmarestia viridis* indicates that gametophytes can still become fertile at a temperature as high as 18°C (NAKAHARA, 1984), instead of 5-10°C in the N Atlantic strains. The lethal temperature of NE Pacific material was identical to that of N Atlantic material (in the 23-25°C interval; LÜNING & FRESHWATER, 1988). The actual distribution area of *Desmarestia viridis* in the N Pacific seems to have a composite southern boundary which is situated along the east coast near the 18°C winter isotherm (a reproduction boundary) and along the west coast near the 24°C summer isotherm (a lethal boundary) (Fig. 2, hatched area in the N Pacific).

The striking difference in the temperature responses of the N Atlantic and N Pacific populations suggest that these populations have been genetically isolated for a rather long period of time. This is unexpected, as both populations are connected by records in the Canadian Arctic (Fig. 2). The temperature responses of the N Pacific and N Atlantic populations of *Desmarestia viridis* have apparently much diverged, albeit within similar lethal temperature limits. 'Ecotypic' variation in temperature responses, when present, mostly pertains to growth and reproduction boundaries, not to lethal bounderies (BREEMAN, 1988).

Desmarestia willei Reinsch is a species from the cold-temperate zone of the Southern Hemisphere (Fig. 2). It is difficult, if not impossible, to differentiate this species from *Desmarestia viridis* (RICKER, 1987). No experimental data are available on its temperature limits. Its northern boundary is situated close to both the 10°C summer isotherm and the 7°C winter isotherm. In Fig. 2, the northern boundary is given as an hypothetical northern reproduction boundary along the 7°C winter isotherm. The 24°C summer isotherm is given as an hypothetical northern lethal boundary. Apparently, *D. willei* is limited towards the equator by lower limiting temperatures than the Northern Hemisphere populations of *D. viridis*.

To explain the amphipolar distribution of certain temperate species such as *Desmarestia viridis* we have invoked in earlier studies (VAN DEN HOEK, 1982a, b) Pleistocene equatorial temperature lowerings of 5°C or more to explain passage across the tropics, especially along the Pacific and Atlantic east coasts where the tropical belts are relatively narrow. In view of recent reconstructions of the sea surface temperatures during the last ice age (CLIMAP, MCIN-

TYRE *et al.*, 1976) a pleistocene temperature decrease even of 5°C in the tropics would seem however improbable. Moreover *D. willei* is limited towards the equator by lower limiting temperatures than the Northern Hemisphere populations of *D. viridis*. The distinct divergence in the temperature responses between the two Northern Hemispere and the Southern Hemisphere populations of '*Desmarestia viridis* in a broader sense' suggests a long genetic isolation between these populations. A recent or subrecent, Pleistocene passage of *Desmarestia viridis* over the equator involving an equatorial temperature lowering of at least 5°C seems therefore unlikely. A passage in a more distant geologic past is more probable.

1.3. *DICTYOSPHAERIA CAVERNOSA* (FORSSK.) BØRG.

The green alga *Dictyosphaeria cavernosa* (order Cladophorales) is representative of the 'disjunct tropical-to-warm-temperate flora' (JOOSTEN & VAN DEN HOEK, 1986). The species forms irregular hollow spheres composed of bubble-like multinucleate cells. The species is widely distributed along the tropical west coasts of the Atlantic and Pacific Oceans, as well as along the coasts of the Indian Ocean. It is lacking on the tropical east coasts of the Atlantic and Pacific Oceans. *Dictyosphaeria cavernosa* shares its highly disjunct W Atlantic- Indo-W Pacific distribution with a considerable number of other tropical species (VAN DEN HOEK, 1982a).

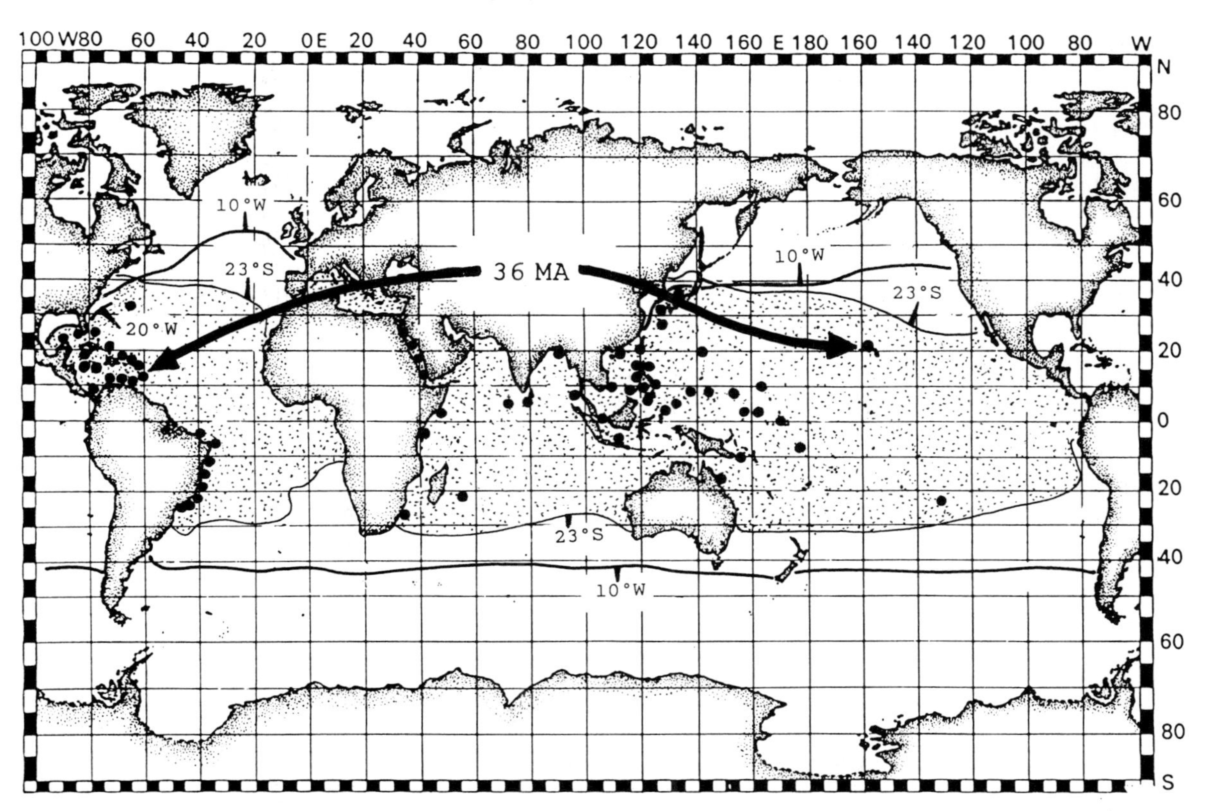

Fig. 3. Distribution of *Dictyosphaeria cavernosa*. Stippled area: area which is thermally suited to *Dictyosphaeria cavernosa*. Arrow gives the intermediate divergence time in million years between widely disjunct populations as tentatively estimated on the basis of DNA-DNA hybridization (see Table 1). See also text to Fig. 1. Distribution according to AKATSUKA 61973?; BØRGESEN (1940); CHIANG, YOUNG-MENG (1973); EGEROD (1952, 1974); FUNAHASHI (1973); HACKETT (1969); ISLAM (1976); JAASUND (1976); KANG JAE WON (1966); LAWSON (1980); DE OLIVEIRA (1977); PAPENFUSS (1968); PHAM-HOANG-HO (1969); PRICE *et al.* (1976); SCHNETTER (1978); SEAGRIEF (1984); SILVA *et al.* (1987); TAYLOR (1960); TRONO (1968); TSENG (1983); TSUDA (1976); TSUDA & TOBIAS (1977); TSUDA & WRAY (1977); WEBER-VAN BOSSE (1928); WEI, TEO LEE & CHIN, WEE YEOW (1983); WOMERSLEY & BAILEY (1970).

The species has a life history with an alternation between isomorphic sporophyte and gametophyte generations. The sporophytes produce quadriflagellate zoospores, the gametophytes biflagellate gametes. In Japanese material, the lower temperature limit for the production of these swarmers is in the 18-23°C temperature interval (ENOMOTO & OKUDA, 1981). This could correspond with a reproduction boundary near the 23°C summer isotherm. No experimental data are available on lethal temperatures, but at its northernmost point in Japan the species is capable of surviving low temperatures corresponding with a mean February temperature of 10°C (SANO *et al.*, 1981). This suggests the 10°C winter isotherms as the lethal boundary.

Both in the NW Pacific and the NW Atlantic the northern boundaries are apparently lethal boundaries. In Japan this boundary is situated close to the 10°C winter isotherm; in Florida close to the 20°C winter isotherm of the open ocean, which corresponds with intermittent cold spells lower than 10°C of the inshore waters (*cf.* CAMBRIDGE *et al.* , 1987). The enormous difference between a mean winter temperature near 20°C and an extreme low temperature near 10°C in southern Florida is due to infrequent cold spells caused by polar continental air masses intruding into this area (WALKER *et al.*, 1982). These cold spells can cause spectacular coral kills and massive fish mortality. They are responsible for an abrupt northward termination of the tropical marine biota in southern inshore Florida. On deep offshore reefs which are not subjected to these cold spells, tropical seaweeds and even some corals penetrate as far north as North Carolina close to where winter temperatures are as low as 10°C (MACINTYRE & PILKEY, 1969; SEARLES, 1984). On the southwest coasts of the Atlantic and Indian Oceans the southern boundaries are apparently reproduction boundaries.

The potential thermal distribution area of *Dictyosphaeria cavernosa* (stippled area in Fig. 3) includes large portions of the Pacific and Atlantic east coasts, including the warm temperate coasts of the Mediterranean. Clearly, the species is not lacking on these coasts because of present-day adverse temperature conditions. A plausible explanation would be that, after the break up of the Tethys Ocean (~ 20 MA ago), populations of *Dictyosphaeria cavernosa* initially persisted on both the east and west coasts of the Atlantic and Pacific Oceans; and that the east coast populations became later extinct by adverse temperature conditions.

It is attractive to invoke the same equatorial temperature lowering used to explain the passage of the equator by amphipolar cold water species such as *Desmarestia viridis*, as the cause of the extinction of *Dictyosphaeria cavernosa*. Extinction of the latter species would have required a temperature lowering of about 5°C (from 28-23°C) down to a temperature where reproduction is no longer possible. This is about the same temperature lowering required for a precarious passage of the tropics by cold water species such as *Desmarestia viridis*. We would like to stress that this scenario implies that the average sea surface temperature of the warmest month at the equator is 23°C (as present day summer temperature of the sea surface near the Canary Islands, *cf.* AFONSO-CARRILLO & GIL-RODRIGUEZ, 1982). Intermittent temperatures as low as 20°C occur at present along eastern equatorial coasts of both the Pacific and Atlantic Oceans, and are caused by local upwelling (LAWSON & JOHN, 1987; GLYNN & STEWART, 1973; LUBCHENKO *et al.*, 1984). These periodic low temperatures could neither cause the extinction of *Dictyosphaeria cavernosa* as the species could sufficiently reproduce during the rest of the year, nor could it allow dispersal of *Desmarestia viridis* across the equator, as transport of plants or disseminules by currents would require a long period of time (in the order of a half to one year) in the course of which temperatures would have reached lethal high values (VAN DEN HOEK, 1987).

We have argued above (section on *Desmarestia viridis*) that available paleoclimatological evidence does not support the idea that the required temperature lowering of tropical waters took place in the late Pleistocene, but probably much earlier. In a recent paper we invoked the Miocene reduction in the complexity and length of the E Atlantic tropical coasts and the concomitant steep cooling trend as the combined causes of the extinction of numerous tropical Tethyan seaweed species along E Atlantic tropical shores (JOOSTEN & VAN DEN HOEK, 1986; PRUD'HOMME VAN REINE & VAN DEN HOEK, 1988). A similar combination of factors may have operated along E Pacific tropical shores .

1.4. *CENTROCERAS CLAVULATUM* (C. AG.) MONT.

The filamentous red alga *Centroceras clavulatum* (order Ceramiales) is widely distributed throughout the tropics and in adjacent warm temperate regions. In contrast to *Dictyosphaeria cavernosa*, the species is also widely distributed along the Atlantic and Pacific east coasts (Fig. 4).

Temperature responses have been investigated in W Atlantic material (BIEBL, 1962; EDWARDS, 1969; VAN DEN HOEK, 1982a). A low lethal temperature lies in the 5-10°C interval (EDWARDS, 1969; BIEBL, 1962), suggesting a lethal boundary near the 10°C winter isotherm (in Florida near the 20°C winter isotherm, see the section on *Dictyosphaeria cavernosa* for an explanation). A low temperature permitting sufficient growth lies in the 15-20°C interval (EDWARDS, 1969), suggesting a growth boundary along the 20°C summer isotherm.

The potential thermal distribution areas of *Dictyosphaeria cavernosa* and *Centrocera clavulatum* (stippled areas in Figs 3 and 4) are quite similar and largely overlap. The differences at a first glance are only marginal: because *C. clavulatum* has a 3-8°C lower limiting temperature for growth/reproduction than *D. cavernosa*, it extends farther towards the poles. However, this difference gives also the most likely explanation for the extinction of *D. cavernosa* and the survival of *C. clavulatum* along the tropical east coasts of the Atlantic and Pacific Oceans. In the geologic past, the sea surface temperatures along the equatorial east coasts of both oceans were probably lowered down to values below the reproduction limit (~23°C) of *D. cavernosa*, but remaining above the growth limit (20-15°C) of *C. clavulatum* . It is likely that *D. cavernosa* is representative for the group of tropical species with a W Atlantic- Indo-W Pacific distribution, while *C. clavulatum* is representative for the large group of species with a pantropical to warm temperate cosmopolitan distribution.

Thus, by testing the temperature responses of representatives of both groups, and those of amphipolar cold water species, it is possible to estimate the degree of temperature lowering in the geologic past along the eastern equatorial coasts of the Atlantic and Pacific Oceans.

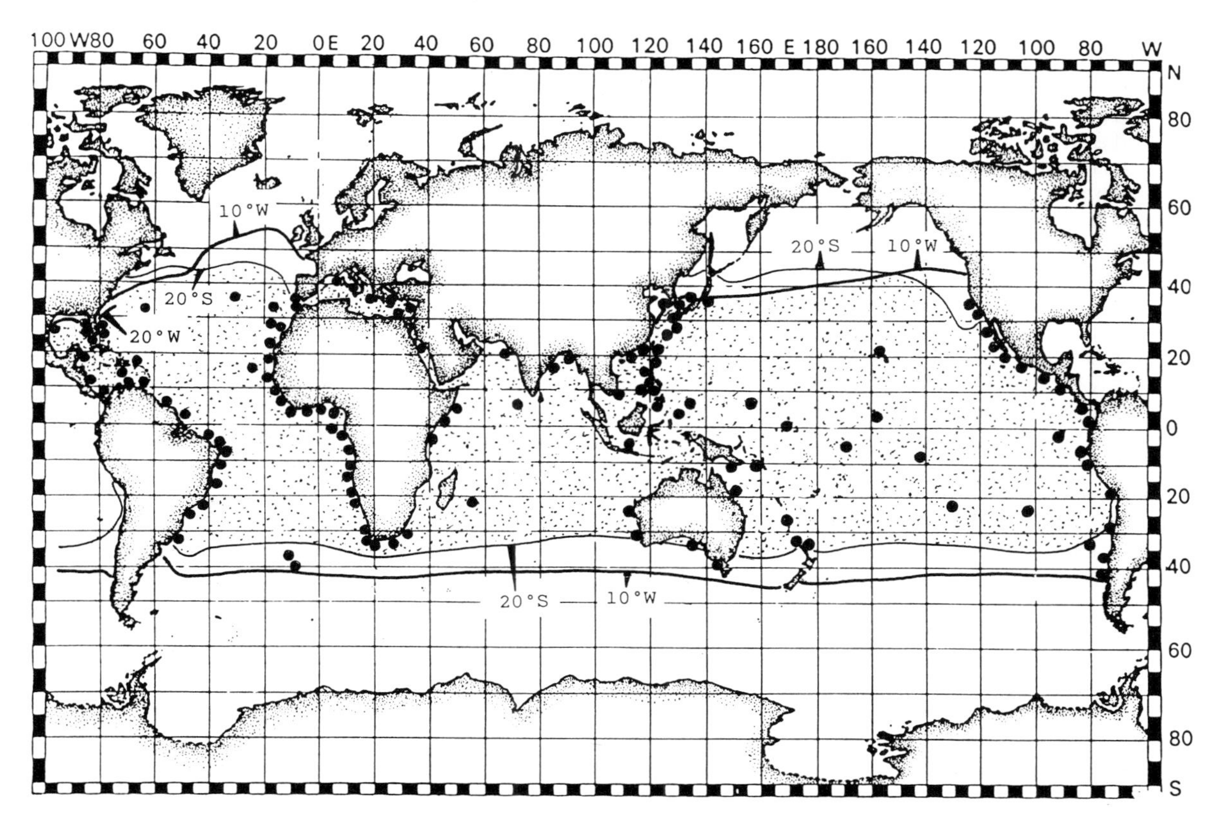

Fig. 4. Distribution of *Centroceras clavulatum* . Stippled area: area which is thermally suited to the species. See also text to Fig. 1. Distribution according to ABBOTT & HOLLENBERG (1976); ABBOTT & SANTELICES (1985); ADAMS & NELSON (1985); AKATSUKA (1973); ALEEM (1978); BAARDSETH (1941); BOLTON & STEGENGA (1987); BOO, SUNG MIN & LEE, IN KYU (1985); BØRGESEN (1934, 1953); CHAMBERLAIN (1965); CHAMBERLAIN *et al.* (1985); CHAPMAN (1977); CHIANG (1973); CRIBB (1983); DAWSON *et al.* (1964); DELLOW (1955); EARLE (1968); ETCHEVERRY (1986); FUNAHASHI (1967); HACKETT (1969); HANISAK & BLAIR (1988); HEYS (1985); VAN DEN HOEK (1982a, and unpublished records from Shark Bay and Rottnest Island, W. Australia); ISLAM (1976); ITONO (1977); JAASUND (1976); KANG JAE WON (1966); KING *et al.* (1971); LAWSON (1980); LAWSON & JOHN (1987); LAWSON *et al.* (1975); LEVRING (1960) MAGRUDER & HUNT (1979); NAKANIWA (1975); DE OLIVEIRA, (1977) PAPENFUSS (1968); PHAM-HOANG-HO (1969); PRICE *et al.* (1986); SCHNETTER & MEYER (1982); SCOTT & RUSS (1987); SEAGRIEF (1984); SETCHELL (1926); SHEPHERD & WOMERSLEY (1971, 1981); STEGENGA (1986); TAYLOR (1945); TRONO, (1968); TSENG (1983); TSENG *et al.* (1982); TSUDA (1976); WEBER-VAN BOSSE (1928); WOMERSLEY & BAILEY (1970); WYNNE (1986).

2. ESTIMATES OF DIVERGENCE TIMES USING THE DNA-DNA HYBRIDIZATION METHOD

We have argued above that, in view of the available paleoclimatological evidence, temperature lowerings along the equatorial W African and W American coasts during the Pleistocene glaciations were probably not large enough to explain the extinction of *Dictyosphaeria cavernosa* and the passage of *Desmarestia viridis*. Also the considerable differences in the temperature responses of populations of the same species in different oceans and hemispheres suggest a relatively long isolation between these populations, and consequently an equatorial temperature lowering in a comparatively distant past.

A crucial question is, of course: how distant is this 'comparatively distant past'. In other words: is it possible to be more specific about the time factor. We tackled this question by estimating genetic divergence between disjunct populations of the same species or species complexes, using the DNA-DNA hybridization method.

According to the molecular clock hypothesis the rate of molecular evolution is approximately constant over time, and is roughly independent on phenotypic evolution. Thus, by determining the degree of genetic divergence between two related organisms, it should in principle be possible to estimate when they diverged (AYALA, 1986). We determined the degree of DNA sequence homology by means of the technique of DNA-DNA hybridization (OLSEN *et al.*, 1987; BOT *et al.*, 1989a, b; STAM *et al.*, 1988). After extraction, the long DNA strands are sheared into fragments approximately 450 base pairs in length. Only the single copy DNA is included in the final procedure (the repeated sequences are included for special purposes). A small quantity of single-stranded radioactive DNA (tracer DNA) from one geographic population of a species is combined with a much larger amount of unlabeled DNA (driver DNA) from another geographic population of the same morphological species, or from another related species, under conditions promoting reassociation into double-stranded hybrid DNA (heteroduplex). The same tracer is also combined with driver from the same material (homoduplex). The reassociation mixture including the hybrid DNA, is loaded on a hydroxyapatite column that binds the double- stranded, but not the single-stranded DNA. The column is subjected to a temperature increase from 60-95°C in steps of 5°C. At each higher temperature step more double stranded DNA will dissociate and the resulting radioactive single stranded DNA will be washed from the column and then counted. In this way the amount of dissociated hetero- or homoduplex material is determined.

Heteroduplexes can be expected to be less stable than homoduplexes, because they are composed of DNA strands that have more or less diverged, so that there is more mismatch between the base pairs. Therefore they dissociate at lower temperatures than homoduplexes. The temperature at which half of the hybrid DNA has dissociated ($T_{m(e)}$) is used for comparing the thermostability of the heteroduplex with that of the homoduplex. The difference between the $T_{m(e)}$ of the heteroduplex and the $T_{m(e)}$ of the homoduplex - the $\delta T_{m(e)}$ -, is a measure of the DNA sequence divergence between the two geographic populations compared.

Estimates of divergence times based on DNA-DNA hybridization can at best be tentative for a variety of reasons (for discussions on their application to seaweeds, see OLSON *et al.*, 1987; STAM *et al.*, 1988; BOT *et al.*, 1989a, b). With this restriction, we estimated divergence time using the following formula: divergence time= $\delta\sigma=_{m(e)}/2$ x 1/average rate of sequence change as % per Ma. We approached the rate of sequence change in two ways. First we used a rate of sequence change of 0.09 % per Ma which was determined as an average of such rates for four different groups of organisms on the basis of a calibration with the fossil record (STAM *et al.*, 1988). Second, we assumed that the divergence time between Atlantic and Pacific populations of *Cladophora albida* is the postulated time of the Miocene closure of the Tethys Ocean in the Middle East (14-20 Ma ago) (RÖGL & STEININGER, 1984). This yielded an average rate of sequence change per Ma of 0.14 - 0.21 % , which seemed to be realistic in comparison to estimated rates of change in the literature of 0.13% (in higher primates) to 0.66% (in Drosophila). In Table 1 we will present $\delta T_{m(e)}$ values of a few examples, and their corresponding divergence time estimates based on the above three values of average rate of basepair change per Ma (0.09, 0.14, and 0,21%).

Our tentative divergence time estimates suggest that in species with a disjunct tropical to warm temperate Pacific-Atlantic distribution the Atlantic and Pacific populations diverged in a comparatively distant past and are not the result of recent dispersal events. Thus, we estimated for *Dictyosphaeria cavernosa*, with a genetic distance between a W Atlantic and a Pacific sample as high as $\delta T_{m(e)}$ =10°C, a divergence time between 24-55 Ma (Early Miocene to Eocene). The divergence time can also be interpreted as a minimum age of the morphological species *Dictyosphaeria cavernosa*, thus indicating an incredibly high morphological conservatism (species ages of flowering plants and mammals are in the order of a few million years or even less than one million years). Long divergence times are also suggested for the *Cladophora pellucida* complex (BOT *et al.*, in prep.) and *Cladophora albida* (Huds.)

Kütz. (Fig. 5). Our interpretation of the above high $\delta T_{m(e)}$ values is that these species had originally a continuous distribution along the coasts of the Tethys Ocean and that N Atlantic and W Pacific populations became finally separated by the Miocene closure of the Tethys Ocean in the Middle East.

In *C. albida* (Fig. 5) NW Pacific (Japan) and SW Pacific (southern Australian) populations diverged much more recently, possibly in Miocene or Pliocene (4-10 Ma). This divergence is difficult to relate to a vicariant event; various possibilities can be invoked (BOT *et al.*, 1989b), but will not be treated here. In the N Atlantic, the NW population (NE America) and the NE population (Europe) possibly diverged in Pliocene (2-4 Ma). We suggest that this divergence reflects the Pliocene intensification of a cold thermal barrier to transoceanic dispersal in the cooling N Atlantic Ocean. NW Atlantic, Icelandic and NE Atlantic plants of the essentially cold temperate species *Cladophora sericea* have *not* genetically diverged (Table 1) (BOT *et al.*, 1989a). This, we think, reflects (sub)recent dispersal of this cold water species across the N Atlantic Ocean, and which was possibly promoted by the reinvasion of coasts uncovered by the retreating Pleistocene glaciers.

In *Cladophora albida* and *Dictyosphaeria caver-*

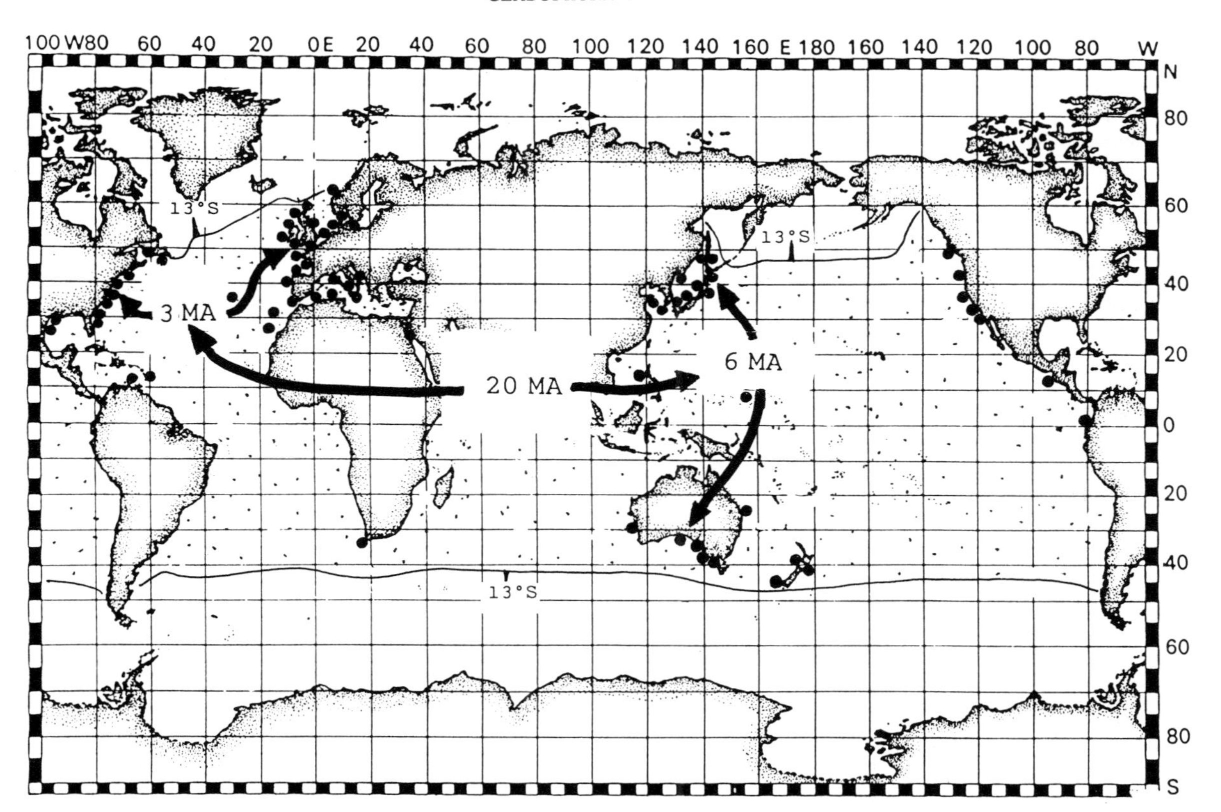

Fig. 5. Distribution of *Cladophora albida* in relation to its thermal boundaries. Stippled area: the potential thermal distribution area. Arrows give intermediate values of tentative divergence times between widely distant populations in millions of years based on DNA-DNA hybridization (see Table 1). *Cladophora albida* is a cosmopolitan warm to cold temperate species with rare records from the tropics. The absence of records from several portions of the world's temperate sea coasts probably reflects lack of local taxonomic investigations on the difficult genus *Cladophora*. In the N Atlantic Ocean, the northern boundary on both sides of the ocean is a growth boundary close to the the 13°C winter isotherm, which corresponds to a growth limit in the 10-15°C interval. The species is capable to grow and survive at tropical temperatures (CAMBRIDGE *et al.*, 1984; BREEMAN, 1988). The temperature responses of *C. albida* from southern Australia are similar to those in the N Atlantic (Cambridge, unpublished results). Distribution according to ABBOTT & HOLLENBERG (1976); CAMBRIDGE *et al.* (1984); CHAPMAN (1956); VAN DEN HOEK (1963, 1982c, and material seen from South Africa); VAN DEN HOEK & WOMERSLEY (1984); KANG, JAE WON (1966); LAWSON *et al.* (1975); OGAWA & MACHIDA (1976); PAPENFUSS (1968); PERESTENKO (1980); SAKAI (1964); SCAGEL (1966); SCHNETTER & MEYER (1982); SEAGRIEF (1984); SILVA *et al.* (1987); TSENG & CHANG (1964); TSUDA & WRAY (1977); VINOGRADOVA (1979).

TABLE 1

Genetic divergence in 3 species between populations from different areas.

Species	*Populations from*	$\delta Tm(e)$ (°C)	*Estimated divergence time (Ma ago)*		
Dictyosphaeria cavernosa	Hawaii and St Croix (Caribbean)	10	24	36	55
Cladophora albida	Rottnest (SW Australia) and Roscoff (W France)	5.7	14	20	32
Cladophora albida	Hokkaido (Japan) and Rottnest (SW Australia)	1.8	4	6	10
Cladophora albida	Connecticut (USA) and Roscoff (W France)	0.8	2	3	4
Cladophora sericea	Labrador (Canada) and Roscoff (W France)	0.1	0		

nosa, populations inhabiting different oceans and/or distant coasts of the same ocean have genetically diverged to a higher or lower degree, but without morphological change and probably also without change of their thermal responses (for thermal responses in *C. albida*, see CAMBRIDGE *et al.*, 1984; BREEMAN, 1988; Cambridge, unpublished results). In other species, for instance *Desmarestia viridis* and *Dumontia contorta* this genetic divergence is parallelled by divergence in thermal responses, and by minute morphological divergence.

3. SOME CONCLUSIONS

One of the purposes of this conference is to formulate plausible predictions about possible temperature changes of the sea surface in the next century and their impact on sealife. The folllowing starting points for the discussions are presented in DE VOOYS (1988;this issue) contribution: a rise in mean summer temperatures of +2 to +4°C, and a change of mean winter temperatures of +2 to -2°C.

On the face of the evidence presented in this contribution, it would seem that temperature changes of an almost similar extent have caused, in a more or less distant geologic past, profound alterations in the composition of the world's seaweed floras.

Thus, passage of cold temperate species like *Dumontia contorta* from the N Pacific to the N Atlantic after the Pliocene opening of the Bering Landbridge (3.5 Ma) would require 6-8°C higher average summer temperatures in the Arctic than the present ones. The apparent genetic divergence in thermal responses between Pacific and Atlantic populations of the same morphological species point to an early and not a recent passage (this paper). This is also suggested by the fact that many cold temperate genera are shared by the N Pacific and N Atlantic Ocean, but that they are represented in both oceans by different species (VAN DEN HOEK, 1984; JOOSTEN & VAN DEN HOEK, 1986; LINDSTROM, 1987).

A precarious passage, across the tropics, of amphipolar cold temperate species like *Desmarestia viridis*, would have necessitated a lowering of the mean temperature of the warmest month with ~5°C along the east coasts of the equatorial Pacific and/or Atlantic Oceans. This same temperature lowering could have caused the extinction of tropical species, as *Dictyosphaeria cavernosa*, needing a reproduction/growth temperature of at least 23°C (average value), but not of warm water species as *Centroceras clavulatum* needing a minimum growth temperature of 20°C (average value). This equatorial temperature lowering is likely to have taken place after the early to mid Miocene disruption of the Tethyan distribution area of tropical to warm temperate species (≈20 Ma), and probably before the Pleistocene. DNA-DNA hybridizations between amphipolar populations of the same morphological species may provide a means to estimate the period of temperature lowering in the equatorial seas.

ACKNOWLEDGEMENTS

Financial support from the Netherlands' Ministry of Housing, Physical Planning and Environmental Hygiene to participate in the workshop on 'Expected effects of climatic changes on marine coastal ecosystems' is gratefully acknowledged.

4. REFERENCES

ABBOTT, I. A. & G.J. HOLLENBERG, 1976. Marine algae of California. Stanford Univ. Press: 1-827.

ABBOTT, I. A. & B. SANTELICES, 1985. The marine algae of Easter Island (Eastern Polynesia).—Proc. 5th Int. Coral Reef Congr. Tahiti **5:** 71-75.

ADAMS, N. M. & W.A. NELSON, 1985. The marine algae of the Three Kings Islands.—Nat. Mus. New Zealand, Miscell. ser. no **13:** 1-29.

AFONSO-CARRILLO, J. & M.C. GIL-RODRIGUEZ, 1982. Aspectos biogeográficos de la flora ficológica marina de las islas Canarias.—Actas il simp. Ibér. estud. bentos mar. III: 41-48.

AKATSUKA, I., 1973. Marine algae of Ishigaki Island and its vicinity in Ryukyu Archipelago.—Bull. Jap. Soc. Phycol. **21:** 39-42.

ALEEM, A. A., 1978. Contributions to the study of the marine algae of the Red Sea.—Bull. Fac. Sci. K. A. U. Jeddah **2:** 99-118.

AYALA, F. J., 1986. On the virtues and the pitfalls of the molecular evolutionary clock.—J. Hered. **77:** 226-235.

BAARDSETH, E., 1941. The marine algae of Tristan da Cunha.—Results of the Norwegian scientific expedition to Tristan da Cunha 1937-1938, No. 9. Oslo, Jacob Dybwad: 174 pp.

BIEBL, R., 1962. Temperaturresistentz tropischer Meeresalgen (verglichen mit jener von Algen in temperierten Meeresgebieten).—Bot. Mar. **4:** 242-254.

BOLTON, J. J. & H. STEGENGA, 1987. The marine algae of Hluleka, Transkei, and the warm temperate/subtropical transition on the east coast of Southern Africa.—Helgoländer Meeresunters. **41:** 165-183.

BOO, SUNG MIN & IN KYU LEE, 1985. Two Korean species of Centroceras Kützing (Ceramiaceae, Rhodophyta).—Korean J. Bot. **28:** 297-304.

BØRGESEN, F., 1934. Some marine algae from the northern part of the Arabian Sea with remarks on their geographical distribution.—Kgl. Danske Vidensk. Selsk. Biol. Medd. **11:** 1-72, pl. I-II.

——, 1940. Some marine algae from Mauritius. I. Chlorophyceae.—Kgl. Dansk. Vidensk. Selsk. Biol. Medd. **15:** 1-81, pl. I-III.

——, 1953. Some marine algae from Mauritius.—Kgl. Dansk. Vidensk. Selsk. Biol. Medd. **21:** 1-62. pl. I-III.

BOT, P. V. M., W.T. STAM, S.A. BOELE-BOS, C. VAN DEN HOEK & W. VAN DELDEN, 1989a. Biogeographic and phylogenetic studies in three North Atlantic species of *Cladophora* (Cladophorales, Chlorophyta) using DNA-DNA hybridization.—Phycologia **28:** 159-168.

BOT, P. V. M., R.W. HOLTON, W.T. STAM, & C. VAN DEN HOEK, 1989b. Molecular divergence between North-Atlantic and Indo-West-Pacific *Cladophora albida* (Cladophorales: Chlorophyta) isolates as indicated by DNA-DNA hybridization.—Mar. Biol. **102:** 307-313.

BREEMAN, A. M., 1988. Factors that keep seaweeds within their geographic boundaries; experimental and phenological evidence.—Helgoländer Meeresunters. **42:** 199-241.

CAMBRIDGE, M. L., A.M. BREEMAN, S. KRAAK & C. VAN DEN HOEK, 1987. Temperature responses of tropical to warm-temperate Cladophora species in relation to their distribution in the North Atlantic Ocean.—Helgoländer Meeresunters. **41:** 329-354.

CAMBRIDGE, M. L., A.M. BREEMAN, R. OOSTERWIJK, & C. VAN DEN HOEK, 1984. Temperature responses of some North Atlantic *Cladophora* species (Chlorophyceae) in relation to their geographic distribution.—Helgoländer Meeresunters. **38:** 349-363.

CHAMBERLAIN, Y. M., 1965. Marine algae of Gough Island.—Bull. Brit. Mus. Nat. Hist. **3:** 175-232, pl. 16-19.

CHAMBERLAIN, Y., M.W. HOLDGATE & N. WACE, 1985. The littoral ecology of Gough Island, South Atlantic Ocean.—Téthys **11:** 302-319.

CHAPMAN, V.J., 1956. The marine algae of New Zealand. Part I. Myxophyceae and Chlorophyceae.—J. Linn. Soc. London Bot. **55:** 333-501, pl. 24-50.

——, 1977. Marine algae of Norfolk Island and the Cook Islands.—Botanica Mar. **20:** 161-165.

CHIANG, Y.-M., 1973. Studies on the marine flora of southern Taiwan.—Bull. Jap. Soc. Phycol. **21:** 97-102.

CRIBB, A.B., 1983. Marine algae of the southern Great Barrier Reef- Rhodophyta. Brisbane, Australian Coral Reef Society, Handbook no. 2, pp. 173, pl. 1-71.

DAWSON, E.Y., C. ACLETO, & N. FOLDVIK, 1964. The seaweeds of Peru. Beih. Nova Hedwigia **13:** 1-111, pl. 1-81.

DELLOW, V., 1955. Marine algal ecology of the Hauraki Gulf, New Zealand.—Trans. Roy. Soc. New Zealand **83:** 1-91.

EARLE, S.A., 1986. Preliminary checklist of marine plants in the vicinity of the Mote marine laboratory, Sarasota, Florida (roneotype).

EDWARDS, P., 1969. Field and cultural studies on the seasonal periodicity of growth and reproduction of selected Texas benthic marine algae.—Contrib. Mar. Sci. **14:** 59-114.

EGEROD, L. EUBANK., 1952. An analysis of the siphonous Chlorophycophyta.—Univ. California Publ. Bot. **25:** 325-454.

EGEROD, L., 1974. Report of the marine algae collected on the fifth Thai-Danish expedition of 1966.—Botanica Mar. **17:** 130-157.

ENOMOTO, S. & K. OKUDA, 1981. Culture studies of Dictyosphaeria (Chlorophyceae, Siphonocladales) 1. Life history and morphogenesis of *Dictyosphaeria cavernosa*.—Jap. J. Phycol. **29:** 225-236.

ETCHEVERRY D.H., 1986. Algas marinas bentonicas de Chile. Montevideo, Unesco, pp. 379.

FUNAHASHI, S., 1967. Marine algae in the vicinity of Noto marine laboratory.—Ann. Rep. Noto Marine Lab. **7:** 15-36.

——, 1973. Distribution of marine algae in the Japan Sea, with reference to the phytogeographical positions of Vladivostok and Noto Peninsula districts.—J. Fac. Sci. Hokkaido Univ. Ser. V (bot.) **10:** 1-31.

GLYNN, P.W. & R.H. STEWART, 1973. Distribution of coral reefs in the Pearl Islands (Gulf of Panama) in relation to thermal conditions.—Limnol. Oceanogr. **18:** 367-379.

HACKETT, H.E., 1969. Marine algae in the Atoll environment: Maldive Islands. Thesis, Duke University. University Microfilms International, Ann Arbor, Michigan, 319 pp.

HANISAK, M.D. & S.M. BLAIR, 1988. The deep-water macroalgal community of the East Florida continental shelf (U.S.A.).—Helgoländer Meeresunters. **42:** 133-163.

HAY, C.H., N.M. ADAMS & M.J. PARSONS, 1985. The marine algae of the Subantarctic Islands of New Zealand.—Nat. Mus. New Zealand Miscell. Ser. No 10, pp. 70.

HEYS, F.M.L., 1985. The seasonal distribution and community structure of the epiphytic algae on *Thalassia hemprichii* (Ehrenb.) Aschers. from Papua New Guinea.—Aquat. Bot. **21:** 295-320.

HOEK, C. VAN DEN, 1963. Revision of the European species of *Cladophora*. Brill, Leiden: 1-248.

——, 1975. Phytogeographic provinces along the coasts of the northern Atlantic Ocean.—Phycologia **14:** 317-330.

——, 1982a. The distribution of benthic marine algae in relation to the temperature regulation of their life histories.—Biol. J. Linn. Soc. **18:** 81-144.

——, 1982b. Phytogeographic distribution groups of benthic marine algae in the North Atlantic Ocean. A review of experimental evidence from life history studies.—Helgoländer Meeresunters. **35:** 153-214.

——, 1982c. A taxonomic revision of the American species of Cladophora (Chlorophyceae) in the North Atlantic Ocean and their geographic distribution.—Verh. Kon. Ned. Akad. W'sch. 2de reeks 78, North-Holland Publishing Cy, Amsterdam, pp 236.

——, 1984. World-wide latitudinal and longitudinal seaweed distribution patterns and their possible causes, as illustrated by the distribution of Rhodophytan genera.—Helgoländer Meeresunters. **38:** 227-257.

——, 1987. The possible significance of long-range dispersal for the biogeography of seaweeds.—Helgoländer Meeresunters. **41:** 261-272.

HOEK, C. VAN DEN & H.B.A. WOMERSLEY, 1984. The genus Cladophora. In: H.B.S. WOMERSLEY. The marine benthic flora of southern Australia. Part I. Woolman, Government Printer, South Australia, pp. 329.

ISLAM, A.K.M. NURUL., 1976. Contribution to the study of the marine algae of Bangladesh. Bibliotheca phycologica 19. Cramer, Vaduz: 253 pp.
ITONO, H., 1977. Studies on the Ceramiaceous algae (Rhodophyta) from southern parts of Japan. Bibliotheca phycologica 35. Cramer, Vaduz: 499 pp.
JAASUND, E., 1976. Seaweeds in Tanzania. University of Tromsø, 160 PP.
JOOSTEN, A.M.T. & C. VAN DEN HOEK, 1986. World-wide relationships between red seaweed floras: a multivariate approach.—Bot. Mar. **29:** 195-214.
KANG, JAE WON., 1966. On the geographical distribution of marine algae in Korea.—Bull. Pusan Fish. Coll. **7** (1,2): 1-125, pl. I-XII.
KENNETT, J.P., 1982. Marine geology. Prentice Hall, Englewood Cliffs, N.J.: 813 pp.
KING, R.J., J. HOPE BLACK & S. DUCKER, 1971. Intertidal ecology of Port Philip Bay with systematic lists of plants and animals.—Mem. Nat. Mus. Vic. **32:** 93-128.
KLEIN, B., 1987. The phenology of *Dumontia contorta* (Rhodophyta) studied by following individual plants *in situ* at Roscoff, northern Brittany.—Bot. Mar. **30:** 187-194.
LAWSON, G.W., 1980. A check-list of East African seaweeds. Dept. Biological Sciences, University of Lagos, Nigeria: 65 pp.
LAWSON, G.W. & D.M. JOHN, 1987. The marine algae and coastal environment of Tropical West Africa (2nd ed.).—Beih. Nova Hedw. **93:** 1-415.
LAWSON, G.W., D.M. JOHN & J.H. PRICE, 1975. The marine algal flora of Angola: its distribution and affinities.—Bot. J. Linn. Soc. **70:** 307-324.
LEVRING, T., 1960. Contribution to the marine algal flora of Chile. Lunds Univ. Arsskr. N. F. (Avd. 2) **56** (10): 1-85.
LINDSTROM, S.C., 1977. An annotated bibliography of the benthic marine algae of Alaska.—Alaska Department of Fish and Game. Data Report No 31: 172 pp.
——, 1987. Possible sister groups and phylogenetic relationships among selected North Pacific and North Atlantic Rhodophyta.—Helgoländer Meeresunters. **41:** 245-260.
LUBCHENKO, J., B.A. MENGE, S.D. GARRITY, P.J. LUBCHENKO, L.R. ASHKENAS, S.D. GAINES, R. EMLER, J. LUCAS & S. STRAUSS, 1984. Structure, persistence and role of consumers in a tropical rocky intertidal community.—J. Exp. Mar. Biol. Ecol. 78 : 23-73.
LÜNING, K., 1985. Meeresbotanik. Thieme, Stuttgart: 1-375.
LÜNING, K. & W. FRESHWATER, 1988. Temperature tolerance of Northeast Pacific marine algae.—J. Phycol. **24:** 310-315.
MACINTYRE, I.G. & O.H. PILKEY, 1969. Tropical reef corals: tolerance of low temperatures on the North Carolina continental shelf.—Science **17:** 374-375.
MAGRUDER, W.H. & J.W. HUNT, 1979. Seaweeds of Hawaii. The Oriental Publ. Cy, Honolulu: 116 pp.
MCINTYRE, A. *et al.*(CLIMAP project members), 1976. The surface of the Ice-age earth.—Science, New York **191:** 1131-1137.
MCKENNA, M.C., 1983. Cenozoic paleogeography of North Atlantic land bridges. In: M.H.P. BOTT, S. SAXOV, M. TALWANI & J. THIEDE. Structure and development of the Greenland Scotland Ridge. Plenum Press, New York: 351-399.
MOROHOSHI, H. & M. MASUDA, 1980. The life history of *Gloiosiphonia capillaris* (Hudson) Carmichael (Rhodophyceae, Cryptonemiales).—Jap. J. Phycol. **28:** 81-91.
NAKAHARA, H., 1984. Alternation of generations of some brown algae in unialgal and axenic cultures.—Scient. Pap. Instit. Algol. Res. Fac. Sci. Hokkaido Univ. **7:** 77-194, pl. 1-12.
NAKANIWA, M., 1975. Marine algae along the coast of Ibaraki Prefecture.—Bull. Jap. Soc. Phycol. **23:** 99-110.
OGAWA, H. & M. MACHIDA, 1976. Marine algae of the Oshika Peninsula.—Tohoku J. Agricult. Res. **27:** 145-153.
OLIVEIRA, E.C. DE, 1977. Algas marinhas bentônicas de Brasil.—Universidado de Sao Paulo, Instituto de Biociências, Thesis, pp. 407.
OLSEN, J.L., W.T. STAM, P.V.M. BOT, & C. VAN DEN HOEK, 1987. scDNA-DNA hybridization studies in Pacific and Caribbean isolates of *Dictyosphaeria cavernosa* (Chlorophyta) indicate a long divergence.—Helgoländer Meeresunters. **41 :** 377-383.
PAPENFUSS, G.F., 1968. A history, catalogue and bibliography of Red Sea benthic algae.—Israel J. Bot. **17** (1,2): 1-118.
PERESTENKO, L.P., 1980. Vodorosli Zaliva Petra Velikogo. Leningrad, Nauka, pp. 232+54.
PHAM-HOANG HO, 1969. Marine Algae of South Vietnam. Trung-Tam Hoc-Lieu Xuat-Ban, pp. 558 (Vietnamese).
POT, R., B. KLEIN, H. RIETEMA, & C. VAN DEN HOEK, 1988. A field study on the growth and development of *Dumontia contorta*.—Helgoländer Meeresunters. **42:** 553-562.
PRICE, I.R., A.W.D. LARKUM & A. BAILEY, 1976. Checklist of marine benthic plants collected in the Lizard Island area.—Aust. J. Plant Physiol. **3:** 3-8.
PRICE, J.H., D.M. JOHN & G.W. LAWSON, 1986. Seaweeds of the western coast of tropical Africa and adjacent islands: a critital assessment. IV. Rhodophyta (Florideae). 1. Genera A-F.—Bull. Brit. Mus. Nat. Hist. Bot. **15:** 1-122.
PRUD'HOMME VAN REINE, W.F., 1988. Phytogeography of seaweeds of the Azores.—Helgoländer Meeresunters. **42:** 165-185.
PRUD'HOMME VAN REINE, W.F. & C. VAN DEN HOEK, 1988. Biogeography of Cape Verdian seaweeds.—Cour. Forsch. Inst. Senckenberg **105:** 35-49.
——, 1989. Biogeography of Macaronesian seaweeds.—Cour. Forsch. Inst. Senckenberg (in press).
RICKER, R.W., 1987. Taxonomy and biogeography of Macqarie Island seaweeds. London, British Museum (Natural History), pp. 344.
RIETEMA, H., 1982. Effects of temperature and photoperiod on *Dumontia contorta* (Rhodophyta).—Mar. Ecol. Progr. Ser. **8:** 187-196.
RIETEMA, H. & A.M. BREEMAN, 1982. The regulation of the life history of *Dumontia contorta* in comparison to that of several other Dumontiaceae (Rhodophyta).—Bot. Mar. **25:** 569-576.
RIETEMA, H. & C. VAN DEN HOEK, 1984. Search for possible latitudinal ecotypes in *Dumontia contorta* (Rhodophyta).—Helgoländer Meeresunters. **38:** 389-399.
RÖGL, F. & F.F. STEININGER, 1984. Neogene Paratethys,

Mediterranean and Indo-pacific seaways. Implications for the paleobiogeography of marine and terrestrial biota.—In: P. BRENCHLEY. Fossils and climate. John Wiley, Chichester: 171-200.

SAKAI, Y., 1964. The species of Cladophora from Japan and its vicinity.—Scient. Pap. Inst. Algol. Res. Hokkaido Univ. **5:** 1-104, pl. I-XVII.

SANO, O., M. IKEMORI & S. ARASAKI, 1981. Distribution and ecology of *Acetabularia calyculus* along the coast of Noto Peninsula.—Jap. J. Phycol. **29:** 31-38.

SCAGEL, R.F., 1966. Marine algae of British Columbia and Northern Washington, Part. I. Chlorophyceae (green algae).—National Museum of Canada Bull. 207, Biological series no. 74, 257 pp..

SCAGEL, R.F., D.J. GARBARY, L. GOLDEN, & M.W. HAWKES, 1986. A synopsis of the benthic marine algae of British Columbia, northern Washington and Southeast Alaska.—Phycological Contribution No 1, Dept. of Botany, University of British Columbia, Vancouver: 444+5 pp..

SCHNETTER, R., 1978. Marine Algen der Karibischen Küsten von Kolumbien. II. Chlorophyceae. Bibliotheca phycologica 42, Cramer, Vaduz: 199 pp.

SCHNETTER, R. & G.M. MEYER, 1982. Marine Algen der Pazifikküste von Kolumbien. Bibliotheca Phycologica 60, Cramer, Vaduz: 287 pp.

SCOTT, F.J. & G.R. RUSS, 1987. Effects of grazing on species composition of the epiphytic algal community on coral reefs of the central Great Barrier Reef.—Marine Ecol. Progr. Ser. **39:** 293-304.

SEAGRIEF, S.C., 1984. A catalogue of South African green, brown and red marine algae.—Mem. Bot. Survey South Africa **47:** 1-72.

SEARLES, R.B., 1984. Seaweed biogeography of the Mid-Atlantic coast of the United States.—Helgoländer Meeresunters. **38:** 259-271.

SETCHELL, W.A., 1926. Tahitian algae.—Univ. California Publ. Bot. **12:** 61-142, pl. 7-22.

SHEPHERD, S.A. & H.B.S. WOMERSLEY, 1971. Pearson Island expedition 1969-7. The subtidal ecology of benthic algae.—Trans. Roy. Soc. S. Aust. **95:** 155-167.

SILVA, P.C., E.G. MEÑEZ & R.L. MOE, 1987. Catalog of the benthic marine algae of the Philippines.—Smithsonian Contrib. Mar. Sci. **27:** 179 pp.

STAM, W.T., P.V.M. BOT, S.A. BOELE-BOS, J.M. VAN ROOY & C. VAN DEN HOEK, 1988. Single copy DNA-DNA hybridizations among five species of *Laminaria* (Phaeophyceae): phylogenetic implications.—Helgoländer Meeresunt. **42:** 252-267.

STEGENGA, H., 1986. The Ceramiaceae (excl. Ceramium) (Rhodophyta) of the South West Cape Province, South Africa. Bibliotheca phycologica 74, Cramer, Vaduz: 149 pp.

TAYLOR, W.R., 1945. Pacific marine algae of the Allan Hancock expeditions to the Galapagos Islands.—Allan Hancock Pacific Expeditions 12, Univ. Southern California Press: 528 pp.

TAYLOR, W.R., 1960. Marine algae of the eastern tropical and subtropical coasts of the Americas. Ann Arbor, Univ. Michigan Press: 870 pp.

TRONO, G., 1968. The taxonomy and ecology of the marine benthic algae of the Caroline Islands. Thesis, University of Hawaii, University Microfilms Inc. Ann Arbor, Michigan: 387 pp.

TSENG, C.K., 1983. Common seaweeds of China. Science Press, Beying: 316 pp.

TSENG, C.K. & C.F. CHANG, 1964. An analytical study of the marine algal flora of the western Yellow Sea coast. II. Phytogeographic nature of the flora.—Oceanologica et Limnologia Sinica **6:** 152-168 (Chinese, English summary).

TSENG, C.K., C.F. CHANG, XIA ENZANG & XIA BANGMEI, 1982. Studies on some marine red algae from Hong Kong. In: C. S. MORTON & C.K. TSENG. Proc. 1st Int. Mar. Biol. Workshop: The marine flora and fauna of Hong Kong and southern China, Hong Kong: 57-84.

TSUDA, R.T., 1976. Some marine benthic algae from Pitcairn Island.—Rev. Algol. N.S. **11:** 325-331.

TSUDA, R.T. & W.J. TOBIAS, 1977. Marine benthic algae from the northern Mariana Islands, Chlorophyta and Phaeophyta.—Bull. Jap. Soc. Phycol. **25:** 67-71.

TSUDA, R.T. & F.O. WRAY, 1977. Bibliography of marine benthic algae in Micronesia.—Micronesica **13:** 85-120.

VINOGRADOVA, K.L., 1979. Opredel'itel' vodoroslej dal'ne vostocnych morej SSSR.—'Nauka', Leningrad, 1-147.

VOOYS, C. DE, 1988. Expected biological effects of long-term changes in temperatures on marine ecosystems in coastal waters around the Netherlands.—NIOZ-rapport 1988-6, Neth. Inst. Sea Res.: 1-38.

WALKER, B.D., H.H. ROBERTS, L.J. ROUSE, & O.K. HUH, 1982. Thermal history of reef-associated environments during a record cold-air outbreak event.—Coral reefs **1:** 83-87.

WEBER-VAN BOSSE, A., 1928. Liste des algues du Siboga. IV. Rhodophyceae. Siboga-expeditie LIX d., Brill, Leiden: 393-533.

WEI, TEO LEE & WEE YEOW CHIN, 1983. Seaweeds of Singapore. Singapore Univ. Press, 123 pp.

WOMERSLEY, H.B.S. & A. BAILEY, 1970. Marine algae of the Solomon Islands.—Phil. Trans. Roy. Soc. London B (Biol. Sci.) **259:** 257-352.

WYNNE, M.J., 1986. Report on a collection of benthic marine algae from the Namibian coast.—Nova Hedwigia **43:** 311-356.

ZANEVELD, J.S., 1966. The occurrence of benthic marine algae under shore fast ice in the western Ross Sea, Antarctica. Proc. 5th Intern. Seaweed Symp. Halifax. Pergamon Press, Oxford: 217-231.

EXPECTED EFFECTS OF CHANGING SEAWATER TEMPERATURES ON THE GEOGRAPHIC DISTRIBUTION OF SEAWEED SPECIES

ANNEKE M. BREEMAN

Department of Marine Biology, Biological Centre, University of Groningen, P.O. Box 14, 9750 AA Haren, (Gn), The Netherlands

ABSTRACT

Seaweeds are generally kept within their geographic boundaries by limiting effects of temperature. Northern boundaries are set by low lethal winter temperatures, or by summer temperatures too low for growth and/or reproduction. Southern boundaries are set by high lethal summer temperatures, or by winter temperatures too high for induction of a crucial step in the life cycle. Characteristic thermal response types, as identified in laboratory experiments, were found to be responsible for characteristic distribution patterns in the North Atlantic Ocean. Changes in seawater temperature have therefore easily predictable effects on the geographic distribution of seaweed species. Locally, species composition (and community structure) will be altered. Apart from latitudinal displacement and regional extinction, changing seawater temperatures may also cause a shift in selection pressure at a boundary, particularly when summer and winter temperatures change to a different extent. For instance, southern boundaries of several cold temperate brown algae in Europe, which are presently set by 'winter reproduction' limits will become 'summer lethal' limits following a rise in summer temperatures of only a few degrees. Reconstruction of thermal regimes during glacial and interglacial periods shows that such shifts in selection pressure have probably occurred more often on eastern than on western Atlantic coasts.

1. INTRODUCTION

Seaweeds are a group of organisms eminently suited for studying the effects of changing seawater temperatures on marine life. Seaweeds are confined to the margins of the world oceans, because insufficient light reaches the deep ocean-bottoms, so they have an essentially linear geographic distribution along the latitudinal gradient where temperature is the main variable. As benthic organisms, growing attached at a fixed location, they are also very directly subjected to changing seawater temperatures. Another reason why seaweeds are such suitable model organisms is that they are relatively easy to culture in the laboratory, which makes it possible to determine their temperature responses experimentally.

Temperature has long been indicated as a key factor for the geographic distribution of seaweed species, but the experimental evidence that might support this conclusion has accumulated only during the past decade (VAN DEN HOEK, 1982a, b; LÜNING, 1985; BREEMAN, 1988). Experimental evidence has now confirmed that temperature effects are indeed responsible for the location of geographic boundaries in the majority of species studied (reviewed by BREEMAN, 1988). Thus, effects of changing seawater temperatures on seaweed distribution can be easily predicted.

Two main types of boundaries can be recognized. Firstly, boundaries can be set by lethal limits. These operate during the adverse season: at the southern boundary during a hot summer, at the northern boundary during a cold winter. Lethal limits are set by the tolerance limits of the hardiest life history stage. This is often a cryptic microthallus or perennating structure instead of the large macrophyte. Some species with perennial macrophytes become seasonal annuals near their geographic limits (BREEMAN, 1988).

Secondly, there are boundaries set by growth or reproduction limits. These operate during the favourable season, at the northern boundary during a cool summer, and at the southern boundary during a warm winter. In species that can persist without regularly completing their sexual cycle, these limits are set by the temperature requirements for growth. In other species, they are set by the temperature requirements for completion of the life history. Often, one or two processes have the most stringent temperature demands. This can be either the induction of gametangia or sporangia, or the development of macrothalli or blades. Temperature demands of all other processes in the life history are then irrelevant

J.J. Beukema et al. (eds.), Expected Effects of Climatic Change on Marine Coastal Ecosystems, 69–76.

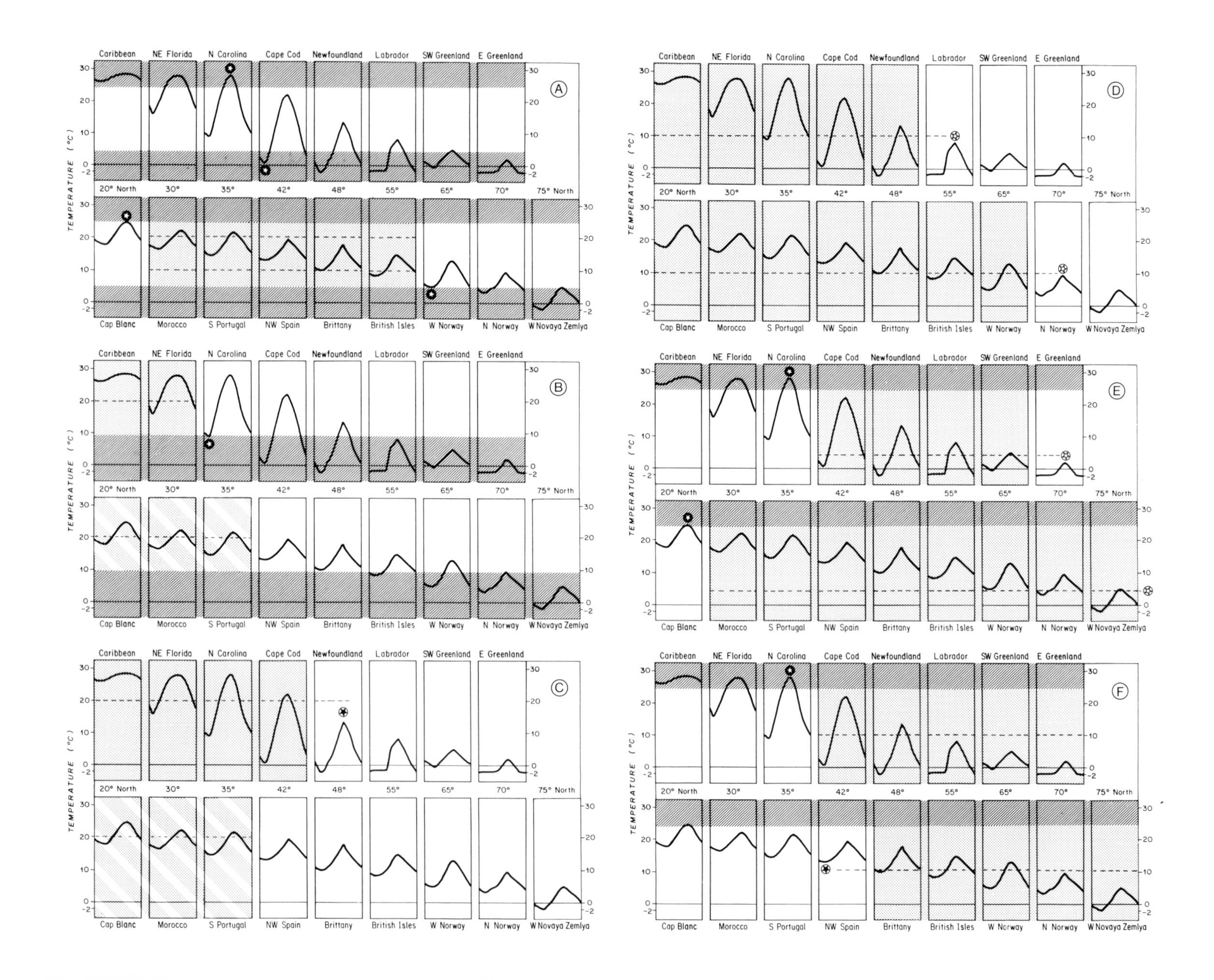
A
B
C
D
E
F
Caribbean
NE Florida
N Carolina
Cape Cod
Newfoundland
Labrador
SW Greenland
E Greenland
20° North
30°
35°
42°
48°
55°
65°
70°
75° North
Cap Blanc
Morocco
S Portugal
NW Spain
Brittany
British Isles
W Norway
N Norway
W Novaya Zemlya
TEMPERATURE (°C)
30
20
10
0
-2

in explaining the geographic distribution. In a few species photoperiodic responses interact with temperature requirements to determine the location of geographic boundaries (Breeman, 1988).

2. THERMAL RESPONSE TYPES OF SEAWEEDS CAUSING CHARACTERISTIC DISTRIBUTION PATTERNS IN THE NORTH ATLANTIC OCEAN

Given the characteristic thermal properties of a species, its potential distribution range will be determined by the annual temperature fluctuations occurring in different parts of the Ocean. Within one latitudinal zone, annual fluctuations in seawater temperature are very dissimilar on the two sides of the Atlantic, with large fluctuations occurring on western and smaller fluctuations on eastern shores (Fig. 1). Consequently, different thermal responses may restrict the distribution on the two sides of the Ocean. Similarly, different limiting factors may set different boundaries on a more local scale (for instance open Atlantic coasts vs. North Sea; see below).

A limited number of thermal response types has been recognized; these result in characteristic distribution patterns in the North Atlantic Ocean, some of which will be discussed below (*cf.* Breeman, 1988; Figs 1a-f):

A) Species endemic to the (warm) temperate eastern Atlantic are relatively stenothermous (Fig. 1a: example *Callophyllis laciniata*). They are characterized by survival ranges between ~5°C and ~25°C. This prevents their occurrence in eastern N. America, where annual temperature fluctuations exceed 20°C everywhere in the (warm) temperate zone. Temperatures above 25°C prevent their occurrence in the tropics (Fig. 1a). In species with isomorphic life histories and without very specific temperature requirements for reproduction, northern and southern boundaries in Eur/Africa are set by lethal limits (Fig. 1a). Species with heteromorphic life histories often require high and/or low temperatures to induce reproduction in one or both life history phases which further restricts the distribution (Breeman, 1988).

B) Species endemic to the tropical western Atlantic are also stenothermous, but they are characterized by survival ranges between ~10°C and ~35°C (Fig. 1b: example *Gracilaria wrightii*). Northern boundaries are set by low, lethal winter temperatures (Fig. 1b). Thermal properties would potentially allow occurrence in the (sub)tropical eastern Atlantic (Fig. 1b: interrupted shading), but adverse temperatures during the glaciations (see below) may have caused extinction. Subsequently, the central Atlantic Ocean must have formed a barrier to dispersal. At present, little experimental evidence is available concerning tropical species with an amphi- Atlantic distribution (Breeman, 1988).

C) Tropical to temperate species endemic to the western Atlantic are more eurythermous, surviving temperatures below zero up to ~35°C (Fig. 1c: example *Lomentaria baileyana*), but they still require high summer temperatures of at least ~20°C for (growth and) reproduction. This determines the location of their northern boundaries (Fig. 1c). Thermal properties would also permit occurrence in the (sub)tropical eastern Atlantic (Fig. 1c: interrupted shading), but along potential 'stepping stones' for dispersal in the northern Atlantic (*i.e.* Greenland, Iceland, NW Europe) summer temperatures would be too low for growth. Adverse temperatures during glaciations (see below), and subsequent failure to disperse across the central Atlantic Ocean may account for their absence from (sub)tropical eastern Atlantic coasts.

D) Amphi-Atlantic (tropical to) temperate species survive at subzero temperatures. They are able to grow and reproduce adequately at temperatures of ~15°C or below. Northern boundaries are generally set by low summer temperatures preventing reproduction (or growth) of the larger life history phase. Species with the more northerly located boundaries have correspondingly lower temperature requirements (Figs 1d, e: examples *Cladophora vagabunda* and *C. rupestris*). Characteristically, species extend further north on European than on American coasts because the Gulf Stream causes summers to be warmer (Figs 1d,e).

E) Amphi-Atlantic (arctic to) temperate species survive at subzero temperatures. In species with isomorphic life histories, not specifically requiring low winter temperatures for reproduction, southern boundaries are set by lethal, high summer temperatures on both sides of the Atlantic (Fig. 1e: example *Cladophora rupestris*). Species with the more southerly located boundaries have correspondingly higher upper lethal limits (Breeman, 1988). None of these species survive temperatures over 28–30°C, which prevents

Fig. 1. Summary of thermal response types of seaweeds responsible for some characteristic distribution patterns in the North Atlantic Ocean. Annual temperature curves are shown for different latitudes in the western (above) and eastern (below) Atlantic. Observed distribution ranges shaded, potential ranges indicated by an interrupted shading. Temperatures beyond survival limits hatched; critical temperatures for growth or completion of the life history indicated by dotted lines. Temperature responses restrictive to distribution indicated as follows: lethal limit (blank asteriks in closed circle), growth limit (blank asteriks in open circle), limit set by life history requirements (black asteriks in open circle). (A) *Callophyllis laciniata*; (B) *Gracilaria wrightii*; (C) *Lomentaria baileyana*; (D) *Cladophora vagabunda*; (E) *Cladophora rupestris*; (F) *Chorda tomentosa*. See text for further explanation. From Breeman (1988).

tropical occurrence. Species with these thermal responses are characterized by distribution patterns in which southern boundaries in Eur/Africa lie further south than those in eastern N. America because of cooler summers (Fig. 1e).

F) In (arctic to) temperate species with heteromorphic life histories (or crustose and erect growth forms) low temperatures are generally required for formation of the macrothalli (either directly or through the induction of sexual reproduction). These species have composite southern boundaries in the north Atlantic Ocean. On American coasts boundaries are set by lethal, high summer temperatures, on European coasts by winter temperatures too high for the induction of macrothalli (Fig. 1f: example *Chorda tomentosa*). Again, differences in lethal or limiting temperatures between species correspond closely with the location of their respective boundaries (BREEMAN, 1988). Species with these thermal responses are characterized by distribution patterns in which the boundaries in Eur/Africa lie further north than those in eastern N. America because of warmer winters (Fig. 1f).

3. TEMPERATURE ECOTYPES IN SEAWEEDS

Thermal ecotypes have evolved in some seaweed species, but not in many others. There is no evidence that populations are always optimally adapted to local conditions; thermal responses rather reflect present (and past) climatic conditions over (part of) the geographical range (BREEMAN, 1988). In general, limits of thermal tolerance are genetically more firmly fixed than temperature effects on reproduction or growth (BREEMAN, 1988; LÜNING & FRESHWATER, 1988). Ecotypic variation in thermal tolerance was mainly found between disjunct tropical and temperate populations of species with a tropical extension, not in temperate species (BREEMAN, 1988; but see GERARD & DU BOIS, 1988). This may reflect the longer evolutionary history of the former group, which is thought to be of Tethyan origin (VAN DEN HOEK, 1984; LÜNING, 1985; JOOSTEN & VAN DEN HOEK, 1986). Evidently, seaweeds are unable to form new, better adapted thermal ecotypes very rapidly. Thus, if temperature conditions at a boundary deteriorate during the season when temperatures used to be limiting, the species will become locally extinct. If temperature conditions improve, the species will be able to extend its distribution range. Following an improvement of conditions, it will take some time before a species reaches the new limits set by its thermal potential.

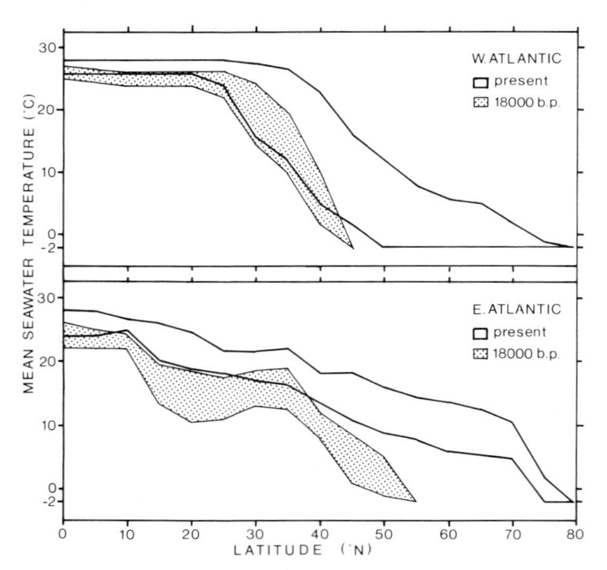

Fig. 2. Annual ranges of sea-surface temperatures between 0° and 80° Northern Latitude on eastern and western Atlantic coasts at present and during the last glaciation, 18000 BP. Ranges represent February and August isotherms (based on U.S. NAVY, 1981; BRADLEY, 1985).

4. LATITUDINAL SHIFTS OF SEAWEED SPECIES WITH DIFFERENT THERMAL RESPONSE TYPES BETWEEN GLACIAL AND INTERGLACIAL PERIODS

In the geologically short Quaternary time span (about 1.6 My) as many as 30 glacial episodes occurred, each of which has been related to extensive ice formation, large scale latitudinal displacement of climatic zones, major fluctuations in oceanic circulation patterns, and sea level oscillations with a range of up to about 100 m (KENNETT, 1982). Some impression of the extent of these climatic changes in the coastal marine environment may be obtained by comparing sea-surface isotherms for August and February (U.S. NAVY, 1981) for the present high interglacial with those reconstructed from paleoclimatic evidence for the maximum of the last glaciation, 18000 years before present (CLIMAP PROJECT MEMBERS, 1976; BRADLEY, 1985) (Fig. 2). Major latitudinal shifts in the geographic distribution of seaweed species must have occurred as a result of these climatic changes (Fig. 3). During the glaciation (CLIMAP PROJECT MEMBERS, 1976; BRADLEY, 1985) there was a considerable southward shift in the position of the Gulf Stream, which ran almost due east from Cape Hatteras to the west coast of Portugal. Landfast ice extended south to ~45 to 55°N, and there was increased upwelling along the coast of northwest Africa. As today, the tropical zone was narrower in the eastern than in the western Atlantic (Fig. 2). Temperatures in the tropics were not much cooler than today (~2–4°C). North of the tropics, there was a marked steepening in the latitudinal gradient, par-

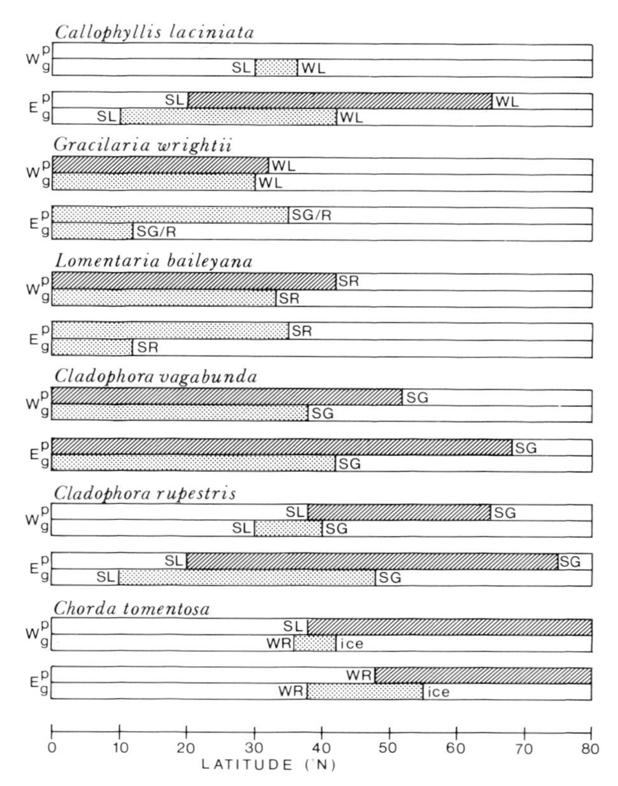

Fig. 3. Present (p) and glacial (g) latitudinal ranges of seaweed species with different thermal response types (Fig. 1) on eastern (E) and western (W) Atlantic coasts. Present ranges are hatched, potential ranges are stippled. Temperature-limits at boundaries indicated as follows: summer lethal (SL), winter lethal (WL), summer growth (SG), summer reproduction (SR), winter reproduction (WR) (based on Figs 1 and 2).

ticularly in the western Atlantic. In the eastern Atlantic summer and winter temperatures were both lowered to the same extent compared with today, so annual temperature ranges were moderate both during glacial and interglacial periods. In contrast, in the western Atlantic, temperatures during the glaciation were lowered especially in summer, much less so in winter (BRADLEY, 1985); so on American coasts glacial conditions were more moderate than they are today (Fig. 2).

Potential distribution ranges during glacial and interglacial times for seaweeds with different thermal response types (*cf.* Fig. 1) are shown in Fig. 3. Because of the sharper latitudinal climate gradient, glacial distribution ranges were narrower than today. Northern boundaries were shifted southward, the latitudinal shifts being larger on eastern than on western Atlantic coasts (Figs 2 and 3). Species now endemic to the eastern Atlantic such as *Callophyllis laciniata* (Figs 1a and 3) would have had a small potential foothold on the other side of the Ocean; so the large annual temperature ranges during the present interglacial rather than extreme conditions during glacial periods (Fig. 2) are responsible for the present absence of stenothermous warm-temperate species in the western Atlantic. Species now endemic to the western Atlantic but with considerable potential distribution ranges in Eur/Africa (Figs 1b and c) would have had these ranges substantially reduced during the glaciation (Fig. 3). This applies not only to stenothermous tropical species like *Gracilaria wrightii* (Figs 1b and 3), but also to more eurythermal ones like *Lomentaria baileyana* (Figs 1c and 3) because these also require high summer temperatures (≥ ~ 20°C) for (growth and) reproduction. Repeated contractions of the tropical belt, with temperature lowerings possibly dating back as far as late Miocene (KENNETT, 1982; see also VAN DEN HOEK *et al.*, this volume), combined with inhospitable sediment coasts (LAWSON & JOHN, 1977; LAWSON, 1982), are probably responsible for the impoverished character of the west African seaweed flora (VAN DEN HOEK, 1982a, b; VAN DEN HOEK, 1984; LÜNING, 1985; JOOSTEN & VAN DEN HOEK, 1986; PRUD'HOMME VAN REINE & VAN DEN HOEK, 1988, 1989; VAN DEN HOEK *et al.*, this volume).

Latitudinal ranges during the glaciation were reduced most for (arctic to) temperate species like *Cladophora rupestris* and *Chorda tomentosa* especially in the western Atlantic (Fig. 3). For both species there was very little latitudinal overlap between glacial and present distribution ranges. Nevertheless, this has not caused a very marked impoverishment of the (arctic to) cold temperate seaweed flora on the American coast (VAN DEN HOEK, 1984; JOOSTEN & VAN DEN HOEK, 1986; LÜNING, 1985; SOUTH & TITTLEY, 1986; SOUTH, 1987). For *Chorda tomentosa* the distribution range in America was further reduced than for *Cladophora rupestris* (Fig. 3) because, during the glaciation, *C. tomentosa* would have met a winter reproduction limit on the American coast before meeting the summer lethal limit that presently sets its southern boundary. This is, again, a result of the reduced annual temperature range in America during the glaciation (Figs 2 and 3). Thus, since the last glaciation, selection pressure for *Chorda tomentosa* at this boundary has shifted to a different thermal capacity (upper tolerance limit instead of low temperature requirement for sexual reproduction) and to another season (summer instead of winter). The same applies to several other cold water species with heteromorphic life histories presently having a summer lethal limit on American coasts (Table 1). On Eur/African coasts, thermal limits at southern boundaries were more similar during glacial and interglacial periods (Table 1, but see below).

TABLE 1

Changing temperature-limits at southern boundaries of heteromorphic brown algae in the eastern and western Atlantic during glacial and interglacial times. Temperature limits at boundaries indicated as follows: summer lethal (SL), winter reproduction (WR). Limits marked with an asteriks are unstable and would revert to the alternative type when temperatures changed by 2°C during summer or winter only. August (A) and February (F) isotherms where survival or reproduction would be limited are indicated (Based on experimental data summarized in Breeman, 1988). Southern ecotype (+).

species	*Limiting isotherms*		*Western Atlantic*		*Eastern Atlantic*	
			present	*18000 B.P.*	*present*	*18000 B.P.*
Laminaria saccharina	16F	20A	SL	SL	SL*	SL*
Scytosiphon lomentaria	+20F	26A	SL	SL+WR*	SL+WR*	SL*
Laminaria hyperborea	13F	18A	--	--	SL+WR*	SL+WR*
Laminaria digitata	11F	20A	SL	SL+WR*	WR*	WR*
Desmarestia aculeata	13F	23A	SL	SL+WR*	WR	WR
Desmarestia viridis	10F	23A	SL	WR*	WR	WR
Chorda filum	13F	26A	SL	WR	WR	WR
Chorda tomentosa	10F	23A	SL	WR	WR	WR

It is clear (Fig. 3) that climatic changes don't cause whole floras to move unaltered to a different latitude. Since local floras are always composed of species with different thermal response types, some of them being near their thermal limits, any change in climatic conditions, even a relatively minor one will alter species composition and thereby community structure.

5. EXPECTED LATITUDINAL SHIFTS OF SEAWEED SPECIES IN EUROPE FOLLOWING A RISE IN SEA-SURFACE TEMPERATURES

In view of the expected rise in air temperatures (GOODESS, this volume; DE BOOIS, this volume), which results from an increased carbon dioxide content of the atmosphere, an elevation of sea surface temperatures may also be expected, but the magnitude of change is difficult to predict (BROECKER, 1987). To my knowledge detailed paleoclimatic reconstructions of sea surface temperatures, such as have been made for the last glaciation (see above: CLIMAP PROJECT MEMBERS, 1976), are not yet available for interglacial periods warmer than today. Following some of the suggestions raised elsewhere in this volume (DE VOOYS, 1990) the effects of an arbitrary rise of 2°C in summer and/or in winter will here be considered. The effects of such a temperature rise on the geographic distribution of seaweed species may be roughly evaluated by comparing the position of the isotherms presently demarcating their geographic boundaries with those representing 2°C colder conditions (Figs 4 and 5).

Many (tropical to) temperate species reach their northern boundaries along the coasts of northwestern Europe and rising seawater temperatures will permit them to extend their distribution ranges. Range extensions of seaweeds have been recorded in the course of this century, and these have been attributed to rising seawater temperatures (LÜNING, 1985: 73).

Some (tropical to) temperate species are relatively eurythermal and tolerate temperatures below zero. Their northern boundaries in northwestern Europe are set by minimum summer temperatures required for growth and/or reproduction (Fig. 4a, example *Cladophora albida*). An increase in summer temperatures would permit northward extension; changing winter temperatures would not affect the distribution as winter temperatures are not limiting.

Many (tropical to) warm-temperate species are more stenothermal and do not tolerate temperatures below about 5°C. They also require high summer temperatures for growth and reproduction (Fig. 4b, example *Cladophora coelothrix*). Their northern boundaries on open Atlantic coasts are set by summer growth and/or reproduction limits, whereas occurrence in the North Sea is prevented by low, lethal winter temperatures. An increase in winter temperatures of about 2°C would not enable extension into the North Sea, but only allow a minor range extension in the English Channel (Fig. 4b). An increase in summer temperatures would permit some northward extension on the west coast of the British Isles, but not very far because low winter temperatures would soon become limiting (Fig. 4b). Instead of a summer growth limit, these species would now meet a winter lethal limit on the Atlantic coast of the British Isles. This is another example where a change in seawater temperatures would shift selection pressure at a boundary to a different thermal capacity and a different season.

The southern boundaries of (arctic to) temperate species in western Europe are set either by summer lethal limits or by winter reproduction limits. In many heteromorphic red and brown algae low winter temperatures are required for completion of the life history and, in general, their southern boundaries are set by winter- reproduction limits (Figs 1f and 5; Table 1). Some species would tolerate high summer temperatures (Fig. 5a, example *Chorda filum*), and the point where they would meet a potential summer lethal limit lies far to the south of the point where high winter temperatures have already become limiting. In these species changing summer temperatures would not affect the distribution, only changing winter temperatures would do so (Fig. 5a).

In many other species, like *Laminaria hyperborea* (Fig. 5b) potential summer lethal and winter 'reproduction' limits (in this case blade initiation is the process critically affected) are located at approximately the same latitude. A change of temperature in either season would affect the location of the boundary. When temperatures would change only during one half of the year, the nature of the boundary would alter (Fig. 5b; Table 1). Such minor changes in the position of summer or of winter isotherms have probably occurred quite often in the course of geological history, even during periods with relatively stable climatic conditions (RUDDIMAN *et al.*, 1970; RUDDIMAN & MCINTYRE, 1981). Thus, for several cold water species selection pressure at southern boun-

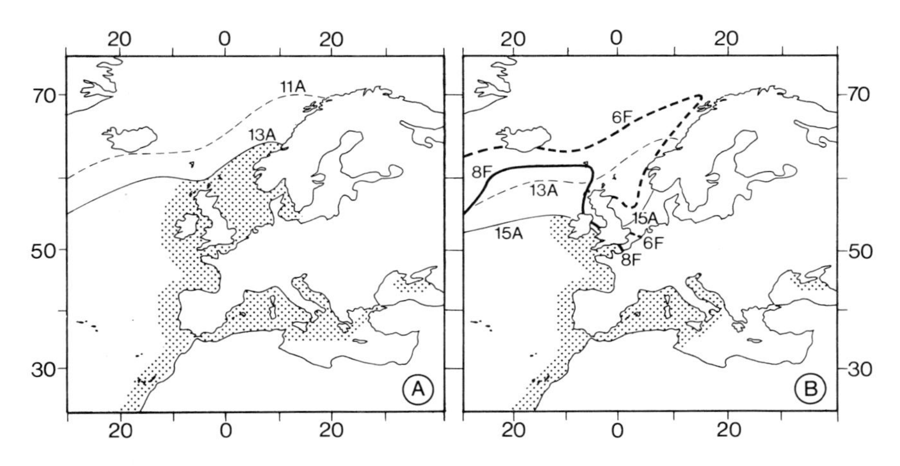

Fig. 4. Expected effects of rising seawater temperatures on the location of northern boundaries in Europe. *Cladophora albida* (a) with summer growth limit at 13° August isotherm; *Cladophora coelothrix* (b) with summer growth limit at 15° August isotherm and winter lethal limit at 8° February isotherm; present distribution ranges are stippled (based on BREEMAN, 1988). See text for further explanation.

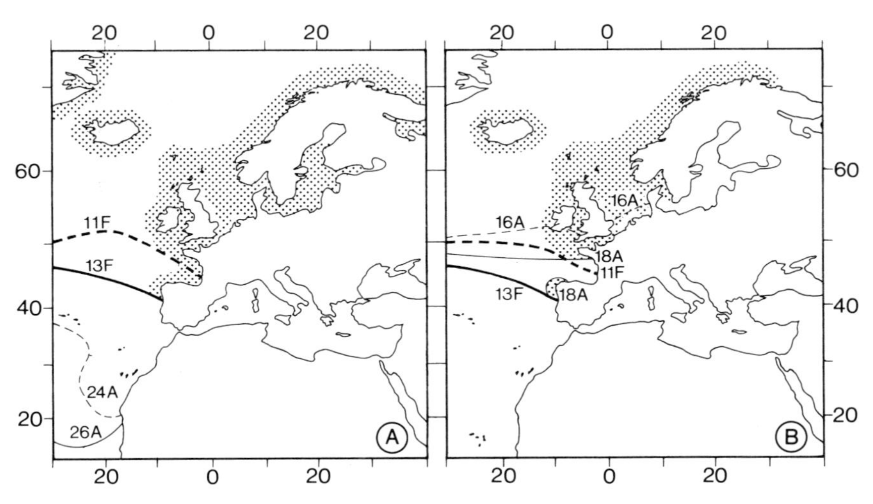

Fig. 5. Expected effects of rising seawater temperatures on the location of southern boundaries in Europe. *Chorda filum* (a) with winter reproduction limit at 13° February isotherm (26° August isotherm represents upper tolerance limit); *Laminaria hyperborea* (b) with winter reproduction limit at 13° February isotherm and summer lethal limit at 18° August isotherm; present distribution ranges are stippled (based on LÜNING, 1985; BREEMAN, 1988). See text for further explanation.

daries in Europe has probably varied through time both during glacial and interglacial periods (Table 1). The same situation existed on American coasts during the glaciation, but it has now changed fundamentally on that coast, all southern boundaries being persistent summer lethal limits (Table 1). In view of the fact that upper lethal limits appear to be genetically more firmly fixed than thermal requirements for reproduction, one might expect a species to adapt these requirements and extend its range, until lethal summer temperatures would also become limiting. This seems to have happened for instance in the brown alga *Scytosiphon lomentaria* (TOM DIECK, 1987), but few data are so far available for other species. The extent to which such ecotypes have evolved may well provide further insight into the evolutionary history and region of origin of a seaweed species.

6. ECOLOGICAL CONSEQUENCES OF SHIFTING GEOGRAPHIC BOUNDARIES OF SEAWEEDS IN EUROPE

Tropical to (warm) temperate seaweed species reaching their northern boundaries in northwestern Europe are mainly smaller red, green, and brown algae. Range-extensions of these species will probably not have much impact. There may be some local enrichment of the flora, but fundamental changes in community structure are not to be expected.

In contrast, much more far-reaching effects are to be expected by northward shifts of southern boundaries of some arctic to cold-temperate seaweeds. For instance, three species of *Laminaria*, *L. hyperborea* (see above), *L. saccharina* and *L. digitata* reach southern boundary in this region. In fact, records of *Laminaria hyperborea* and *Laminaria saccharina* near their southern boundaries in the Iberian Peninsula have become more scarce since the early part of this in century (LÜNING, 1985: 70). Following a rise in summer and/or winter temperatures, marked northward shifts of their southern boundaries are to be expected. In the extreme case of summer temperatures rising by 4°C (*c.f.* DE VOOYS, this volume), these *Laminaria* species would disappear from the Iberian Peninsula, the Atlantic coast of France, the southern parts of the British Isles, the North Sea, and southern Norway. As major canopy forming algae, they determine community structure in subtidal kelp forests, and their extinction would undoubtedly cause major changes in these ecosystems.

Acknowledgements—I thank O.A. Singelenberg, C. van den Hoek and I. Novaczek for critical comments on the manuscript. Financial support from the Ministry of Housing, Physical Planning and Environmental Hygiene to participate in a workshop on 'Expected effects of climatic changes on marine coastal ecosystems' is gratefully ackowledged.

7. REFERENCES

BRADLEY, R.S., 1985. Quaternary paleoclimatology. Allan and Unwin, London: 1-472.

BREEMAN, A.M., 1988. Relative importance of temperature and other factors in determining geographic boundaries of seaweeds: experimental and phenological evidence.—Helgoländer Meeresunters. **42:** 199-241.

BROECKER, W.S., 1987. Unpleasant surprises in the greenhouse?—Nature **328:** 123-126.

CLIMAP PROJECT MEMBERS, 1976. The surface of the ice-age earth.—Science **191:** 1131-1137.

DIECK, I. TOM, 1987. Temperature tolerance and daylength effects of several geographical isolates of *Scytosiphon lomentaria* (Scytosiphonales, Phaeophyceae) of the North Atlantic and North Pacific Ocean.—Helgoländer Meeresunters. **41:** 307-321.

GERARD, V.A. & K.R. DU BOIS, 1988. Temperature ecotypes near the southern boundary of the kelp *Laminaria saccharina*.—Mar. Biol. **97:** 575-580.

HOEK, C. VAN DEN, 1982a. Phytogeographic distribution groups of benthic marine algae in the North Atlantic Ocean. A review of experimental evidence from life history studies.—Helgoländer Meeresunters. **35:** 153-214.

——, 1982b. The distribution of benthic marine algae in relation to the temperature regulation of their life histories.—Biol. J. Linn. Soc. **18:** 81-144.

——, 1984. World-wide latitudinal and longitudinal seaweed distribution patterns and their possible causes, as illustrated by the distribution of Rhodophytan genera.—Helgoländer Meeresunters. **38:** 227-257.

JOOSTEN, A.M.T. & C. VAN DEN HOEK, 1986. World-wide relationships between red seaweed floras: a multivariate approach.—Botanica Marina **24:** 195-214.

KENNETT, J., 1982. Marine Geology. Prentice-Hall, Inc., Englewood Cliffs, N.J.: 1-813.

LAWSON, G.M. & D.M. JOHN, 1977. The marine flora of the Cap Blanc peninsula: its distribution and affinities.—Bot. J. Linn. Soc. **75:** 99-118.

LAWSON, G.M. & D.M. JOHN, 1982. The marine algae and coastal environment of tropical West Africa. Cramer, Vaduz: 1-455.

LÜNING, K., 1985. Meeresbotanik. Thieme Verlag, Stuttgart: 1-375.

LÜNING, K. & W. FRESHWATER, 1988. Temperature tolerance of northeast Pacific marine algae.—J. Phycol. **24:** 310-315.

PRUD'HOMME VAN REINE, W.F. & C. VAN DEN HOEK, 1988. Biogeography of Cape Verdian seaweeds.—Cour. Forsch. Inst. Senckenb. **105:** 35-49.

——, 1989. Biogeography of Macaronesian seaweeds.—Cour. Forsch. Inst. Senckenb.: in press.

RUDDIMAN, W.F., D.S. TOLDERLUND & A.W.H. BÉ, 1970. Foraminiferal evidence of a modern warming of the North Atlantic Ocean.—Deep-Sea Research **17:** 141-155.

RUDDIMAN, W.F. & A. MCINTYRE, 1981. Ocean mechanism for amplification of the 23,000-year ice-volume cycle.—Science **212:** 617-627.

SOUTH, G.R., 1987. Biogeography of the benthic marine algae of the North Atlantic Ocean - an overview.—Helgoländer Meeresunters. **41:** 273-282.

SOUTH, G.R. & I. TITTLEY, 1986. A checklist and distributional index of the benthic marine algae of the North Atlantic Ocean.—Br. Mus. (Nat. Hist.), London: 1-76.

U.S. NAVY, 1981. Marine Climatic Atlas of the World. Vol. 9. World-wide means and standard deviations. U.S. Government Printing Office, Washington.

EXPECTED BIOLOGICAL EFFECTS OF LONG-TERM CHANGES IN TEMPERATURES ON BENTHIC ECOSYSTEMS IN COASTAL WATERS AROUND THE NETHERLANDS

C.G.N. DE VOOYS

Netherlands Institute for Sea Research, P.O. Box 59, 1790 AB Den Burg, Texel, The Netherlands

ABSTRACT

A comparison between the course of air temperatures during the last century (either on a global scale or at De Bilt) with temperatures of Dutch coastal waters shows that the temperatures in Dutch coastal waters are not only determined by air temperatures; hydrographical alternations on the North Atlantic Ocean will have had a clear influence too.

A comparison of the macrobenthic fauna of the Wadden Sea with those of the Seine estuary (mean temperatures 2°C higher) and the Gironde estuary (mean temperatures 4°C higher) leads to the conclusion that at a rise in temperature of 2°C an enrichment of ~20% of the number of species would occur, whereas at a rise of 4°C an enrichment of ~30% might be expected.

In the warmest period of the Eemien (Pleistocene) the annual mean water temperature in the Eemien Sea was about 14–15°C, compared with ~10°C for the present Dutch Wadden Sea. All species of molluscs known from the Eemien still occur in Europe, and a good resemblance exists between the molluscs species of the Dutch Eemien and the recent ones of Arcachon Bay. This supports the above conclusion.

1. INTRODUCTION

The extensive use of fossil fuel resources for energy production causes an increasing accumulation of carbon dioxide in the atmosphere. From this accumulation and that of some other gaseous substances a rise of temperature of the earth's atmosphere can be predicted. However, so far no precise predictions of future climate developments can be made.

This study was carried out to assess the possible effects of a supposed rise in coastal seawater temperatures: a rise in mean summer temperatures of +2°C to +4°C in the coming 50–100 years, compared with the mean summer temperatures at present.

First, changes in coastal water temperatures in the past 100 years will be discussed. Next, a comparison is made between the present bottom faunas of the Dutch Wadden Sea with bottom faunas of southern European estuaries with higher mean annual water temperatures. Also a comparison is made with the Mollusc bottom fauna of the Eemien Sea. On the basis of the results of these comparisons, the biological effects of the assumed rise in coastal water temperatures will be discussed.

2. LONG-TERM GLOBAL CHANGES IN AIR TEMPERATURES VS CHANGES IN THE DUTCH COASTAL WATER TEMPERATURES.

Temperature measurements in the air as well as in Dutch coastal waters during the last 90–120 years are available. The changes over that period can be compared with global changes in air temperature to see whether the Dutch coastal water temperatures ran parallel with the global ones.

JONES *et al.* (1986) published global mean air temperatures for each of the years between 1861 and 1984 combined for 15 regions on land as well as on sea. The data from the 2 hemispheres have been combined to produce comprehensive estimates of global mean air temperatures (Fig. 1). The results show only little trend before 1900, a clear warming till about 1940, a stabilization (or slight decrease) till the mid-seventies, followed by a rapid warming during the last 10 to 15 years. They conclude that a long-term warming trend occurs, with the overall change in the right direction and of the correct magnitude, compared with the change expected from the CO_2 accumulation hypothesis. The relative stability of temperatures between 1940 and the mid-seventies presumes the existence of some compensatory factor.

In Dutch coastal waters long-term series of temperature measurements of surface waters are available from Den Helder - 't Horntje (1861–1980) and the Oosterschelde (1894–1980) (VAN DER HOEVEN, 1982, 1983). To straighten out variations in the mean yearly temperatures, means for groups of 10 years have

J.J. Beukema et al. (eds.), Expected Effects of Climatic Change on Marine Coastal Ecosystems, 77–82.

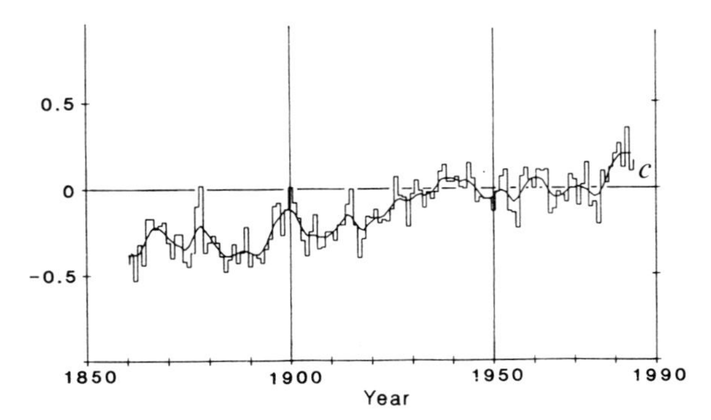

Fig. 1. Variations of global mean air temperature since 1861, based on sea-surface air temperature data. The smooth curve shows 10-yr gaussion filtered values (from JONES *et al.*, 1986).

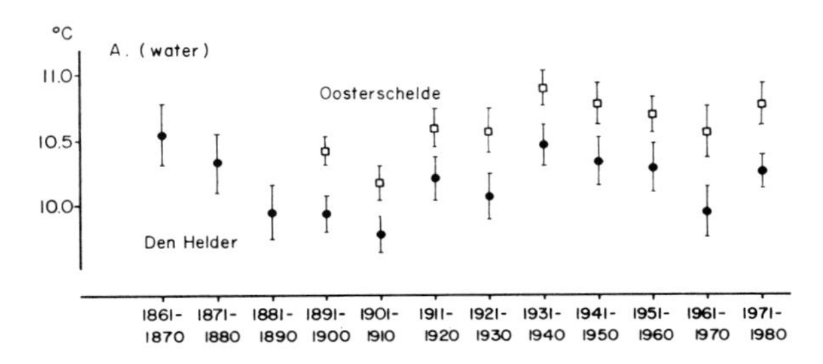

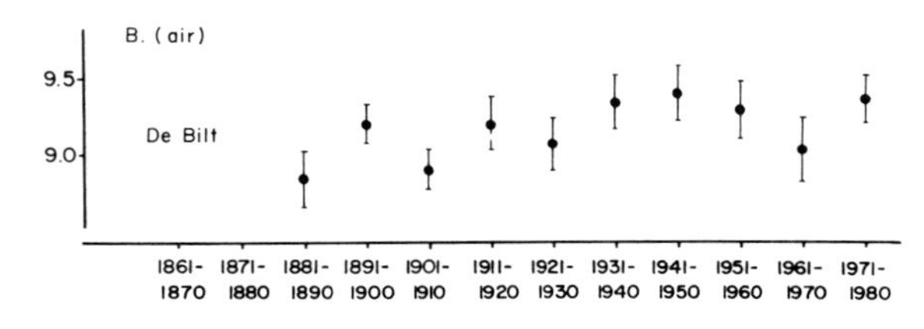

Fig. 2. Mean temperatures (10-yr averages ±1 s.e.) for:
a. water temperatures at Den Helder (westernmost entrance of the Wadden Sea) (solid points) and in the Oosterschelde (S.W. Netherlands) (open points)
b. air temperature at De Bilt (central part of the Netherlands).

been calculated, with standard errors. The results are shown in Figs 2a and b.

In Fig. 2a (Den Helder) two periods of temperature decrease can be distinguished, from 1861–1910 and 1941–1970. A warming occurred between 1910 and 1940 and after 1970. In the Oosterschelde (Fig. 2a) about the same temperature changes can be observed, but they were smaller, with smaller standard errors. In the water temperatures there is no clear evidence for a steady trend of temperature increase from 1900 to 1980. In fact, at Den Helder (Fig. 2a), mean temperatures from 1861 to 1870 were higher than from 1971 to 1980.

When a comparison is made between the changes in global air temperature as determined by JONES *et al.* (1986) and the measurements in Dutch coastal waters, especially with the longest series Den Helder (Fig. 2a), a significant difference is observed between 1860 and 1910 (little trend *vs* significant decrease), a correspondence occurs between 1910 and 1941, and a slight difference between 1941 and 1970 (stabilization *vs* slight, non-significant decrease). There is again a correspondence during the last 10 or 15 years (increase).

When air temperatures at De Bilt over the period 1880–1980 (VAN ENGELEN, 1983) are plotted in the same way as the seawater temperatures, results are obtained as given in Fig. 2b. A comparison with Fig. 2a (seawater temperatures) shows a clear difference in the trend of temperature development over the period 1880–1910, but in the period 1910–1980 the trend is the same.

This means that the temperature of the Dutch coastal waters will not only be determined by air temperature, but also by hydrographic changes on the North Atlantic Ocean, for example the course of the Gulf stream.

Thus, temperatures in Dutch coastal waters do not exactly follow the global and local air temperature changes. However, the recent warming up starting around 1970 and the lowering trend after 1940 as well as the rising trend between 1920 and 1940 in the global air temperatures were reflected in the Dutch coastal seawater temperatures.

3. SHORT-TERM WATER TEMPERATURE CHANGES AND EXPECTED MEAN TEMPERATURES

In the shallow parts of the North Sea as well as in the Wadden Sea and the Oosterschelde, where tidal currents prevent the development of summer stratification, the water follows the air temperature with a short time lag of 6 days for the Wadden Sea and 35 days for the North Sea. This means that the North Sea water will usually show an appreciable time lag with respect to seasonal meteorological conditions. The sea follows the annual course more closely the more shallow the water is. Also, the amplitudes are smaller where the water is deeper (Fig. 3).

Seawater temperatures in relatively shallow and sheltered basins such as the Dutch Wadden Sea (in the north) and the Oosterschelde (in the south) will in both summer and winter deviate from those of offshore coastal water. Since in the Wadden Sea and in the Oosterschelde the amplitudes are closer to those of the air, these areas are warmer in spring and summer and colder in autumn and winter. Mean monthly temperatures in the Oosterschelde basin in the period 1971–1980 are given in Fig. 4. Also, for the

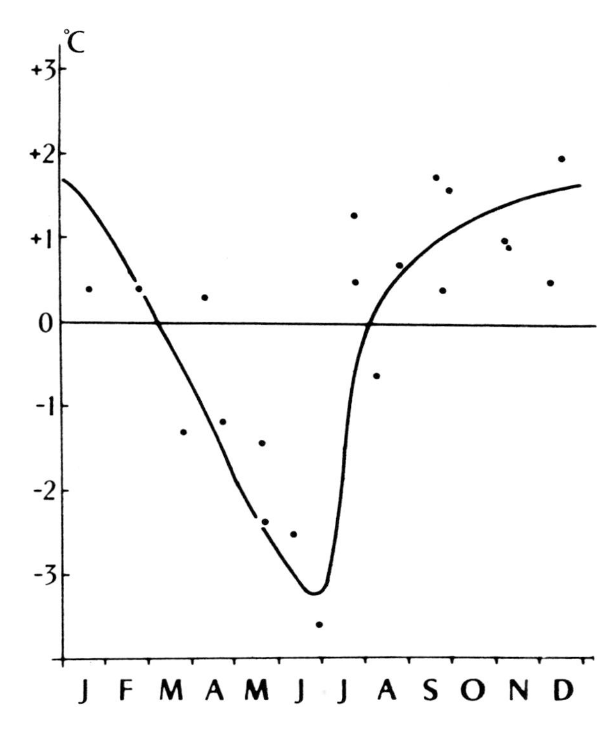

Fig. 3. Temperature differences between North Sea and Wadden Sea (Ameland watershed) in course of the year (from POSTMA, 1983).

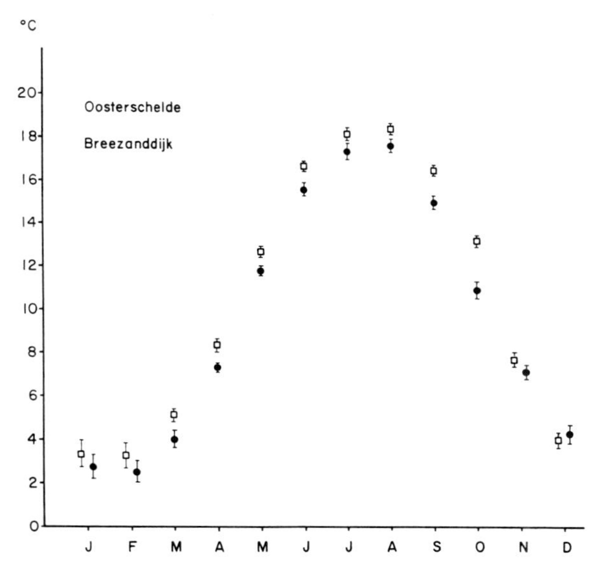

Fig. 4. Annual course of mean seawater temperature (monthly averages with 1 s.e. for the 1971-1980 period) at 2 coastal stations, viz. in the Wadden sea (Breezanddijk: solid points) and in the S.W. of the Netherlands (Oosterschelde: open points).

Wadden Sea, mean temperatures at Breezanddijk, near the Afsluitdijk, are shown.

To indicate summer temperatures, the mean monthly temperatures of August (the warmest month) are taken, and the highest and lowest mean temperatures recorded during the 1971–1980 period are given to show the variation. Mean summer temperatures in the Oosterschelde, ~18.5°C in August, were somewhat higher than those of the Wadden Sea, ~17.5°C. The difference in mean winter temperatures between the Oosterschelde and the Wadden Sea was ~1°C. In the Oosterschelde mean February temperatures were ~3.5°C, and in the Wadden Sea ~2.5°C.

Offshore water temperatures along the Dutch North Sea coast were obtained from two light vessels, Noord Hinder (before the Flemish coast) and Haaks-Texel (before Texel Island). Generally, mean summer temperatures were only slightly lower than those of the Wadden Sea and the Oosterschelde, but mean winter temperatures were higher by 2 to 3°C. To get an impression of the variations in the coastal water temperatures, and the extreme temperatures which can occur in various locations, mean seawater temperatures in the warmest month (August) and in the coldest winter month (February) are given in Table 1.

If summer temperatures were to rise by 2°C (or 4°C), mean summer water temperatures would become in the Oosterschelde basin ~20.5° (or 22.5°C), and in the Wadden Sea ~19.5°C (and 21.5°C), respectively.

If winter temperatures were to rise by 2°C, mean temperatures during the coldest month would become 5°C in the Oosterschelde, and 4°C in the Wadden Sea. If winter temperatures were to decrease by 2°C, mean temperatures would become ~1°C in the Oosterschelde basin and ~0°C in the Wadden Sea.

Not only mean summer temperatures and mean winter temperatures are important for life in the Dutch coastal waters, but so are extreme temperatures in summer and winter. When the highest and the lowest mean monthly temperatures are plotted against the mean yearly temperature, a linear relation appears. The mean temperature of the coldest month clearly

TABLE 1

Monthly averages of summer (August) and winter (February) seawater temperatures (in °C) in various Dutch coastal areas. Also shown: ranges, standard deviation of mean and number of observations (n).

	Summer			*Winter*			
	mean	*range*	*st.dev.*	*mean*	*range*	*st.dev.*	*n*
Haaks-Texel 1948-1976	17.3	15.7 to 19.5	0.91	4.6	-0.1 to 6.2	1.38	29
Noord Hinder 1948-1981	16.8	15.7 to 18.5	0.73	6.5	1.3 to 8.6	1.34	33
Wadden Sea 1951-1980	17.4	15.4 to 20.2	0.98	2.0	-1.1 to 5.0	1.58	29
Oosterschelde 1951-1980	18.6	17.3 to 21.4	0.95	2.9	-1.4 to 5.4	1.86	29

becomes higher when the annual mean temperature rises. Also, but less clearly, the mean temperature of the warmest month rises when the mean year temperature is higher. Generally, an alternation in the mean year temperature goes with a considerable alternation in the mean winter temperature, and a lesser alternation in summer temperature.

The temperature of the water on the tidal flats at high tide can be higher or lower than the mean water temperature in the entire Oosterschelde or the Wadden Sea. Solar radiation during daytime warms up the shallow water on the tidal flats. During the night the reverse occurs and the water cools by losing radiant heat. The shift of the lunar tide through the day causes a pulse in the daily water temperature cycle with a period of about 2 weeks (14.76 days) and a variation of the daily mean temperature with the same period. This periodicity is caused by the fact that for one week tidal flats are emerged during the mid of the day and directly exposed to solar radiation and the next week submerged during the day. The temperature of the water on the tidal flats at high tide differs from the mean temperature by ~0.5°C with a standard deviation of ±0.1°C (VUGTS & ZIMMERMAN, 1975, 1985). In stagnant pools on drained tidal flats, water temperature can rise to ~30°C (KRISTENSEN, 1957).

When the flats are emerged during the day and exposed to solar radiation, the sediment warms up. In spring and summer the surface sediment often considerably surpasses the air temperature, whereas during autumn and winter the surface sediment generally stays below the maximum temperature of the air (DE WILDE & BERGHUIS, 1979). According to these authors, temperatures up to 29°C have been measured on the surface of the sediment. If the temperature of the coastal water were to rise 2 or 4°C, temperatures of emerged sediment would become even higher because of a higher start temperature.

Temperature variations in the sediment differ clearly from those in the water column, and as a consequence the bottom fauna also lives in temperature ranges different from those in the water column.

4. A COMPARISON BETWEEN THE PRESENT BOTTOM FAUNAS IN THE DUTCH WADDEN SEA AND THOSE OF THE ESTUARIES AT LOWER LATITUDES (SEINE AND GIRONDE)

Effects of a possible rise in seawater temperature in coastal waters of 2°C and 4°C can be investigated by comparison with more southernly areas with higher mean temperatures and a fauna similar to that of the Dutch Wadden Sea.

The mean yearly seawater temperature in the Marsdiep (Dutch Wadden Sea) is 10.1°C. In the Seine estuary, a mean yearly temperature of 12.2°C occurs (DESPREZ, 1981). The difference of about 2°C is found the whole year round (fig. 1 of BEUKEMA & DESPREZ, 1986). In the Gironde estuary, a mean yearly seawater temperature of 13.8°C is found (BACHELET, 1979), nearly 4°C higher than that of the Marsdiep. Again such differences occur throughout the year.

The marine bottom fauna of the Dutch Wadden Sea has been listed by WOLFF & DANKERS (1983); an inventory of the marine fauna of the Seine estuary is given by DESPREZ (1981), and an inventory of that of the Gironde estuary has been made by BACHELET (1979).

A comparison will be attempted between the macrozoobenthos of these regions, on the basis of two assumptions. The first is that the knowledge of the number of macrozoobenthos species in the Wadden Sea is nearly complete, *i.e.* that the chance that new macrozoobenthos species are discovered in the Wadden Sea is small. The second assumption is that sampling is aselect, *i.e.* the chance that a Seine or Gironde sample contains macrozoobenthos species occurring in the Wadden Sea is the same as for species that do not. This assumption is necessary, because the number of species found in a restricted sampling programme, as in the Seine or Gironde estuary, is much smaller than the number of species known from the Wadden Sea (compiled from records of a high number of authors). The species lists given by the two French authors are very incomplete. Hydrographical and geographical conditions of French estuaries roughly correspond with those in the Dutch Wadden Sea.

For a comparison with a macrobenthic fauna living at a mean temperature of nearly 4°C higher than that of the Wadden Sea, the numbers of macrozoobenthos species belonging to 8 taxonomic groups from the Gironde estuary as given by BACHELET (1979) were compared with the numbers of Wadden Sea species belonging to the same taxonomic groups

TABLE 2A

A comparison between numbers of benthic animal species observed in the Gironde estuary (by Bachelet, 1979) and the total number ever found in the Wadden Sea (Wolff & Dankers, 1983).

Taxonomic group	Gironde	Wadden Sea	Nr of Gironde sp. not found in the Wadden Sea
Gastropoda	4	47	2
Bivalvia	4	35	0
Polychaeta	12	105	4
Mysidaceae	2	8	0
Cumaceae	2	5	2
Isopoda	4	14	2
Amphipoda	6	62	1
Decapoda	3	23	0
total	37	299	11 (=30%)

TABLE 2B

A comparison between numbers of benthic animal species observed in the Seine estuary ((by Desprez, 1981) and the total number ever found in the Wadden Sea (Wolff & Dankers, 1983).

Taxonomic group	Seine	Wadden Sea	Nr of Seine sp. not found in the Wadden Sea
Bivalvia	6	35	0
Polychaeta	30	105	8
Isopoda	4	14	0
Amphipoda	17	62	5
Decapoda	2	23	0
total	59	239	13 (=22%)

(WOLFF & DANKERS, 1983). For each taxonomic group the number of species was noted that occurred in the Gironde estuary but not in the Wadden Sea. Among the 8 taxonomic groups considered, a total of 299 species are mentioned for the Wadden Sea and 37 for the Gironde estuary. Among the 37 species mentioned for the Gironde estuary, 11 (or 29.7% of the total) have not been found in the Wadden Sea (Table 2a).

For a comparison with a macrobenthos fauna living at a mean temperature of 2°C higher than that of the Wadden Sea, the number of macrozoobenthos species belonging to 5 taxonomic groups (DESPREZ, 1981) were compared in the same way with the number of Wadden Sea species belonging to the same taxonomic groups (WOLFF & DANKERS, 1983). For each taxonomic group the number of species was noted that occurred in the Seine estuary but not in the Wadden Sea. Totally, of the 5 taxonomic groups considered, 239 species are mentioned in the Wadden Sea and 59 in the Seine estuary. Of the 59 species mentioned to occur in the Seine estuary, 13 (or 22% of the total) have not been found in the Wadden Sea (Table 2b).

When the differences in macrozoobenthos species between the Gironde estuary, the Seine estuary and the Wadden Sea are compared, the difference found between the Seine estuary and the Wadden Sea lies between the difference found between the Gironde estuary and the Wadden Sea. This can be expected when temperature difference is a dominant factor in determining the occurrence of macrozoobenthos species.

An increase in the mean water temperature of Dutch coastal waters by 2°C or 4°C may eventually result in an increase in the number of macrobenthos species by ~22 and ~30%, respectively.

No information can be given on the possible rate of dispersal of macrobenthos in a northern direction. Neither is it possible to indicate which species would vanish if an increase in the mean coastal water temperature of 2°C or 4°C would occur.

5. A COMPARISON BETWEEN THE PRESENT BOTTOM FAUNAS IN THE DUTCH WADDEN SEA AN THOSE OF THE EEMIEN SEA (PLEISTOCENE)

A comparison of the Wadden Sea bottom fauna with a similar one at higher mean water temperatures can be made also by studying the fauna of a relatively recent geological period, the Eemien (the last interglacial).

During the Eemien a shallow sea, which may be compared with the present Wadden Sea (but it was somewhat deeper), existed in the western part of the Netherlands. The members of the fossil fauna of mollusc shells are nearly all species that still occur in European coastal waters.

TABLE 3

Recent distribution of Eemien Lamellibranch species (from Spaink, 1958).

Area	*number of species*
North Sea	45 = 34.6%
Mediterranean	37 = 28.5%
Atlantic coast from Scotland to Gibraltar	100 = 76.9%
Norwegian coast up to Finmark	5 = 3.8%

When the highest sea level was reached during the Eemien, a moderate climate reigned which was warmer than the present-day climate in the Netherlands. Mollusc species from the Eemien fauna are now dispered along the Atlantic coast from Gibraltar to the Channel, and even in the Mediterranean. The recent distribution over various coastal regions of mollusc species that also occurred in the Eemien is shown in Table 3.

VAN STRAATEN (1956) found a high similarity between the fossil Eemien mollusc fauna and the recent mollusc fauna of the Bay of Arcachon (situated in SW France, to the West of Bordeaux). Of the Dutch fossil Eemien fauna, 63.3% of the species are now found in the Bay of Arcachon.

The mean yearly seawater temperature in the Bay of Arcachon is 14.6°C, against the Wadden Seas (Marsdiep) 10.1°C, a difference of 4.5°C. The similarity of the bottom fauna of the Dutch Eemien and that of the present-day Bay of Arcachon, and the difference between the Eemien fauna and that of the present Wadden Sea indicate that the Eemien Sea was warmer than the Wadden Sea at present. The temperature of the Eemien Sea during its warmest period is thought to have been 14–15°C on the basis of faunal comparison. The diatom assemblage of the Eemien has not yet been studied sufficiently intensively to make any independent estimation of the possible temperature of the Eemien Sea.

From the fossil records of the Eemien the conclusion may be drawn that an increase in the mean temperature of the Dutch coastal waters by ~4°C could result in a change in the bottom fauna which will probably become similar to the present-day bottom fauna in Arcachon Bay.

6. REFERENCES

BACHELET, G., 1979. Dynamique de la macrofauna benthique et production des Lamellibranches de l'estuaire de la Gironde. Thèse, Université Pierre et Marie Curie: 163 pp.

BEUKEMA, J.J. & M. DESPREZ, 1986. Single and dual growing seasons in the tellinid Bivalve *Macoma balthica* (L.).—J. exp. mar. Biol. Ecol. **102:** 35-45.

DESPREZ, M., 1981. Etude du macrozoobenthos intertidal de l'estuaire de la Seine. Thèse, Univ. de Rouen: 186 pp.

ENGELEN, A.F.V. VAN, 1983. Quantielklassificatie en presentatie van maand-, seizoen- en jaargemiddelde temperaturen van De Bilt over het tijdvak

1881-1982.—K.N.M.I. Technische Rapporten TR-38: 8 pp, 17 tabellen.

HOEVEN, P.C.T. VAN DER, 1982. Watertemperatuur en zoutgehalte waarnemingen van het Rijksinstituut voor visserijonderzoek (RIVO): 1860-1981.—K.N.M.I. Scientific Report W.R. 82-8.

——, 1983. Observations of surface watertemperature and salinity in the Easterscheldt 1894-1982.—K.N.M.I. Scientific Report W.R. 83-12.

JONES, P.D., T.M.L. WIGLEY & P.B. WRIGHT, 1986. Global temperature variation between 1861 and 1984.—Nature **322:** 430-434.

KRISTENSEN, I., 1957. Differences in density and growth in a cockle population in the Dutch Wadden Sea.—Archs néerl. Zool. **12:** 351-453.

POSTMA, H., 1983. Hydrography of the Wadden Sea movements and properties of water and particulate matter. In: W.J. WOLFF. Ecology of the Wadden sea. Vol 1, part 2. A.A. Balkema, Rotterdam: 1-75.

SPAINK, G.,1958. De Nederlandse Eemlagen.1. Algemeen overzicht.—Wet. Meded. KNNV no **29:** 25-33.

STRAATEN, I.M.J.U. VAN, 1956. Composition of shell beds formed in tidal flat environment in The Netherlands and in the Bay of Arcachon (France).—Geologie Mijnb. **18:** 206-226.

VUGTS, H.F. & J.T.F. ZIMMERMAN, 1975. Interaction between daily heat balance and the tidal cycle.—Nature, Lond. **225:** 113-117.

VUGTS, H.F. & J.T.F. ZIMMERMAN, 1985. The heat balance of a tidal flat area.—Neth. J. Sea Res. **19:** 1-14.

WILDE, P.A.W.J. DE & E.M. BERGHUIS, 1979. Cyclic temperature fluctuations in a tidal mud-flat. In: E. NAYLOR & R.G. HARTNOLL. Cyclic phenomena in marine plants and animals. Proc. 13th EMBS. Pergamon Press, Oxford: 435-441.

WOLFF, W.J. & N. DANKERS, 1983. Preliminary checklist of the zoobenthos and nekton species of the Wadden Sea. In: W.J. WOLFF. Ecology of the Wadden Sea. Vol. 1, Part 4. A.A. Balkema, Rotterdam: 24-60.

EXPECTED EFFECTS OF CHANGES IN WINTER TEMPERATURES ON BENTHIC ANIMALS LIVING IN SOFT SEDIMENTS IN COASTAL NORTH SEA AREAS

J.J. BEUKEMA

Netherlands Institute for Sea Research, P.O.Box 59, 1790 AB Den Burg, Texel, The Netherlands

ABSTRACT

Dynamics of macrobenthic animals were studied quantitatively during a 20-year period on tidal flats in the Wadden Sea. During this period, 4 winters were very cold (mean temperature more than 2°C below the long-term average), whereas also a period with 8 too mild winters in succession occurred. No less than 10 out of the total of 28 species studied in detail were found to be sensitive to cold winters. Their overwinter survival and numbers found after a severe winter were lower than after normal or mild winters. The 4 polychaete and 4 bivalve species among these 10 sensitive species probably died from low temperatures, whereas the 2 crustaceans moved to deeper off-shore waters.

Species numbers and biomass were relatively low after cold winters, particularly at the lower tidal flats where sensitive species were numerous. At high tidal flats the share of sensitive species was low; individuals of sensitive species living there suffered more from cold winters than those at lower intertidal levels.

Recovery after cold winters was rapid (within 1 or 2 years) in most species. Total-biomass values increased even faster as a consequence of generally highly successful recruitment after a cold winter. Because most invertebrate predators were scarce on the tidal flats after a severe winter, densities in several prey species could rapidly increase during the summers following cold winters.

It is predicted that higher winter temperatures in the Wadden Sea area will cause higher species richness and a more stable biomass of the macrozoobenthic fauna living on the tidal flats.

1. INTRODUCTION

Though an increasing trend in temperatures appears to emerge now on a global scale (JONES *et al.*, 1986), local trends may deviate (compare also DE VOOYS, this issue, and GOODESS & PALUTIKOF,this issue). In particular in several coastal areas, including NW Europe, temperature trends during the last few decades appear to be decreasing rather than increasing. Moreover, a period with relatively high mean annual temperatures in Europe was characterized by relatively low and variable winter temperatures over most of Europe (LOUGH *et al.*, 1983).Thus, to predict the possible impact of future temperature changes on coastal ecosystems, effects of both higher and lower than average temperatures should be considered.

Several reports have been published on high rates of mortality in various macrozoobenthic species during severe winters in the North Sea area, both on tidal flats (BLEGVAD, 1929; SMIDT, 1944; CRISP *et al.*, 1964; HANCOCK & URQUHART, 1964; BEUKEMA, 1979, 1985; REICHERT & DÖRJES, 1980) and in the shallow sublitoral (CRISP *et al.*, 1964; ZIEGELMEIER, 1964, 1978; DÖRJES, 1980; BEUKEMA *et al.*, 1988). On the other hand, few reports appear to be available reporting on such effects of extreme summer temperatures. Maximum lethal temperatures reported for tidal-flat species appear to be high relative to water or air temperatures reached in summer. For *Macoma balthica*, a species with a northern area of distribution, RATCLIFFE *et al.* (1981) found ~35°C, which temperature will be reached only rarely and for short periods at the surface of the sediment of tidal flats in the North Sea area. Neither am I aware from my own experience (a 20-year sampling programme) of any soft-bottom species in this area sensitive to high summer temperatures. Therefore, the following evaluation of temperature effects on macrozoobenthos in North Sea coastal areas will be restricted to a discussion of influences of fluctuations in winter temperatures.

Data have been used both from the literature and from an extensive sampling programme in the westernmost part of the Wadden Sea.

2. MATERIAL AND METHODS

At Balgzand, a tidal-flat area of 50 km^2 in the westernmost part of the Wadden Sea, 15 permanent

J.J. Beukema et al. (eds.), Expected Effects of Climatic Change on Marine Coastal Ecosystems, 83–92.

stations were sampled during 20 years (1969–1988) at least annually in a uniform way. Among these 15 stations, 3 were square plots of 900 m^2 and these were sampled 2 to 5 times per year. The samples of August and March have been used to calculate over-winter survival in each of the 20 years. Additional data are available from 12 transects of 1 km length, sampled at least annually in March and yielding an estimate of densities at the end of each winter (together with the March data of the 3 plots). These transects were also sampled in August-September during the periods 1978-1988 and (for shallow-living molluscs only) 1969-1977.

Details of the sampling procedure can be found in earlier papers (BEUKEMA, 1974, 1979). In short: cores were taken and sieved in the field on 1 mm mesh screens. All samples were sorted in the laboratory while the animals were still alive. The total area sampled at each sampling was 0.9 m^2 along the transects and 1.0 to 1.7 m^2 at the plots. Abundance estimates are expressed both in numbers per m^2 and in $g.m^{-2}$ AFDW (ash-free dry weight).

The 15 stations are scattered over the area and cover nearly the full tidal range (almost 1.5 m) and sediment types (soft muds to clean sands).

Monthly means of water temperatures (as measured in the tidal inlet) range from 3°C for the coldest to 18°C for the warmest month (VAN DER HOEVEN, 1982; compare fig.1 in BEUKEMA & DESPREZ, 1986 and fig. 4 in DE VOOYS, this issue). Mean air temperatures follow a similar pattern. On the tidal flats, bottom surface temperatures will be affected by both water and air temperatures. As the latter are more readily available (monthly weather reports issued by KNMI, de Bilt, the Netherlands), these will be used for grading the winters and studying correlations with rates of overwinter survival in the various species. In 28 species sufficient data were available to correlate late-winter abundance (and in nearly all of these species overwinter survival too) with winter temperatures (data were used from the nearby weather stations at Den Helder and De Kooy, *viz.* mean air temperatures during the January-February-March periods of each year, compare Fig. 1a).

Winters in the western Wadden Sea differ widely in character (Fig. 1). In most winters mild westerly winds prevail, temperature never drops below −10°C and only during less than 10 days the maximum temperature is below 0°C. Out of the 20 winters during the period of observation, 13 were mild, *i.e.* with mean temperatures above the long-term average. Only 4 were severe (see Fig. 1a) with records below −10°C on ~7 days and with 15 to 32 days showing below-zero temperatures all day. The other 3 winters were too cold, but not really severe, with 0 to 2 days with

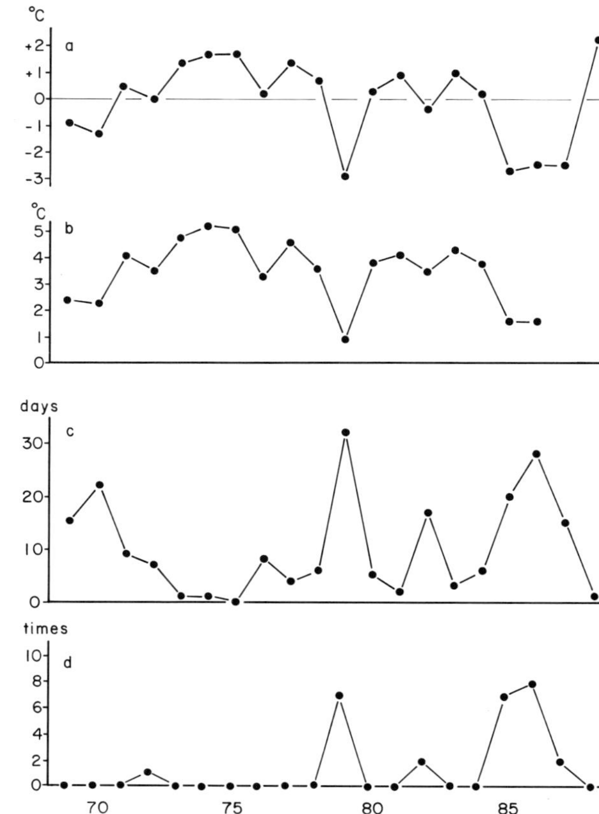

Fig. 1. Some parameters of winter character in the western part of the Wadden Sea during the 1969-1988 period: (a) Deviation of mean air temperature (°C) in winter (Jan-Febr-Mar) from the long-term average at the Den Helder weather station (b) Mean seawater temperature (°C) at the Marsdiep tidal inlet in winter (Jan-Febr-Mar) (c) Numbers of days with maximum temperatures below 0°C for each (entire) winter (d) Numbers of days with a minimum temperature below −10°C for each (entire) winter. Data from monthly weather reports issued by KNMI, De Bilt.

a minimum temperature below −10°C and 15 to 22 days with maximum temperatures below 0°C.

During the period of observation of the tidal-flat fauna, the coldest winter was 1978-79 (Fig. 1). Winters as severe as this one have been observed in this area only about 10 times per century. During the 1984-87 period 3 cold winters occurred in succession, whereas a series of 8 mild winters occurred during the seventies. In particular the changes in the fauna observed during these periods may indicate what will happen during possible future periods with consistent temperature changes.

Correlations were evaluated statistically by the Spearman rank correlation test (usually with n=20).

TABLE 1

Data showing which macrobenthos species living on tidal flats in the western part of the Wadden Sea appear to be sensitive to low winter tempeatures (mean air temperatures during the Jan-Febr-Mar period at a nearby weather station).
(a) correlation between overwinter survival and winter temperature (Spearman r, n=20, 20-year data set from 3 square plots at Balgzand);
(b) correlation between late-winter density ($n.m^{-2}$) and winter temperature (Spearman r, n=20, 20-year data set from 15 sampling stations on Balgzand);
(c) overwinter survival during the coldest winter (1978-79) of the period of observation;
(d) mean overwinter survival during 3 other cold winters (1984-85, 1985-86 and 1986-87, compare Fig. 1);
(e) mean overwinter survival during 5 average to mild winters (1979-84 period);
(f) publications stating winter sensitivity in the species concerned.
The statistical significance is indicated by *, ** or *** for p<0.05, 0.01 and 0.001, respectively. In 1 species (*H. lunulata*) the summer densities were in most years too low to enable an estimate of winter survival.

species	*corr. coeff.*		*prop. surviving*			*references*[1]
	surv. *a*	*dens.* *b*	*coldest* *c*	*cold* *d*	*mild* *e*	*f*
Lanice conchilega	.65**	.89***	.00	.00	.12	1,2,6,12,13,17
Harmothoe lunulata		.68**				
Abra tenuis	.85***	.50*	.00	.02	.19	1
Mysella bidentata	.84**	.49*	.00	.15	.57	1,16
Antinoella sarsi	.66**	.70**	.00	.09	.22	1
Nemerteans	.76**	.57*	.00	.22	.44	
Angulus tenuis	.51*	.43	.00	.25	.12	1,3,17
Cerastoderma edule	.55*	.40	.02	.08	.59	1,2,4,7,9,10,11,14,16
Nephtys hombergii	.86***	.79***	.02	.13	.52	1,2,3,8,13,15
Crangon crangon	.63*	.64**	.01	.03	.05	
Carcinus maenas	.73**	.40	.07	.05	.10	5

1) 1=Beukema(1979), 2=Beukema(1984), 3=Beukema *et al.*(1988), 4=Blegvad(1929), 5=Crisp (1964), 6=Crisp *et al.*(1964), 7=Dörjes(1980), 8=Hamond(1966), 9=Hancock & Urquhart (1964), 10=Kristensen(1957), 11=Kristensen(1959), 12=Newell(1964), 13=Reichert & Dörjes (1980), 14=Smidt(1944), 15=Vader(1962), 16=Wolff(1973), 17=Ziegelmeier(1964).

3. RESULTS

3.1. RESPONSES OF SPECIES TO LOW TEMPERATURES

Several species have been reported to show sensivity to low winter temperatures either by higher-than-average overwinter mortality during severe winters and/or by lower-than-normal numbers (including complete absence) after such winters. For the species found to be sensitive at Balgzand, these references have been included in Table 1. Out of the 28 species studied in detail at the tidal flats of Balgzand, no less than 10 showed survival rates to be correlated significantly with winter temperatures (first column in Table 1). One further species (the polychaete *Harmothoe lunulata)* is added to this list, because it was found only after mild winters (it is a commensal of the highly sensitive species *Lanice conchilega*). The sensitive species listed in Table 1 are arranged according to the proportions (of the summer numbers) surviving during the coldest winters (as compared to these proportions during milder winters: 5th column in Table 1). In more than half of these species no living specimens at all were found on the tidal flats after the coldest winter (3rd column in Table 1).

In most of these 10 or 11 sensitive species, a significantly positive relationship was observed between the densities found at the end of a winter and the mean winter temperature (second column in Table 1). This correlation was in several cases less perfect than the one for the survival rate, because the density of a species at the end of a winter of course also depends on its abundance before that winter. Examples of the dynamics and winter-temperature dependence are shown in two sensitive species, *viz.* the polychaetes *Lanice conchilega* and *Nephtys hombergii* (Fig. 2b). Note the similarity in their fluctuation patterns, the low numbers after the cold winters of 1978-79 and again in 1985-87 and the high abundance during the period with mild winters in 1973-75. In both species overwinter survival rates were high during mild winters only, but *Nephtys* appears to be less sensitive than *Lanice* (Fig. 2c). In both species numbers increased significantly with temperatures of the foregoing winter (Fig. 2d). Compare Table 1 for the correlation coefficients calculated from the data shown in Fig. 2.

Most of the species listed in Table 1 appear to die directly by the low temperatures prevalent during cold winters. KRISTENSEN (1957) showed a direct influence of temperatures below −2°C on survival in *Cerastoderma edule*. Species (such as *Lanice con-*

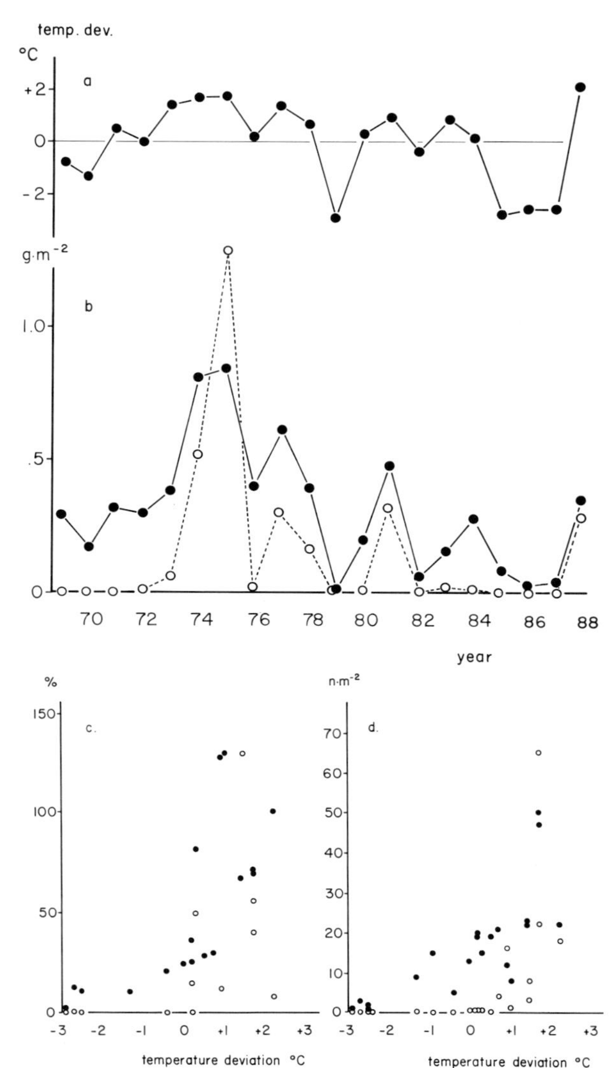

Fig. 2. Responses of the polychaete worms *Nephtys hombergii* (●) and *Lanice conchilega* (○) to winter temperatures: (a) Deviations of mean air temperatures in winter from long-term averages (as Fig. 1a); (b) Mean biomass values (g.m^{-2} AFDW) as observed in the late-winter/early-spring period of each year at 15 stations on Balgzand; (c) Relationship between the temperature deviation (shown in a) and the mean overwinter survival (March numbers as a % of numbers found in August) in the two species (data from 3 stations at Balgzand); (d) Relationship between the temperature deviation of the preceding winter (as shown in a) and the mean numerical density (n.m^{-2}) in the two species in late winter/early spring (15 stations on Balgzand).

chilega, Nephtys hombergii and *Angulus tenuis*), which show low cold-winter survival even in subtidal areas (ZIEGELMEIER, 1964; BEUKEMA *et al.*, 1988), will die in severe winters almost surely by direct effects of low temperatures (slightly below 0°C). In some highly mobile species (such as the shrimp *Crangon crangon*, the shore crab *Carcinus maenas* and probably also the polychaete *Antinoella sarsi*) enhanced migration during cold winters to the deeper and less cold water of the open North Sea may have been the cause of their relatively low abundance at tidal flats immediately after cold winters (causing apparently low survival rates in these areas in such winters). In *Crangon*, the sexually mature individuals always leave the Wadden Sea in late autumn, but juveniles remain during mild winters to leave the area almost completely during severe winters only (BODDEKE, 1976).

Decreased numbers by ice scouring and transport by floating ice during cold winters appears to be significant only in mussels (*Mytilus edulis*). In this species, high mortalities during cold winters were rare in the western part of the Wadden Sea and were limited to certain intertidal areas. In more eastern and northern parts of the Wadden Sea, high mussel mortalities by ice scouring have been observed more often on tidal flats (BLEGVAD, 1929; DÖRJES, 1980) and also in the shallow subtidal (BLEGVAD, 1929). After the severe winter of 1978-79 (and not after several milder winters) high numbers of mussels were observed scattered on the North Sea bottom off the Frisian Islands. These had most probably been transported from the Wadden Sea to this area by floating and melting ice. A similar observation of living *Mytilus* reaching in this way even Helgoland was made during the cold winter of 1923 (HAGMEIER & KÄNDLER, 1927).

Suffocation (by lack of oxygen or sulfide poisoning) under extensive ice fields completely covering vast tidal-flat areas may be another cause of death during cold winters. On Balgzand, I observed it only once, *viz.* in a population of the polychaete *Scoloplos armiger* during the 1986-87 winter (see fig. 4 in BEUKEMA, 1989). In most cold winters ice fields in the western part of the Wadden Sea are soon broken up by tidal action. In the colder Danish Wadden Sea, however, suffocation of cold-resistant species (such as *Mya arenaria, Macoma balthica* and *Arenicola marina*) appears to occur at a larger scale (BLEGVAD, 1929).

Temperatures are more extreme (and during longer periods so) at high than at low tidal flats. This difference is caused by the mitigating influence of flood water (POSTMA, 1983). As to be expected, the response of sensitive species to cold winters was stronger at high than at low tidal flats. Because not all sensitive species occur at a wide range of levels, this differential response could be proven (BEUKEMA, 1985) only in 3 species (the polychaetes *Nephtys hombergii* and *Lanice conchilega* and the bivalve *Cerastoderma edule*). In all of these species survival

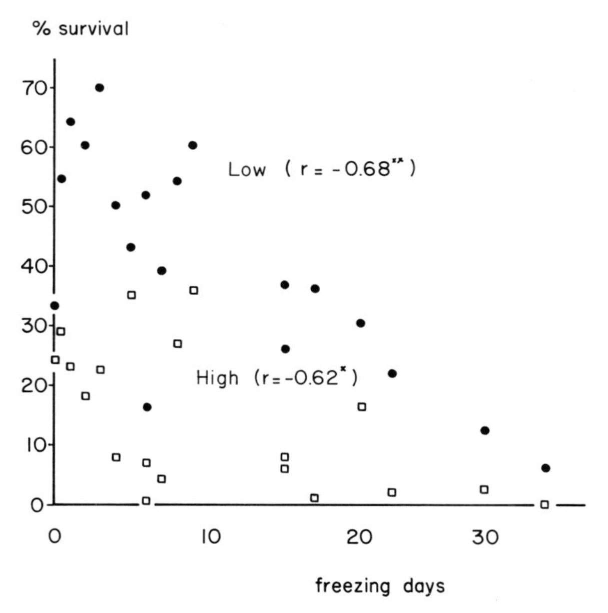

Fig. 3. The relationship between overwinter survival in the cockle *Cerastoderma edule* (in % of the numbers observed in August) and the character of the winter (expressed in number of freezing days, see Fig. 1c), separately for two parts of Balgzand: high (□) and low (●) tidal flats.

was better both in milder winters (at all levels) and at lower intertidal levels (in all winters). As an example the data concerning *Cerastoderma edule* are shown (Fig. 3). During the cold winter of 1962-63, all *Lanice* were killed between the tide marks in Britain, but they partly survived at subtidal levels (Crisp *et al.*, 1964). In the Danish Wadden Sea, Thamdrup (1935) observed some overwinter survival in *Nephtys hombergii* at his lowest intertidal station only, whereas all of these worms disappeared at higher stations during winter.

Winters are colder at higher latitudes. Along the Dutch and German North Sea coast, winter temperatures decline in a SW to NE direction (Huhn, 1973; compare table 1 of de Vooys, this issue). The sensitive species distributed over all of this area will suffer most from cold winters in the NE part of this area. For several species this could indeed be proven (Beukema *et al.*, 1988).

3.2. RESPONSES OF TIDAL-FLAT COMMUNITIES TO COLD WINTERS

The incidence of cold- sensivity in more than a third of the macrozoobenthic species of the tidal flats of the Wadden Sea must cause differences in some community parameters after mild and cold winters. Total numbers of species observed during the late-winter samplings were indeed relatively low (~28) after cold winters as compared to mild winters (~33, see Fig. 4b). A similar difference was observed (Fig. 4c) for the 'species densities', *i.e.* the mean number of species per areas sampled (0.9 m²). The two indices of species richness were positively correlated with temperatures during the preceding winter (r= 0.64 and 0.45, $p<0.01$ and <0.05, respectively). These correlations are not very high because two factors spoil the effect of winter temperature on species richness somewhat: there is a time lag of 1

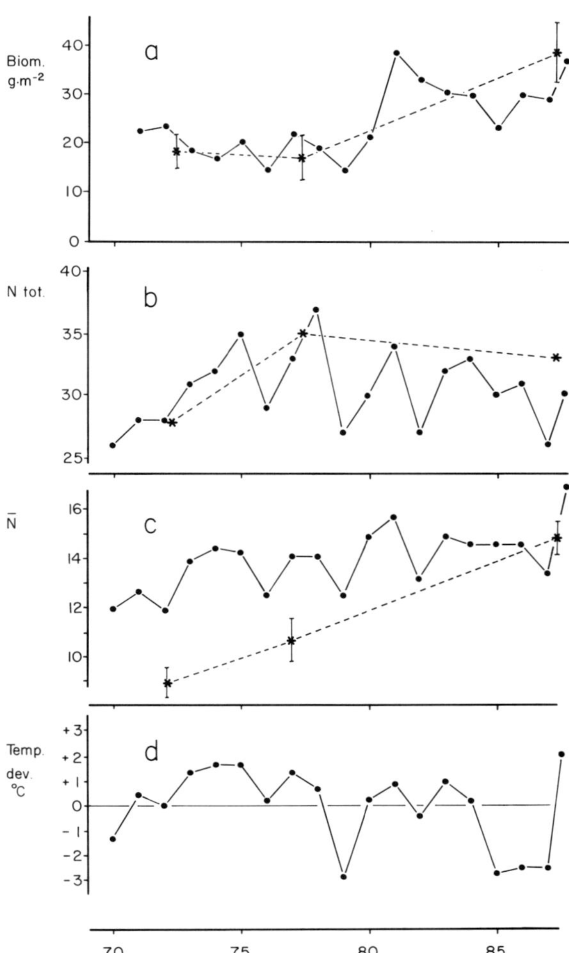

Fig. 4. Long-term trends in biomass and species richness of macrozoobenthos on tidal flats in the western half of the Dutch Wadden Sea. Annual data from late-winter/early-spring sampling at Balgzand (means of 15 stations, full points and solid lines) and less frequent late-summer/autumn data (broken lines) from tidal flats outside Balgzand (means of 26 stations with 1 s.e.). (a) Mean biomass in $g.m^{-2}$ AFDW; (b) Total number of species found (at 13.5 m² on Balgzand and 12 m² on other tidal flats); (c) Mean number of species per sampling station (0.9 m² on Balgzand and 0.45 m² on other tidal flats); (d) Deviation (in °C) of mean winter temperature from long-term average (as in Fig. 1a).

or 2 years in the restoration of species number after a cold winter and there is a general upward trend in species number during the 20-year period of observation (BEUKEMA, 1989). Biomass values (Fig. 4a) were also about 10 to 20% lower after cold winters as compared to precedent or subsequent mild winters. The biomass difference was only slight, because nearly all sensitive species contribute only little to total biomass (the bivalve *Cerastoderma edule* being the only exception).

Nearly all of the sensitive species observed at Balgzand showed their maximum abundance at the lower tidal flats (well below MTL). The only exception was the bivalve *Abra tenuis*. Individuals of this species were found almost exclusively above mean-tide level (where they contributed generally no more than a few percent to total biomass). Thus the group of sensitive species accounted for relatively high proportions of total biomass only at low tidal flats. Therefore, only on such tidal flats was total biomass seriously reduced after severe winters (as compared to foregoing years with milder winters). After the 1978-79 winter, biomass values were roughly halved at the lowest stations, whereas no reductions were found at the highest stations (see fig. 4 in BEUKEMA, 1985). If anything, biomass values even were relatively high after the 1978-79 winter at the highest stations. This may have been due to a relatively low predation pressure during this winter as a consequence of low predator numbers and deeper-than-normal burrowing in some prey species, as observed in the lugworm *Arenicola marina* and the ragworm *Nereis diversicolor*. BLEGVAD (1929) also observed abnormally deep burrowing in *Arenicola* during a severe winter.

3.3. RECOVERY AFTER SEVERE WINTERS

Within 1 or 2 years after a severe winter, tidal-flat zoobenthos restored the original biomass and species richness (Fig. 4). Individual species showed great differences in their ability to recover from low densities after a cold winter. Some species showed higher-than-average numbers already during the summer immediately following the severe winter. This appears to be the rule in several species of bivalves, including *Cerastoderma edule, Mytilus edulis, Mya arenaria* and *Macoma balthica* (see BEUKEMA, 1982 for a review). This group contains both sensitive and hard species. On the other hand, there was also a bivalve species that hardly recovered within the 6-year period between the cold winters of 1978-79 and 1984-85, *viz. Angulus tenuis* (Fig. 5). This species appears to need a warm summer to reproduce succesfully in the North Sea area (STEPHEN, 1938; ANONYMOUS, 1984). Most sensitive species, however, reached about-normal levels of abundance within 1 or 2 years (*e.g. Nephtys hombergii* and *Lanice conchilega*, see Fig. 2).

Conditions for rapid recovery appear to be favourable immediately after a cold winter. Though the tidal flats are far from deserted, more space is available than after other winters (at least at the lower tidal flats). A more important factor will be the low abundance of both epibenthic and infaunal predators during spring and early-summer periods after cold winters. Examples are shrimps (*Crangon crangon*) and shore crabs (*Carcinus maenas*) and the predatory worm *Nephtys hombergii* (Table 1).

3.4. CHANGES DURING LONG PERIODS WITH MILD OR COLD WINTERS

During the 1969-1988 period of observation of the tidal-flat fauna in the western Wadden Sea, two periods occurred with a succession of either mild (1971-1978 and in particular 1973-1975) or cold (1985-1987) winters (Fig. 1). The faunal changes observed during these periods might provide some information to be used for a prediction of faunal changes during longer periods with consistent climatic changes. Therefore, the relevant data for the ends of these periods (*i.e.* 1975, 1978 and 1987) have been summarized in Table 2 (together with the same

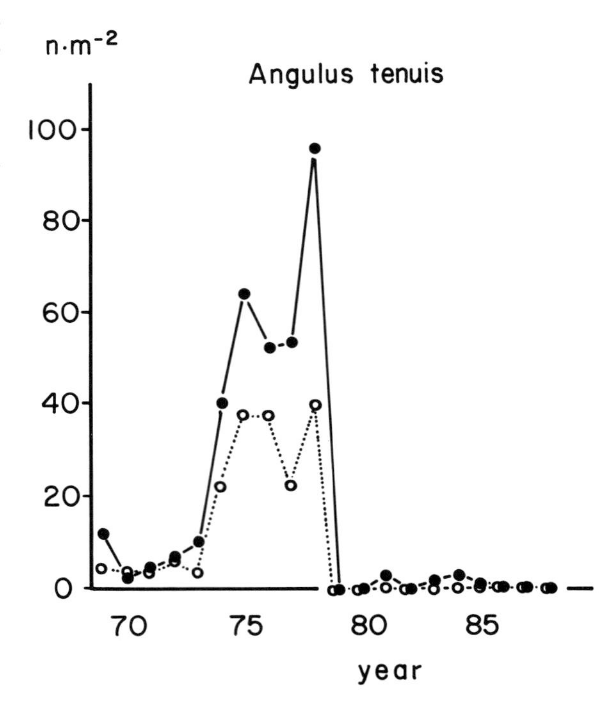

Fig. 5. Numerical densities (n.m^{-2}) of the bivalve *Angulus tenuis* as observed at 2 low tidal-flat stations (● and ○) on Balgzand in each of the 20 late-winter/early-spring sampling periods of 1969 upto and including 1988.

TABLE 2

Mean numerical densities ($n.m^{-2}$) and biomass values ($g.m^{-2}$ AFDW) in 11 sensitive (see Table 1) and 17 hard species observed at 15 tidal-flat stations in the western part of the Wadden Sea in March of 4 different years: after a series of mild winters (1975), a mild winter (1978) preceding an extremely cold winter (1979) and a series of cold winters (1987). Low biomass values (<0.05 $g.m^{-2}$) have been omitted.

	numbers ($n.m^{-2}$)				*biomass ($g.m^{-2}$)*			
	mild		*cold*		*mild*		*cold*	
	1975	*1978*	*1979*	*1987*	*1975*	*1978*	*1979*	*1987*
sensitive species:								
Lanice conchilega	64	4	0	0	1.3	0.2		
Harmothoe lunulata	1	1	0	0				
Abra tenuis	2	18	0	0				
Mysella bidentata	3	0	0	0				
Antinoella sarsi	1	0	0	1				
Nemerteans	1	1	0	0				
Angulus tenuis	7	9	0	0				
Cerastoderma edule	9	55	10	25	1.3	9.1	2.2	0.5
Nephtys hombergii	47	21	1	2	0.8	0.4		
Crangon crangon	1	2	0	1				
Carcinus maenas	3	1	1	1	0.2			
Total sensitive	139	112	12	30	3.8	9.9	2.2	0.5
hard species:								
Macoma balthica	76	103	104	188	1.1	1.6	2.0	5.4
Mya arenaria	3	4	9	25	3.1	2.0	2.0	12.1
Mytilus edulis	25	12	21	8	6.2	0.4	1.8	0.7
Scobicularia plana	1	1	1	1	0.2	0.1	0.2	0.2
Littorina littorea	0	1	1	2				
Hydrobia ulvae	756	444	4148	1448	0.2	0.1	1.2	0.4
Retusa alba	1	1	1					
Arenicola marina	19	14	21	26	4.0	3.0	3.5	4.9
Nereis diversicolor	22	24	55	151	0.5	0.6	1.0	2.4
Heteromastus filiformis	22	41	42	213	0.2	0.4	0.3	1.3
Scoloplos armiger	19	20	14	164	0.2	0.3	0.2	0.7
Scolelepis foliosa	1	1	1	4	0.1	0.1	0.1	0.1
Anaitides spec.	10	6	10	9				
Eteone longa	2	4	6	4				
Magelona papillicornis	2	1	1	1				
Corophium volutator	738	934	459	518	0.6	0.8	0.4	0.5
Bathyporeia spec.	1	1	2	1				
Total hard species	1698	1612	4895	2796	16.4	9.4	12.7	28.7
Total all species	1837	1724	4907	2826	20.2	19.2	14.8	29.2
Share sens. sp. (%)	7.6	6.5	0.2	1.1	19	52	15	2
Species number (total)	35	37	27	26				
Species nr per 0.9 m^2	15	14	12	13				

data for 1979, *i.e.* after the coldest winter of the entire period of observation).

As to be expected from the above discussion of the responses of species and communities to cold winters, the species richness was higher at the end of the mild-winter periods than at the end of the cold period. The difference in the total number of species (encountered on ~15 m^2 of tidal flats with widely varying environmental conditions) after such periods amounted to ~30% (see bottom lines of Table 2). The share of sensitive species amounted to ~7% of the total numbers of macrobenthic animals at the end of the mild periods and to only ~1% at the end of the cold period. The shares of total biomass were more variable as a consequence of the strongly fluctuating amounts of *Cerastoderma edule*, but showed the same trend (Table 2: from as high as 52% at the end of a mild period to as low as 2% at the end of the cold period).

The value for total biomass was relatively low in 1979, but not in 1987. By the end of the period of observation, several hard species showed higher densities than in earlier years, both after cold and mild winters. Such changes should not be attributed particularly to the character of the winters but to a general increase of the productivity of the zoobenthos in the western part of the Wadden Sea during recent years, probably as a consequence of eutrophication (BEUKEMA & CADÉE, 1986). These increases had started already before the cold winters of 1985-1987, and partly already before the severe one of 1979. It is hard to tell, which part of the increases during 1985-87 is to be attributed to eutrophication and which part to the series of cold winters.

Faunal changes during the mild-winter period 1971-1978 were described earlier by BEUKEMA *et al.* (1978) for a much larger area than Balgzand, *viz.* the whole western half of the Dutch Wadden Sea. They also observed a high species richness at the end of this period, without a clear change in total biomass. The share of sensitive species (such as *Lanice conchilega, Nephtys hombergii* and *Angulus tenuis*) increased, whereas a few hard species were found to decrease (*viz. Mya arenaria, Arenicola marina* and *Macoma balthica*). In the latter two of these species, the decrease was explained by a differential dispersal in winter, which is supposed to be activated by low temperatures in these species and might have been insufficient during mild winters. This would have affected particularly their abundance in areas far away from the nurseries which are situated at high coastal tidal flats (*Arenicola*: BEUKEMA & DE VLAS, 1979; FARKE *et al.*, 1979; *Macoma*: Beukema, 1989).

Note that no species disappeared during the period with mild winters as contrasted to the periods with one or more cold winters (Table 2).

4. DISCUSSION

The character of the preceding winter exerts a significant influence on the composition and richness of the tidal-flat fauna in the Wadden Sea. The immediate and direct effect of a cold winter is a (mostly strong) reduction of the abundance in the numerous cold-sensitive species. The abundance of these species increases during periods with mild winters, whereas no species disappear during mild winters. The expected effect of a general (but moderate) warming, therefore, would be an increase in species richness, because the sensitive species would thrive and the cold-adapted species would remain.

It is less easy to predict the development of the total biomass of the benthic fauna during a prolonged period with higher temperatures. Total-biomass values during the period with mild winters in the seventies were relatively stable (Fig. 4a). This stability will have been favoured by the relatively low rates of mortality and moderate recruitment success. Severe winters, on the other hand, cause at first an immediate decrease in biomass over most of the tidal flats (*viz.* those below MTL), quickly followed by an overcompensation in several important species due to highly successful recruitment (BEUKEMA, 1982). Thus, fluctuations in total biomass become heavier if the frequency of severe winters would increase. On the other hand, a prevalence of mild winters will stabilize the community at an intermediate biomass level, a high species number and strong interspecific interactions.

This prediction is in accordance with the behaviour of similar macrobenthic faunas in more northern areas where winters frequently are severe with long periods of ice cover. Along the west coast of Sweden, MÖLLER (1986) usually observed strong fluctuations in the numbers of several important species. Only during short periods with (rare) mild winters good survival of already present stocks was followed by low recruitment of a new generation and this resulted in a relatively stable zoobenthic biomass with strong interspecific competition. As in the Wadden Sea, the transformation of an instable winter-struck fauna to a more stable mild-weather fauna was rapid.

The impact of cold winters was more serious on low than on high tidal flats, just the reverse situation as in the individual sensitive species. At high tidal flats the community consisted almost exclusively of hard species, whereas at low tidal flats sensitive species accounted for a significant share of the total community. Only a moderate reduction of the sensitive species would, therefore, affect biomass and species density at low intertidal levels, whereas even total extinction of sensitive species would hardly affect the community at high levels. More generally, eurythermic species prevail at high intertidal levels where better adaptations to the more extreme (seasonal and daily) temperature ranges are indeed needed. This appears to be true also for high summer temperatures (EVANS, 1948).

Rapid recovery after a severe winter will be propagated by the better survival of nearby populations in deeper water. Nearly all of the sensitive species listed in Table 1 also occur abundantly in subtidal parts of the Wadden Sea (DEKKER, 1989) and in a coastal zone of the North Sea (DÖRJES, 1976; own observations). Several of these species even show maximum densities in such subtidal areas: *Lanice conchilega, Harmothoe lunulata, Angulus tenuis, Mysella bidentata,* and *Nephtys hombergii*. At tidal flats, these species are more or less restricted to lower intertidal levels (DANKERS & BEUKEMA, 1983). Such areas are drained only during short daily periods and therefore resemble subtidal bottoms.

A notable exception is *Abra tenuis*, which distribution appears to be limited to high tidal flats, above mean-tide level. Exactly this one species (and none of the other sensitive species listed in Table 1) is limited to the western part of the Wadden Sea and has not been observed in the German and Danish Wadden Sea (WOLFF & DANKERS, 1983).

All other sensitive species find a nearby refuge for too low temperatures in deeper and more off-shore waters and can recolonize the Wadden Sea after a severe winter either by mobile larval stages (bivalves and polychaetes) or by swimming of adults (crabs and shrimps). Particularly along the more southern and western parts of the Dutch North Sea coast, the water remains relatively warm in winter, allowing relatively high survival rates in sensitive species (BEUKEMA *et al.*, 1988). Net water transport along the Dutch mainland coast is in a northern direction, thus facilitating a rapid colonization or recolonization with 'southern' species. This may be an explanation for the relatively high share of winter-sensitive species among the tidal-flat fauna of the Wadden Sea. Despite their frequent annihilation, these species are far from rare inhabitants of the tidal flats during most years.

Cold winters occur simultaneously over vast areas. Over all of the Wadden Sea (and even larger areas) the same winters were the coldest of the century (*e.g.* 1929, 1941, 1947, 1963, 1979). That is why the populations of sensitive species fluctuate in a synchronized way over all of the Wadden Sea (BEUKEMA & ESSINK, 1986). Even some winter-hard species show such fluctuation patterns as a consequence of highly successful reproduction after severe winters, *e.g. Macoma balthica* (fig. 5 of BEUKEMA & ESSINK, 1986).

Such parallel fluctuations cause temporarily uniform conditions over vast areas and may, therefore, affect the success of populations of predators which use the stocks of benthic animals in the Wadden Sea as an indispensable food source, such as small plaice (BERGMAN *et al.*, 1988) and many species of birds (SMIT, 1983). In particular for several species of migratory birds, the tidal flats provide a vital source of food for replenishing their reserves during migration. All of such species will meet a relative scarcity of food on these feeding grounds immediately after a severe winter but a superfluous supply during the subsequent summer. Zoobenthic biomass on the tidal flats of the Wadden Sea is at a minimum every year in late winter and at a maximum in summer (BEUKEMA, 1974). Most fishes leave the shallow parts of the area in winter (FONDS, 1983) and need less food in winter than in summer. Several species of wading birds, however, have to restrict their feeding almost completely to tidal flats and need more food per day in winter than in summer. A milder climate would guarantee these predators a more stable (but probably never very high) supply of food in all seasons. In mild winters, food supply would not only be higher than in cold winters, but would be more accessible too by less ice cover and shallower burrying (see also GOSS-CUSTARD, this issue).

Some indirect shifts in the composition of the tidal-flat macrozoobenthos may be expected if (nearly) all winters would become mild. The increase of the density of the predatory worm *Nephtys hombergii* would be coupled with decreases in its prey, the worms *Scoloplos armiger* and in particular *Heteromastus filiformis* (BEUKEMA, 1987). Increased abundances of crabs and shrimps and their earlier appearance on the tidal flats in spring would suppress reproductive success in several other benthic species, including the important bivalves (JENSEN & JENSEN, 1985; Dekker, pers.comm.). Other interspecific and intraspecific interactions would also intensify and might contribute to a stabilization, *e.g.* by suppressing high recruitments. Bivalves would be affected negatively in yet another way: they would loose higher proportions of their body weights by a higher metabolism during warmer winters (own observations). This could unfavourably affect the number and/or quality of their eggs and thereby reproductive success. Such additive effects of higher predator and adult abundance and lower quality of young stages after mild winters might at least partly explain the generally low recruitment in bivalves after mild winters. Mild winters certainly are not purely favourable for all components of the benthic community living on tidal flats of the Wadden Sea.

5. REFERENCES

ANONYMOUS, 1984. Adaptive ecology in the inshore benthos.—Scott. Mar. Biol. Ass., Ann. Rep. 1984: 26-36.

BERGMAN, M.J.N., H.W. VAN DER VEER & J.J. ZIJLSTRA, 1988. Plaice nurseries: effects on recruitment.—J. Fish Biol. **33** (Suppl.A): 201-218.

BEUKEMA, J.J., 1974. Seasonal changes in the biomass of the macro-benthos of a tidal flat area in the Dutch Wadden Sea.—Neth. J. Sea Res. **8:** 94-107.

——, 1979. Biomass and species richness of the macrobenthic animals living on a tidal flat area in the Dutch Wadden Sea: effects of a severe winter.—Neth. J. Sea Res. **13:** 203-223.

——, 1982. Annual variation in reproductive success and biomass of the major macozoobenthic species living in a tidal flat area of the Wadden Sea.—Neth. J. Sea Res. **16:** 37-45.

——, 1985. Zoobenthos survival during severe winters on high and low tidal flats in the Dutch Wadden Sea. In: J.S. GRAY & M.E. CHRISTIANSEN. Marine biology of polar regions and effects of stress on marine organisms. John Wiley, Chichester: 351-361.

——, 1987. Influence of the predatory polychaete *Nephtys hombergii* on the abundance of other polychaetes.—Mar. Ecol. Prog. Ser. **40:** 95-101.

——, 1989. Tidal-current transport of thread-drifting postlarvae juveniles of the bivalve *Macoma balthica* from the Wadden Sea to the North Sea.—Mar. Ecol. Prog. Ser. **52:** 193-200.

BEUKEMA, J.J. & G.C. CADÉE, 1986. Zoobenthos responses to eutrophication of the Dutch Wadden Sea.—Ophelia **26:** 55-64.

BEUKEMA, J.J. & M. DESPREZ, 1986. Single and dual growing seasons in the tellinid bivalve *Macoma balthica* (L.).—J. exp. mar. Biol. Ecol. **102:** 35-45.

BEUKEMA, J.J. & K. ESSINK, 1986. Common patterns in the fluctuations of macrozoobenthic species living at different places on tidal flats in the Wadden Sea.—Hydrobiologia **142:** 199-207.

BEUKEMA, J.J. & J. DE VLAS, 1979. Population parameters of the lugworm, *Arenicola marina*, living on tidal flats in the Dutch Wadden Sea.—Neth. J. Sea Res. **13:** 331-353.

BEUKEMA, J.J., W. DE BRUIN & J.J.M. JANSEN, 1978. Biomass and species richness of the macrobenthic animals living on the tidal flats of the Dutch Wadden Sea: long-term changes during a period with mild winters.—Neth. J. Sea Res. **12:** 58-77.

BEUKEMA, J.J., J. DÖRJES & K. ESSINK, 1988. Latitudinal differences in survival during a severe winter in macrozoobenthic species sensitive to low temperatures.—Senckenbergiana marit. **20:** 19-30.

BLEGVAD, H., 1929. Mortality among animals of the littoral region in ice winters.—Rep. Dan. biol. Stn **35:** 49-62.

BODDEKE, R., 1976. The seasonal migration of the brown shrimp *Crangon crangon*.—Neth. J. Sea Res. **10:** 103-130.

CRISP, D.J., 1964. Mortalities in marine life in North Wales during the winter of 1962-63. In: D.J. CRISP. The effects of the severe winter of 1962-63 on marine life in Britain. J. Anim. Ecol. **33:** 190-197.

CRISP, D.J., J. MOYSE & A. NELSON-SMITH, 1964. General conclusions. In: D.J. CRISP. The effects of the severe winter of 1962-63 on marine life in Britain.—J. Anim. Ecol. **33:** 202-210.

DANKERS, N. & J.J. BEUKEMA, 1983. Distributional patterns of macrozoobenthic species in relation to some environmental factors. In: W.J. WOLFF. Ecology of the Wadden Sea. Vol.1, part 4. Balkema, Rotterdam: 69-103.

DEKKER, R., 1989. The macrozoobenthos of the subtidal western Dutch Wadden Sea. I. Biomass and species richness.—Neth. J. Sea Res. **23:** 57-68.

DÖRJES, J., 1976. Primärgefüge, Bioturbation und Makrofauna als Indikatoren des Sandversatzes im Seegebiet vor Norderney (Nordsee). II. Zonierung und Verteilung der Makrofauna.—Senckenbergiana marit. **8:** 171-188.

——, 1980. Auswirkung des kalten Winters 1978/1979 auf das marine Makrobenthos.—Natur u. Mus. **110:** 109-115.

EVANS, R.G., 1948. The lethal temperatures of some common British littoral molluscs.—J. Anim. Ecol. **17:** 165-173.

FARKE, H., P.A.W.J. DE WILDE & E.M. BERGHUIS, 1979. Distribution of juvenile and adult *Arenicola marina* on a tidal mud flat and the importance of nearshore areas for recruitment.—Neth.J.Sea Res. **13:** 354-361.

FONDS, M., 1983. The seasonal distribution of some fish species in the western Dutch Wadden Sea. In: W.J. WOLFF. Ecology of the Wadden Sea. Balkema, Rotterdam. Vol. 2, part 5: 42-77.

HAGMEIER, A. & R. KÄNDLER, 1927. Neue Untersuchungen im nordfriesischen Wattenmeer und auf den fiskalischen Austernbänken.—Wiss. Meeresunt. NF 16, Abt. Helgoland, H 2, Nr **6:** 1-90.

HAMOND, R., 1966. The Polychaeta of the coast of Norfolk.—Cah. Biol. Mar. **7:** 383-436.

HANCOCK, D.A. & A.E. URQUHART, 1964. Mortalities of edible cockles (*Cardium edule* L.) during the severe winter of 1962-63. In: D.J. CRISP. The effects of the severe winter of 1962-63 on marine life in Britain.—J. Anim. Ecol. **33:** 176-178.

HOEVEN, P.C.T. VAN DER, 1982. Observations of surface watertemperature and salinity, State Office of Fishery Research (RIVO): 1860-1981.—Scient. Rep. W.R.82-8, KNMI, De Bilt: 118 pp.

HUHN, R., 1973. On the climatology of the North sea. In: E.D. GOLDBERG. North Sea Science. MIT Press, London: 183-236.

JENSEN, K.TH. & J.N. JENSEN, 1985. The importance of some epibenthic predators on the density of juvenile benthic macrofauna in the Danish Wadden Sea.—J. exp. mar. Biol. Ecol. **89:** 157-174.

JONES, P.D., T.M.L. WIGLEY & P.B. WRIGHT, 1986. Global temperature variation between 1861 and 1984.—Nature **322:** 430-434.

KRISTENSEN, I., 1957. Diffences in density and growth in a cockle population in the Dutch Wadden Sea.—Archs néerl. Zool. **12:** 351-453.

——, 1959. The coastal waters of the Netherlands as an environment of molluscan life.—Basteria (Suppl.) **23:** 18-46.

LOUGH, J.M., T.M.L. WIGLEY & J.P. PALUTIKOF, 1983. Climate and climate impact scenarios for Europe in a warmer world.—J. Climate appl. Meteorol. **22:** 1673-1684.

MULLER, P., 1986. Physical factors and biological interactions regulating infauna in shallow boreal areas.—Mar. Ecol. Prog. Ser. **30:** 33-47.

NEWELL, G.E., 1964. The south-east coast, Whitstable area. In: D.J. CRISP. The effects of the severe winter of 1962-63 on marine life in Britain.—J. Anim. Ecol. **33:** 178-179.

POSTMA, H., 1983. Salinity, temperature and density. In: W.J. WOLFF. Ecology of the Wadden Sea. Balkema, Rotterdam. Vol. 1, part 2: 24-33.

RATCLIFFE, P.J., N.V. JONES & N.J. WALTERS, 1981. The survival of *Macoma balthica* in mobile sediments. In: N.V. JONES & W.J. WOLFF. Feeding and survival strategies of estuarine organisms. Plenum Press, New York: 91-108.

REICHERT, A. & J. DÖRJES, 1980. Die Bodenfauna des Crildumersieler Wattes (Jade, Nordsee) und ihre Veränderung nach dem Eiswinter 1978/1979.—Senckenbergiana marit. **12:** 213-245.

SMIDT, E.L.B., 1944. Das Wattenmeer bei Skallingen. 3.The effects of icewinters on marine littoral faunas.—Folia geogr. dan. **2:** 1-36.

SMIT, C.J., 1983. Production of biomass by invertebrates and consumption by birds in the Dutch Wadden Sea. In: W.J. WOLFF. Ecology of the Wadden Sea. Balkema, Rotterdam. Vol. 2, part 6: 290-301.

STEPHEN, A.C., 1938. Production of large broods in certain marine lamellibranchs with a possible relation to weather conditions.—J. anim. Ecol. **7:** 130-143.

THAMDRUP, H.M., 1935. Beiträge zur Oekologie der Wattenfauna auf experimenteller Grundlage.—Medd. Komm. Danm. Fisk.-Havunders., Ser. Fisk., **10:** 1-125.

VADER, W.J.M., 1962. Wadwetenswaardigheden. 1. Enkele aantekeningen over de uitwerking van de vorstperiode van eind-december 1961 op de wadfauna.—Zeepaard **22:** 83-85.

WOLFF, W.J., 1973. The estuary as a habitat. An analysis of data on the soft-bottom macrofauna of the estuarine areas of the rivers Rhine, Meuse and Scheldt.—Zool. Verh., Leiden **126:** 1-242.

WOLFF, W.J. & N. DANKERS, 1983. Preliminary checklist of the zoobenthos and nekton species of the Wadden Sea. In: W.J. WOLFF. Ecology of the Wadden Sea I(4). Balkema, Rotterdam: 24-60.

ZIEGELMEIER, E., 1964. Einwirkung des kalten Winter 1962/63 auf das Makrobenthos in Ostteil des Deutschen Bucht.—Helgoländer wiss. Meeresunters. **10:** 276-282.

——, 1978. Macrobenthos investigations in the eastern part of the German Bight from 1950 to 1974.—Rapp. P.-v. Réun. Cons. perm. int. Explor. Mer **172:** 432-444.

EFFECTS OF TEMPERATURE CHANGES ON INFAUNAL CIRCALITTORAL BIVALVES, PARTICULARLY *T. TENUIS* AND *T. FABULA*

JAMES G. WILSON

Environmental Sciences Unit, Trinity College, Dublin 2, Ireland

ABSTRACT

The limits of thermal tolerance are associated closely with the latitudinal and local variations found within an animals' range, yet in most situations the absolute limits of tolerance are seldom reached. Under these conditions the survival of a species is related more to its ability for capacity adaptations (*e.g.* growth rate, reproduction) than to its absolute tolerances (resistance adaptations).

Exceptions to the above can occur in the littoral zone, with the effect being proportional to the height up to the shore: here both elevated temperatures during the summer and lower temperature, during the winter especially those low enough to result in ice, can lead directly to mortality.

The most pronounced sub-lethal effects are those on growth and reproduction. Under conditions of elevated temperatures growth, especially in the initial stages, is faster. Reproduction is earlier in warmer waters and may occur more than once or continuously throughout the summer. This may lead to them being in poorer condition compared to northern populations which spawn once in spring for overwintering and hence to decreased long term survival.

1. INTRODUCTION

Temperature is one of the most important of the physical variables in the aquatic environment. On land, organisms have evolved various mechanisms, both physiological (*e.g.* sweating) and behavioural (*e.g.* sunning) to modify its effect, but in the water no such adaptations are possible (although some littoral species do show remarkably similar activities!). However, there are compensations to living in the sea, in that the absolute range of temperatures is much less than on land, variation in temperature is very much less and changes generally operate on a much longer time-scale. An exception to this is the littoral zone which is uncovered by the tide, and here as a consequence, conditions may be as extreme and as variable as on the land itself.

This paper will examine the effects of temperature on infaunal bivalves and in particular the closely related *Tellina tenuis* and *Tellina fabula* as representative marine organisms.

2. RESPONSE TO TEMPERATURE

The response of a bivalve or indeed any organism to temperature can be summarised as a function of the species involved, its past thermal history, the temperature to which it is exposed and the length of time for which it is exposed. The response itself can present at different levels, with different functions displaying different sensitivities. These generalisations can be depicted graphically as in Fig. 1. In Fig. 1a there are absolute limits to the thermal tolerance, but within these, there are zones in which some functions are impaired: the degree of impairment and indeed the extent of these zones is still largely a matter of conjecture.

The limits themselves can be modified; a history of higher temperatures raises the limit (Fig. 1b) although even here there are limits both to the long term temperatures under which acclimatisation can take place and to the short term exposure when the response is measured.

3. LETHAL LIMITS

The distribution of different bivalve species is well correlated with their thermal tolerances, yet the published values for lethal limits may yield little in the way of information as to the role of temperature in a particular location. This is partly a function of the methods by which limits were obtained. For example, early work by HENDERSON (1929) in which the temperature was raised by 1°C every 5 min gave lethal temperatures in excess of 40°C for *Macoma balthica* and *Mya arenaria* while KENNEDY & MIHURSKY (1971) with a 24-h exposure at fixed temperatures reported limits some 10°C lower. Likewise ANSELL *et al.*, (1980a; 1980b) found that a 96-h exposure resulted in

J.J. Beukema et al. (eds.), Expected Effects of Climatic Change on Marine Coastal Ecosystems, 93–97.

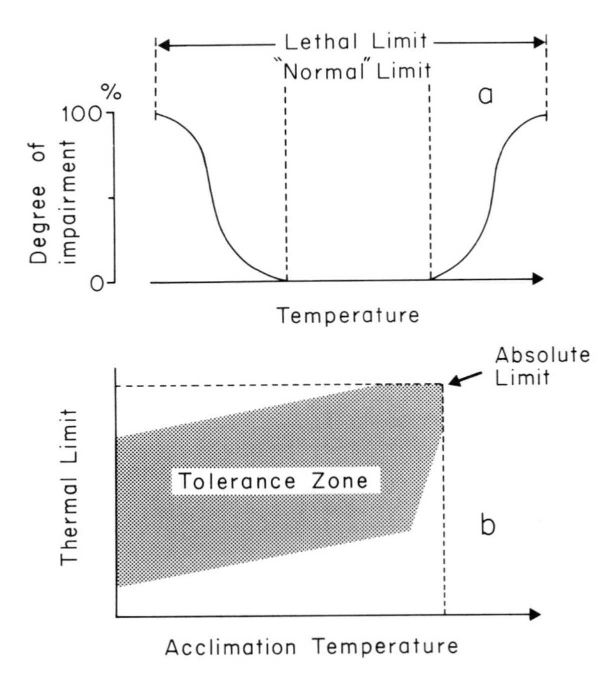

Fig. 1. Effect of temperature changes on an organism: (a) relationship of function with temperature; (b) modification of response with past history of exposure (after FRY, 1947).

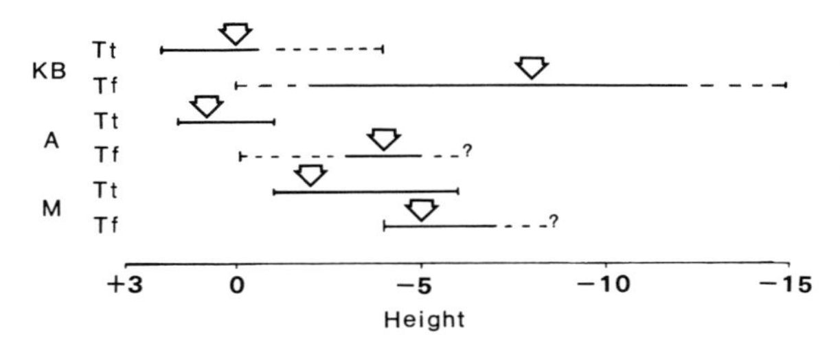

Fig. 2. Distribution of *T. tenuis* (Tt) and *T. fabula* (Tf) with shore height at Kames Bay (KB), Arcachon (A) and Marseille (M). Solid line, usual range; dotted line exceptional range: arrow indicates zone of maximal abundance.

upper thermal limits 2 to 3°C lower than a 24-h exposure. Clearly then such data must be regarded as relative rather than absolute and for comparative purposes rather than as a final truth.

Used in this way, such data can provide useful indications as to the effect of temperature on species distribution (*e.g.* WILSON, 1978, 1981): southern species have higher limits than northern, shallow burrowers higher limits than deep burrowers and littoral species higher limits than sublittoral (although within a single species shore height has only a slight effect on thermal tolerance). WILSON (1976) has even suggested that thermal limits may recapitulate evolutionary history and thus indicate species of northern or southern origin in our waters.

The majority of studies on thermal tolerance relate to upper limits: comparatively little has been done on lower limits. Those data which are available (*e.g.* KANWISHER, 1955) suggest that northern Europe bivalve species can survive temperatures well below that at which sea water freezes, and therefore under field conditions the problem is more one of isolation by ice than of direct thermal stress.

The conclusion from these types of observations must be that, while there is a link between the thermal tolerance and the distribution in the great majority of situations temperature is not acting directly but indirectly, such that by impairment of some function the survival or viability of the organism is itself impaired.

4. SUB-LETHAL EFFECTS

In Fig. 1a the zone of impairment indicates a realm in which the individuals performance is sub-optimal right up to the stage of heat coma but this does not neccessarily imply merely a delaying of the death point nor even that death itself is inevitable. The animal may be stressed, but recovery is perfectly possible when the right conditions return. Burrowing for example, was impaired at temperatures a few degrees lower than the upper lethal limits and the upper temperature limit for burrowing decreased with increasing exposure time (after the initial thermal shock) as did the lethal limit (ANSELL *et al.*, 1980a; 1980b).

However, short-term experiments such as these can only serve as indicators of what responses we may expect, and it is necessary to look at long-term responses in the context of the environmental changes we are considering. Three sources of data may be examined: the first is data on the life styles of populations along a latitudinal gradient from north to south, the second is to look at the changes induced by local large scale thermal deformations caused by cooling water effluents, and the third to relate long term changes in temperature and populations.

Fig. 2 shows the distribution of the two Tellins *T. tenuis* and *T. fabula* at 3 locations: Kames Bay (KB) in Scotland, Arcachon (A) on the Atlantic coast of France and Marseille (M) in the Mediterranean. The temperature gradient from Kames Bay to Marseille is in the order of ~10 °C and it can be seen from the figure that there is a tendency for the distribution to move down the shore as the ambient temperature increases. The Mediterranean is (virtually) tideless, so perhaps this is an extreme example, but there is still a difference between Kames Bay and Arcachon. For example the upper limit of the distribution of *T. tenuis* in Kames Bay is at HWN (High Water Neaps) equivalent to ~5 hours emersion per tidal cycle (STEPHEN, 1928) while at Arcachon the limit is +1.52

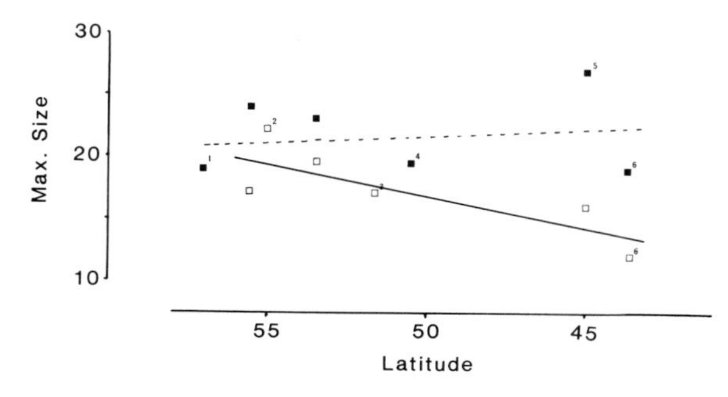

Fig. 3. Maximum size of *T. tenuis* (solid symbols, dotted line) and *T. fabula* (open symbols, solid line) with latitude (see also text). Reference 1 - McINTYRE (1971); 2 - SALZWEDEL (1979); 3 - WARWICK *et al.* (1978); 4 - HOLME (1949); 5 - SALVAT (1967); 6 - MASSÉ (1972); otherwise Wilson (unpublished data).

m, representing 3 to 4 hours emersion per tidal cycle (SALVAT, 1967).

TREVALLION (1971) suggested that burrowing depth in *T. tenuis* was correlated with shore height, and it could be argued that by burrowing deeper, littoral species could ameliorate temperature stress while the tide is out. Certainly, burrowing depth has been proposed as a factor in species thermal tolerance (McMAHON & WILSON, 1981). However, WILSON (1976, 1979) found no such correlations of burrowing and shore height, but rather suggested that burrowing depth was controlled by a combination of siphon extensibility and sediment penetrability and therefore no such amelioration mechanisms would operate.

T. fabula is the more thermally sensitive species, even in the Mediterranean where both species are sublittoral (ANSELL *et al.*, 1980a; WILSON, 1978) yet it is worth noting that the distribution of both *T. fabula* and *T. tenuis* seems to be similarly affected from one location to the other (Fig. 2). However, there does seem to be a difference between the response of the two species in growth. Fig. 3 shows the maximum size in a number of populations from the north of Scotland to the Mediterranean. The maximum size in *T. tenuis* did not change over the range studied ($r=-0.16$, $P>0.1$) but there was a significant correlation between latitude and size in *T. fabula* ($r=0.8$, $0.02<P<0.05$), with the northern populations attaining larger maximum sizes. There is also a difference in growth rates and longevity between northern and southern populations. MASSÉ (1972) noted that Mediterranean specimens of *T. tenuis* attained a shell length in one year that Scottish animals took two years to reach, and that their life span did not seem to exceed 2 years, compared to 5 on the Atlantic coast of France and 8 in northern Scottish populations. Similarly Mediterranean *T. fabula* seem to live only one year (MASSÉ, 1972) compared to around 5 years in Scotland (STEPHEN, 1932) and the German Bight (SALZWEDEL, 1979). BEUKEMA & MEEHAN (1985) have discussed this situation in *M. balthica* and have shown for example that in American populations growth rate and maximum size increased continuously to the south, whereas maximal growth and size occurred at intermediate latitudes in Europe. It is interesting here to note the observations of BARNETT (1971) on a population of *T. tenuis* in the vicinity of a cooling water outfall where temperatures were raised 2 to 4°C above ambient. This population had a faster initial growth rate, but the final size did not differ from unaffected populations nearby.

Of course increased production need not go to somatic growth but can go instead to reproduction, an event which in itself may be triggered by temperature. The *T. tenuis* population mentioned above (BARNETT, 1971) near the cooling outfall showed comparatively early and prolonged breeding and MASSÉ (1972) implied two breeding periods in Mediterranean *T. tenuis* populations compared to one in the north Atlantic. A similar situation exists in *M. balthica* (*e.g.* BACHELET, 1980). STEPHEN (1932) linked good recruitment of both *T. tenuis* and *T. fabula* to warm summers and BEUKEMA (1979) has investigated winter effects. In general, however, long-term data of this kind adequate to identify trends are lacking.

One consequence of multiple or prolonged breeding can be energy imbalance, especially if food is scarce, and imbalance can also result directly from temperature changes unless both input and output adjust accordingly. NEWELL & BRANCH (1980) have reviewed this subject at some length and have suggested that in most bivalves the feeding rate (consumption and assimilation efficiency) adjusts to temperature while the metabolism may or may not adjust according to the species involved. Intertidal species at least, they suggest, may be temperature independent (*i.e.* adjust) over part of the range.

Fig. 4 shows the respiration of *T. tenuis* and *T. fabula* from 3 locations - Dublin Bay, Ireland (summer and winter), Arcachon, France (summer) and Marseille, France (summer). From the data it is apparent that there is little evidence of short term regulation or adjustment of respiration except in the Marseille specimens. However, it is noticeable that both species demonstrate reverse acclimation (Dublin, winter versus summer) which has been put forward as a mechanism to conserve energy when food is scarce. In *T. tenuis* the Arcachon respiration was lower than that of the Dublin specimens (summer or winter) over the range 5 to 20°C, suggesting that the metabolism has been adjusted, but in *T. fabula* the Arcachon values were always higher, suggesting no adjustment and this difference in response support NEWELL & BRANCH (1980) (above) on adjustment in intertidal species. The Marseille data (Fig. 4) would seem to indicate some adjustment

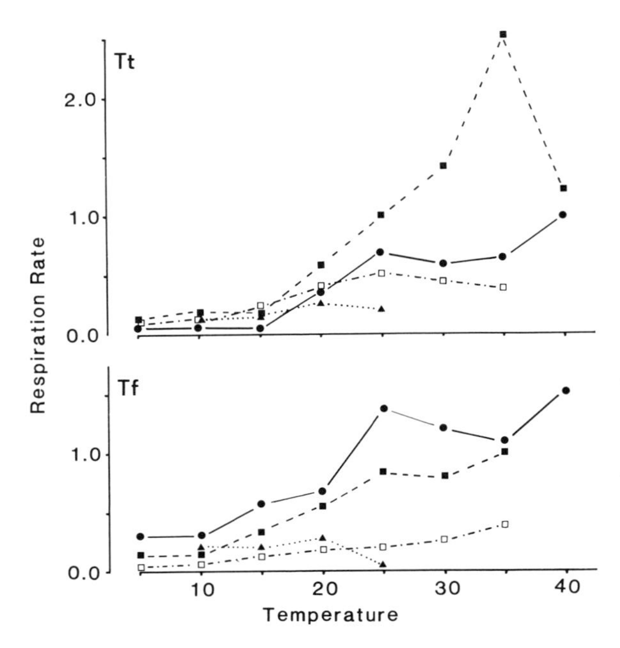

Fig. 4. Respiration rate (mm^3 O_2.mg^{-1}.h^{-1}) with temperature (°C) of *T. tenuis* (Tt) and *T. fabula* (Tf) from Dublin Bay, Ireland (summer-conditioned ■---■ and winter-conditioned □-·-□), Arcachon, France (●——●) and Marseille, France (▲····▲).. From McMahon & Wilson (1981) and Wilson (1985 and unpublished data).

in both species, but these populations are near the limit of their tolerance (especially *T. fabula*) and so the responses may not reflect the true pattern overall. In fact the *T. fabula* populations around Marseille have declined noticeably in the past few years, although it is uncertain whether this is due to high temperatures to some other cause (*e.g.* pollution).

Beukema (Beukema *et al.*, 1985; Beukema & Desprez, 1986) has suggested that it is food availability and not temperature *per se* which controls energy balance in *M. balthica*, and of course the effect of temperature on a species cannot be considered in isolation, but the rest of the system must be examined also. Three major interactions need to be investigated: food, as mentioned above, predators (the next link in the chain) and competitors. In southern waters predatory gastrodops are much more abundant than in northern, where the major predation on *Tellina* may be siphon cropping by juvenile flatfish (Trevallion, 1971). Temperature changes may also permit the establishment of immigrant species (Barnett & Hardy, 1984), or otherwise alter the competitive balance by selectively affecting species (Fig. 3).

Finally, it must be remembered that other biological conditions will also change: food type and pattern of availability, predator type and pressure, immigration of new species etc., and all these will affect the balance of indigenous communities.

What is apparent is that while it may be possible at this stage to estimate roughly how and to what extent many of the prominent species around our coasts may be affected by temperature changes, such predictions are necessarily made in isolation. It therefore seems necessary to encourage larger scale investigations (both temporal and spatial) so that individual reactions may be put into context.

5. REFERENCES

Ansell, A.D., P.R.O. Barnett, A. Bodoy & H. Massé, 1980a. Upper temperature tolerances of some European molluscs. I. *Tellina fabula* and *Tellina tenuis*.—Mar. Biol. **58:** 33-39.

——, 1980b. Upper temperature tolerances of some European molluscs. II. *Donax vittatus, D. semistriatus* and *D. trunculus*.—Mar. Biol. **58:** 41–46.

Bachelet, G., 1980. Growth and recruitment of the Tellinid bivalve *Macoma balthica* at the southern limit of its geographical distribution.—Mar. Biol. **59:** 105–117.

Barnett, P.R.O., 1971. Some changes in intertidal sand communities due to thermal pollution.—Proc. R. Soc. Lond. (B) **177:** 353–364.

Barnett, P.R.O. & B.L.S. Hardy, 1984. Thermal deformations. In: O. Kinne. Marine Ecology. Volume 5.4. John Wiley & Sons, London: 1769-1963.

Beukema, J.J., 1979. Biomass and species richness of the macrobenthic animals living on a tidal flat area in the Dutch Wadden Sea: effects of a severe winter.—Neth. J. Sea Res. **13:** 203–223.

Beukema, J.J. & M. Desprez, 1986. Single and dual animal growing seasons in the tellinid bivalve *Macoma balthica* (L.).—J. exp. mar. Biol. Ecol. **102:** 35–45.

Beukema, J.J., E. Knol & G.C. Cadée, 1985. Effects of temperature on the length of the annual growing season in the tellinid bivalve *Macoma balthica* (L.) living on the tidal flats in the Dutch Wadden Sea.—J. exp. mar. Biol. Ecol. **90:** 129-144.

Beukema, J.J. & B.W. Meehan, 1985. Latitudinal variation in linear growth and other shell characteristics of *Macoma balthica*.—Mar. Biol. **90:** 27–33.

Fry, F.E.J., 1947. Effects of the environment on animal activity.—Univ. Toronto Stud. Biol. Ser. **55:** 5–62.

Henderson, J.T., 1929. Lethal temperatures of Lamellibranchiata.—Contr. Canad. Biol. Fish. N.S. **4:** 397–411.

Holme, N.A., 1949. The fauna of sand and mud banks near the mouth of the Exe estuary.—J. mar. biol. Ass. U.K. **28:** 189–237.

Kanwisher, J.W., 1955. Freezing in intertidal animals.—Biol. Bull. **109:** 56–63.

Kennedy, V.S. & J.A. Mihursky, 1971. Upper temperature tolerance of some estuarine bivalves.—Chesapeake Sci. **12:** 193-204.

Massé, H., 1972. Contribution à l'étude de la macrofaune de peuplements des sables fins infralittoraux des côtes de Provence. VI. - Données sur la biologie des espèces.—Téthys **4:** 63-84.

McIntyre, A.D., 1971. The range of biomass in intertidal sand with special reference to the bivalve *Tellina tenuis*.—J. mar. biol. Ass. U.K. **50:** 561–575.

McMAHON, R.F. & J.G. WILSON, 1981. Seasonal respiratory responses to temperature and hypoxia in relation to burrowing depth in three intertidal bivalves.—J. therm. Biol. **6:** 267- 277.

NEWELL, R.C. & G.M. BRANCH, 1980. The influence of temperature on the maintenance of energy balance in marine invertebrates.—Adv. mar. Biol. **17:** 329–396.

SALVAT, B., 1967. Mollusques des plages océaniques et semi-abritées du Bassin d'Arcachon.—Bull. Mus. natn. Hist. nat. (2e série) **39:** 1177–1191.

SALZWEDEL, H., 1979. Reproduction, growth, mortality and variations in abundance and biomass of *Tellina fabula* (Bivalvia) in the German Bight in 1975/76.—Veröff. Inst. Meeresforsch. Bremerh. **18:** 111–202.

STEPHEN, A.C., 1928. Notes on the biology of *Tellina tenuis* da Costa.—J. mar. biol. Ass. U.K. **15:** 327–341.

——, 1932. Notes on the biology of some lamellibranchs in the Clyde area.—J. mar. biol. Ass. U.K. **18:** 51–68.

TREVALLION, A., 1971. Studies on *Tellina tenuis* da Costa. III. Aspects of general biology and energy flow.—J. exp. mar. Biol. Ecol. **7:** 95–122.

WARWICK, R.M., C.L. GEORGE & J.R. DAVIES, 1978. Annual macrofauna production in a *Venus* community.—Estuar. coast. mar. Sci. **7:** 215–241.

WILSON, J.G., 1976. Abundance and distribution of British Tellinidae. Ph.D. Thesis, University of Glasgow.

——, 1978. Upper temperature tolerance of *Tellina tenuis* and *T. fabula*—Mar. Biol. **45:** 123–128.

——, 1979. The burrowing of *Tellina tenuis* da Costa and *Tellina fabula* Gmelin in relation of sediment characteristics.—J. Life Sci. R. Dubl. Soc. **1:** 91–98.

——, 1981. Temperature tolerance of circatidal bivalves in relation to their distribution.—J. therm. Biol. **6:** 279–286.

——, 1985. Oxygen consumption of *Tellina fabula* Gmelin in relation to temperature and low oxygen tension.—Soosiana **13:** 27–32.

EXPECTED EFFECTS OF TEMPERATURE CHANGES ON ESTUARINE FISH POPULATIONS

M.J. COSTA

Dep. Zoologia e Antropologia, Bloco C–2, 3º Piso, Campo Grande, 1700 Lisboa, Portugal

ABSTRACT

Because the temperature of coastal waters will probably rise in the next 50–100 years, some possible effects of this temperature change on fish populations in coastal areas, particularly the Wadden Sea, are discussed.

This discussion is based on a comparison of some fish populations existing in the Wadden Sea and in two Portuguese estuaries (Tagus and Mira), both having higher average temperatures than the Wadden Sea.

1. INTRODUCTION

Estuarine fish are generally eurythermal, *i.e.* they can live within the limits of a considerable temperature range. Nevertheless as all other fish, they are dependent on the temperature of the water where they live. Temperature affects such processes as metabolic rate, spawning, migration and also rate of development.

This paper considers the expected long-term effects of temperature rise on the Wadden Sea fish, mainly based on a comparison with the fish fauna of two Portuguese estuaries - the Tagus and the Mira.

2. AREA DESCRIPTION

The Tagus is a large estuary with an area of ~320 km². Its tidal flats represent 3.7% of the total area. It is subject to intense disturbance (domestic sewage from Lisbon, industrial pollution and fishing activity). In 1987, the mean monthly water temperature reached 22.8°C in summer (August) and was 13.75°C in winter (February).

The Mira estuary is much smaller (about 2 km²), but may prove to be one of the best in Europe in terms of water quality, since the only source of pollution is derived from domestic sewage. It possesses *Zostera* beds like the Wadden Sea with two species - *Z. noltii* and *Z. marina*. In 1987 the mean monthly temperatures were 12.2°C and 22.5°C in winter and summer, respectively.

In Fig. 1 monthly mean water temperatures of the Tagus, the Mira and the Wadden Sea are shown.

3. SOME BIOLOGICAL EFFECTS OF TEMPERATURE ON FISHES

3.1. TEMPERATURE TOLERANCE AND DEATH

Temperature tolerance is genetically controlled and the optimal temperature to which a fish is adapted is species characteristic.

The range of temperature at which fishes can live varies considerably from one species to another and within the same species, for different life stages. Rates of temperature decrease are at least 20 times more lethal than temperature increases (BRETT, 1960). Systematic differences between species of known history have demonstrated that the lethal temperature may be used as a taxonomic tool (FRY, 1957).

Death from extreme temperatures has been well documented. But temperature can be involved with other environmental factors such as salinity and dissolved oxygen. Also the toxic action of various substances upon the fish varies with temperature changes.

3.2. FOOD

For any given species a rise in temperature usually leads to an increase in the digestion rate. The rate of food conversion and the primary process of digestion are highly temperature dependent, increasing by a factor of two or three for a 10° change (NIKOLSKY, 1963). Because of the economic interest of flatfish and the importance of the Wadden Sea for the survival of these populations, the main food items of these fish and the expected effects of temperature change on food availability will be dealt with in the following.

a) Flounder *(Platichthys flesus)*

According to DOORNBOS & TWISK (1984) the food of

J.J. Beukema et al. (eds.), Expected Effects of Climatic Change on Marine Coastal Ecosystems, 99–103.

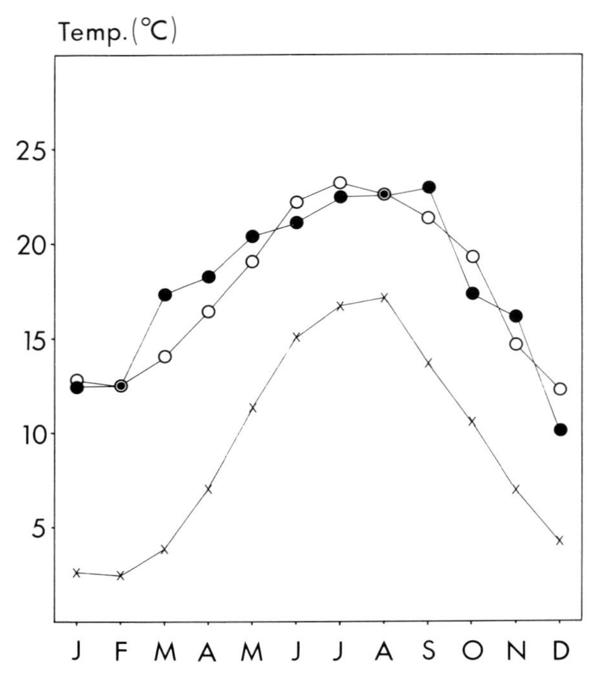

Fig. 1. Mean monthly water temperatures in the Tagus estuary (1987,●), the Mira estuary (1987,○) and the Wadden Sea (average 1960–1980, x, from BEUKEMA & DESPREZ, 1986).

the flounder is composed mainly of polychaete worms, particulary *Arenicola marina*, but also *Nereis virens* and *Nephtys hombergii*. Molluscs are hardly eaten by flounders. Decapod Crustaceans (*Crangon crangon* and *Carcinus maenas)* also play an important role in the food of this species in Lake Grevelingen. In the Tagus estuary I found the polychaetes *Lanice conchilega* and *Nereis diversicolor* to be the preferential prey. Crustaceans (especially *C. crangon*) were secondary prey, while molluscs and *Pomatoschistus minutus* were eaten only occasionally (COSTA, 1988). The rarity of molluscs in the Tagus could be the reason for this.

In the Wadden Sea polychaetes are preyed upon by flounder of all age groups. In this area, flounders also consume large numbers of amphipods *(Corophium* sp.) and *C. crangon* (DE VLAS, 1979).

b) Plaice (*Pleuronectes platessa*)

The diet of plaice, *Pleuronectes platessa*, in the Wadden Sea is mainly composed of molluscs and polychaete worms, including *Arenicola marina, Nereis diversicolor, Nephtys hombergii* and *Heteromastus filiformes* (DE VLAS, 1979). Crustaceans are of little importance, except for *C.crangon*.

In Lake Grevelingen, DOORNBOS & TWISK (1984) found polychaetes, especially *Arenicola marina*, Nereidae (*Nereis virens* and *N. diversicolor*), *Nephtys hombergii* and *Capitella capitata* in the stomachs of plaice. Crustaceans, particularly *C. crangon* and *Carcinus maenas*, are eaten throughout the year, while predation on molluscs is restricted to summer.

c) Sole (*Solea solea*)

In the Tagus estuary, sole preyed mainly on polychaetes, especially on *Nereis diversicolor* and *Pectinaria korenii,* molluscs (especially *Venerupis decussata)* as well as crustaceans (*C. crangon*) (COSTA, 1988). BRABER & DE GROOT (1973) studied the food of the sole in the southern part of the North Sea and show that polychaetes are the principal food of sole, but crustaceans and molluscs are also fed upon.

It is apparent that flatfishes feed on a wide variety of organisms, but prefer to prey upon polychaete worms. The difference in food preference from place to place must be related to differences in macrobenthos fauna. For instance, it was noted that at the mouth of the Tagus estuary, soles fed mainly on *Pectinaria korenii,* which were abundant in that area.

What could happen to these food species if a rise in temperature would occur?

According to BEUKEMA *et al.* (1978) and BEUKEMA (this issue) less extreme winter conditions in the Wadden Sea would result in a higher species diversity and a relatively stable biomass in the macrozoobenthos.

If there is a rise in temperature in winter we can expect that most species that play an important role in food of fishes will be favoured. This is the case for species susceptible to low temperatures such as *Nephtys hombergii* and *Lanice conchilega*. This latter species is abundant in the Tagus estuary. *Crangon crangon*, the brown shrimp, which populations use estuaries as nurseries, would increase. *C. crangon* is the most abundant invertebrate in the Tagus estuary; it is preyed upon by all demersal fish species.

FONDS & SAKSENA (1977) studied the food intake of young sole at 5 constant temperatures from 10 to 26°C and showed that food intake in this species was maximal at the higher temperatures.

If, however, the mean winter temperature would decrease by 2°C and if macrozoobenthos would be seriously affected, a dramatic lack of food available to fishes may be expected.

3.3. GROWTH

Growth rate is one of the essential parameters in the

assessment of population productivity. Together with salinity and photoperiod, temperature is the most important environmental factor influencing growth.

Growth rates of plaice and flounder show an approximately linear relation to temperature.

The growth of sole in the Tagus estuary is very fast. Its monthly growth constant K has a value of 0.0392 which is close to the maximum value of about 0.04 recorded for North Sea soles (FONDS, 1975).

At a temperature of 22°C, FONDS (1979) obtained a maximum length of 21.1 mm for 30-day old soles. From the daily growth patterns observed in the otoliths of Tagus soles, it can be concluded that specimens sampled in May with a length between 57 and 63 mm were ~100 days old.

So, by comparison with data for the Tagus, it can be predicted that, at least in the case of sole and flounder, higher temperatures will be beneficial by favouring the rates of food digestion and growth.

If mean water temperature in winter would decrease, the growth of soles will also be affected. Experiments by FONDS (1975) show that the growth rate of young soles will cease at ~4-5°C. He also shows that with an unlimited food supply, growth appears to be a function of the ambient water temperature.

3.4. SEXUAL MATURITY AND SPAWNING

Temperatures appear to confine spawning to a narrower range than the majority of other functions (KINNE, 1963). Most fishes require a specific temperature range for spawning, sometimes preceded by a temperature increase or decrease. Only a few fish can breed in estuaries and most of them are small and have short life cycles.

In the Tagus estuary, the resident species are *Engraulis encrasicolus, Syngnathus abaster, Gobius niger, Gobius paganellus* and *Pomatoschistus minutus*. These species and their spawning periods are shown in Fig. 2.

They have a long spawning period in the estuary. *S. abaster* even spawns all year-round.

P. minutus, which lives also in the Wadden Sea, is a species with a very important role in the food webb. In the Tagus its broad spawning period extends from January to July with a peak in February. During that period, mean water temperatures varies from 12.5°C to 18.2°C. According to FONDS & VAN BUURT (1974) the eggs of *P. minutus* are not adapted to development at high temperatures (25°C).

In the southern North Sea, shells with eggs of *P. minutus* are found at temperatures between 8 and 15°C (FONDS & VAN BUURT, 1974). SWEDMARK (1957) noticed a spawning period between March and July near Roscoff at the French coast. In the Ythan estuary the spawning period extends from April to June (HEALEY, 1971).

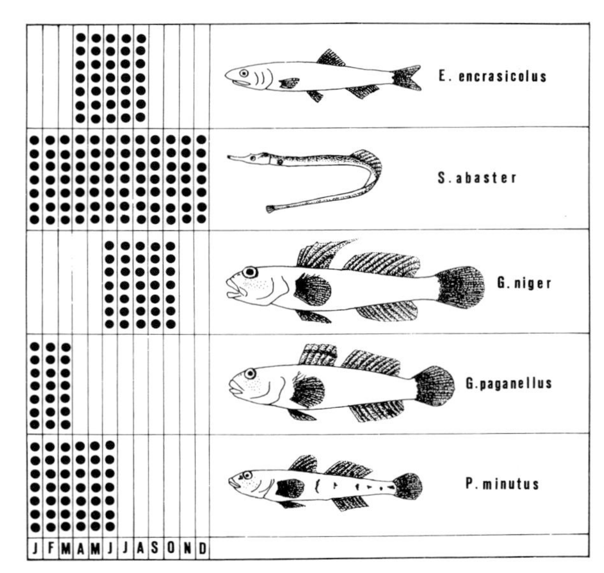

Fig. 2. Spawning periods of resident species in the Tagus estuary.

If the temperature increases, a broadening of the spawning period for this species, and maybe also for all other resident species in the Wadden Sea, can be expected.

Because it is assumed that food availability will increase with temperature and since gonadal development is determined by food supply, I think that a rise in temperature will stimulate spawning. On the other hand, a decrease in winter temperatures could be catastrophic for these species.

3.5 NURSERY AREAS

The Wadden Sea represents perhaps the most important nursery of the Dutch coast for two commersial fish species, *viz.* plaice and sole (ZIJLSTRA, 1972). According to the above cited experiments by FONDS & SAKSENA (1977), high summer temperatures and a rich food supply can be the explanation for this. Young soles spend the summers of their first two years in the coastal nurseries of the Wadden Sea (CREUTZBERG & FONDS, 1971; FONDS, 1975). They leave the shallow coastal waters in the autumn (November, December) when the water temperature falls below ~10°C, to return again in spring.

Young soles enter the Tagus estuary in April-May, colonizing the shallow zones near Vila Franca de Xira with highly productive tidal flats where muddy substrate and feeding conditions are best. The mean monthly temperature in this nursery zone reaches 18.5°C in April and a maximum of 22.4°C in August.

For 0-group sole the highest observed density in the nursery zone was 143.2 individuals per 1000 m^2 in May which corresponds to a biomass of 221.3 g per 1000 m^2.

If the temperature rises by 2 to 4°C, an optimization of nursery grounds for sole in the Wadden Sea can be expected. With higher temperatures at the spawning grounds, we can expect a higher recruitment. At low temperatures, which would cause high mortality in 0-group sole (TIEWS, 1971), this species would disappear. The even more extreme temperatures occurring on mud-flats would not affect soles, because they do not accomplish intertidal feeding migrations.

Zostera beds act as very important spawning and nursery grounds for the fishfauna. In the Mira estuary, where beds of both *Z. marina* and *Z. noltii* occur near the mouth, the overall density of fishes attains 462 per 1000 m^2 (ALMEIDA, 1988).

Because the ecological races of *Zostera marina* are considered stenothermic, an increase of 2 to 4°C in mean temperature may result in mass destruction of this species. Its ecological role as nursery will then be greatly affected. This may be minimized by the introduction of other races more tolerant of higher temperatures.

3.6. COMMUNITY STRUCTURE

With changing temperatures community structure of the estuarine fish will show certain changes. If we assume that the spawning period of resident fishes will vary, we can also assume that a rise in the temperature of the North Sea will change the spawning period of species that are estuarine-dependent. Their permanency in the Wadden Sea may change and consequently, the community structure in the course of the year will change.

If the mean temperature would increase we also predict the appearrence of some fish species from warmer waters. During the recent period of warming, subtropical invaders appeared from southerly waters in the seas round the British isles. This also happened in southern Portuguese estuaries, where the subtropical *Sphoeroides spengleri* has established itself in recent years.

If winter temperatures decrease some species may disappear and their ecological niche will eventually be occupied by others more eurythermic and resistent to low temperatures.

4. CONCLUSIONS

In conclusion, a rise in temperature will increase the feeding rate and promote growth in some Wadden Sea flatfish species. Some other resident species may show a broadening of the spawning period.

The process of acclimatization, allowing increased tolerance to high temperatures, is accompanied by a decreased resistance to low temperatures and a reduction in the tolerance range, leaving the organisms more vulnerable to cold. During cold winters many fishes, especially soles, die either due to direct effects of the temperature or because they are an easy prey to fishermen.

In the above, the effects of temperature have been considered in isolation. Synergic action with other environmental factors such as salinity, dissolved oxygen and chemical pollutants have been disregarded.

5. REFERENCES

ALMEIDA, A., 1988. Estrutura, dinâmica e produçã o da macrofauna acompanhante dos povoamentos de *Zostera noltii* e *Zostera marina* do estuário do rio Mira. Faculdade de Ciências, Universidade de Lisboa (Tese): 1-363.

BEUKEMA, J.J. & M. DESPREZ, 1986. Single and dual growing seasons in the tellinid bivalve *Macoma balthica* (L.).—J. exp. mar. Biol. Ecol. **102:** 35-45.

BEUKEMA, J.J., W. DE BRUIN & J.J.M. JANSEN, 1978. Biomass and species richness of the macrobenthic animals living on the tidal flats of the Dutch Wadden Sea: long-term changes during a period with mild winters.—Neth. J. Sea Res. **12:** 58-77.

BRABER, L. & S.J. DE GROOT, 1973. The food of five flatfish species (Pleuronectiformes) in the southern North Sea.—Neth. J. Sea Res. **6:** 163-172.

BRETT, J.R., 1960. Thermal requirements of fish—three decads of study, 1940-1970. In: C.M. TARZWELL. Biological problems in water pollution. Transactions 1959 Seminar. U.S. Depart. Health, Education, Welfare. R.A. Taft San. Eng. Center, Cincinnati, Ohio: 110-117 (Techn. Rep. W 60-3).

COSTA, M.J., 1988. Ecologie alimentaire des poissons de l'estuaire du Tage.—Cybium **12:** 301-320.

CREUTZBERG, F. & M. FONDS, 1971. The seasonal variation in the distribution of some demersal fish species in the Dutch Wadden Sea.—Thalassia Jugosl. **7:** 13-23.

DOORNBOS, G. & F. TWISK, 1984. Density, growth and annual food consumption of plaice (*Pleuronectes platessa* L.) and flounder (*Platichthys flesus* (L.)) in Lake Grevelingen, the Netherlands.—Neth. J. Sea Res. **18:** 434-456.

FONDS, M., 1975. The influence of temperature and salinity on growth of young sole *Solea solea* L. In: G. PERSOONE & J. JASPERS. Proc. 10th Europ. Symp. Mar. Biol., Ostende, Belgium **1:** 109-125.

—, 1979. Laboratory observations on the influence of temperature and salinity on development of the eggs and growth of the larvae of *Solea solea* (Pisces).—Mar. Ecol. Prog. Ser. **1:** 91-99.

FONDS, M. & G. VAN BUURT, 1974. The influence of temperature and salinity on development and survival of goby eggs (Pisces, Gobiidae).—Hydrobiol. Bull. **8:** 110-116.

FONDS, M. & V.P. SAKSENA, 1977. The daily food intake of young soles (*Solea solea* L.) in relation to their size and the water temperature.—Actes Colloques C.N.E.X.O. **4:** 51-58.

FRY, F.E.J., 1957. The lethal temperature as a tool in taxonomy.—Année biol. **33:** 205-219.

HEALEY, M.C., 1971. Gonad development and fecundity in the sand goby *Gobius minutus* Pallas.—Trans. Am. Fish. Soc. **100:** 520-526.

KINNE, O., 1963. The effects of temperature and salinity on marine and brackish water animals. I. Temperature.—Oceanogr. mar. Biol. Ann. Rev. **1:** 301-340.

NIKOLSKY, G.V., 1963. The ecology of fishes. Academic Press, New York: 1-329.

RAUCK, G. & J.J. ZIJLSTRA, 1978. On the nursery-aspects of the Wadden Sea for some commercial fish species and possible long-term changes.—Rapp. P.-V. Réun. Cons. int. Explor. Mer. **172:** 266-275.

SWEDMARK, M., 1957. Variation de la croissance et de la taille dans différentes populations du téléostéen *Gobius minutus*.—Année biol. **33:** 163-170.

TIEWS, K., 1971. Weitere ergebnisse von Langzeitbeobachtungen über das Auftreten von Beifangfischen und -krebsen in den Fängen der deutschen Garnelen fischerei (1961-1967).—Arch. Fisch. Wiss. **22:** 214-255.

VLAS, J. DE, 1979. Annual food intake by plaice and flounder in a tidal flat area in the Dutch Wadden Sea, with special reference to consumption of regenerating parts of macrobenthic prey.—Neth. J. Sea Res. **13:** 117-153.

ZIJLSTRA, J.J., 1972. On the importance of the Wadden Sea as a nursery area in relation to the conservation of the southern North Sea fishery resources.—Symp. Zool. Soc. Lond. **29:** 233-258.

SEA-LEVEL CHANGES AND TIDAL FLAT CHARACTERISTICS

WINFRIED SIEFERT

Hydrol. Res. Group; Hamburg Port Authority, Lentzkai, 2190 Cuxhaven, F.R.Germany

ABSTRACT

When the greenhouse effect would result in an acceleration of the mean sea-level rise, it is of crucial importance for the coastal areas of the southern North Sea whether or not the mean height of the tidal flats will rise accordingly. If not, coastal protection in areas behind tidal flats will have to be adapted, since wave energy will no longer be dissipated in front of the coastline and will attack the protective dikes heavily.

An analysis of long-term data of tidal flat heights and a comparison with the sea-level rise have been carried out, including data from ~ 200 continuous and temporary tide-recording stations along the Dutch, German, and Danish coast. The results demonstrate a spatially more or less horizontal mean sea level (MSL). Its rise during this century varied between 10 and 20 cm per 100 y. Not any acceleration of this rise during the last decade has been recognized. The tidal range, however, did increase during the last two decades.

A significant height change of the tidal flats parallel to the rising sea level has not taken place. If the rate of global sea-level rise would increase rapidly in the future, the mean tidal flat height might even be reduced, due to both increasing turbulence and higher wave and current energy with rising water levels.

1. INTRODUCTION

The German North Sea Coast is relatively short (370 km). Its shape is extremely irregular with estuaries and extensive tidal flats, partly protected by sandy islands with deep tidal channels in between (Fig. 1).

In order to gain more detailed information and sufficient knowledge about the processes in this region, the German Coastal Engineering Board (KFKI, Kuratorium für Forschung im Küsteningenieurwesen) has initiated an extensive programme, 'MORAN' (Morphological Analysis North-Sea-Coast). The purpose and the first ideas of the programme were presented by SIEFERT & BARTHEL (1980). Investigations are still under way.

It is important to know whether the mean height of the tidal flats will rise, if the greenhouse effect results in an acceleration of the rate of mean sea-level rise. If the flats would not keep pace, coastal protection in areas behind tidal flats will have to be changed completely, as wave energy then will no longer be dissipated in front of the coastline and will attack the dikes heavily.

An analysis of data on tidal flat characteristics over long periods and a comparison of these sea level data over the same periods may give first indications to the future behaviour of this important area in times of faster rising water level.

2. TIDES IN GENERAL

The relative motions of the earth, moon, and sun cause a number of periodic tide-producing forces. The period of each constituent can be obtained from astronomical studies; however, the amplitude and phase of the tidal response to each constituent are difficult to express in terms of the causitive forces. At any point the resultant tide is composed of a finite number of constituents, each with its own periodicity, phase angle, and amplitude. All constituents are simple-harmonic in time and are mutually independent.

With sufficient data available (~370 days of tidal records), it is possible to determine the characteristics of the various constituents composing the resultant tide. In most cases, the measured tide can be represented with reasonable accuracy if about 10 constituents are considered. In areas with relatively small tidal differences the mean tidal conditions can already be identified with a record of ~ 100 tides.

All the tidal characteristics, such as height and time, depend on a number of astronomical magnitudes, *viz.* the true solar time of the moon's meridian passage, as well as the horizontal parallax and declination of the moon and the sun respectively. Such tidal characteristics can be derived and tabulated for each place provided observation series are sufficient long, *i.e.* covering a period of 19 years. Inversely, it is possible to calculate in advance the

J.J. Beukema et al. (eds.), Expected Effects of Climatic Change on Marine Coastal Ecosystems, 105–112.

tides for a place with the aid of such tables and the precalculated astronomical magnitudes which are regularly published in astronomical yearbooks.

3. HYDROLOGICAL AND SEDIMENTOLOGICAL CONDITIONS (BASED ON REINECK, 1978).

On the average, the tidal flats of the southern North Sea extend 5 to 7 km seaward, reaching a maximum of 10 to 15 km. They are intersected by estuaries and numerous tidal channels.

The tides in this area are semidiurnal. Their range varies regionnaly and locally from ~1.5 to 3.7 m. Spring and neap influence are of minor importance. The tidal currents over the flats reach velocities of 30 to 50 $cm.s^{-1}$, with a maximum (at storm surges) of 150 $cm.s^{-1}$. In gullies the current velocities are less than or about 100 $cm.s^{-1}$ and in large channels up to 150 $cm.s^{-1}$.

Waves are a very important factor. They come from various directions and with changing strength. The predominant wind direction is from the west. The outer parts of the tidal flats are extensive breaker zones. The wave movement over the tidal flats is the major factor influencing erosion and sedimentation and the resulting distribution of sediments.

Tidal currents and wave movements combine to transport large quantities of suspended material over the tidal flats. Measurements at the Neuwerk flat showed concentrations of suspended matter ranging from 30 $mg.dm^{-3}$ for a calm tide to 300 $mg.dm^{-3}$ on a storm tide. If the shifting direction and volume with which the tide flows over the flats is taken into account, the transport of sediments during storm surges rises by a facor of 2 to 3 above normal.

The present day tidal flats receive very little new sediment from outside. Most sediment changes take place within the flats in the form of redistribution. Hence, not the sediment balance should be taken as a main parameter of redistribution, but the turnover rate, as will be shown later.

The sand on the tidal flats is fine with a mean diameter of ~0.1 mm. On the beaches subject to wave action, a fine-medium grained to medium grained sand with a mean diameter of ~0.2 mm prevails. The commonest mineral is quartz (80%), next in order come feldspat, mica and carbonate.

Gravel is mostly found in the channels, partly protecting the bed. The flora and fauna on the tidal flats enhance the threshold of beginning sand transport.

The position of the larger channels changes only slowly - except when they suddenly break through sand bars or even barrier islands during storm surges. Small channels meander considerably (up to 100 m per year) and form bank cutting faces and point bars.

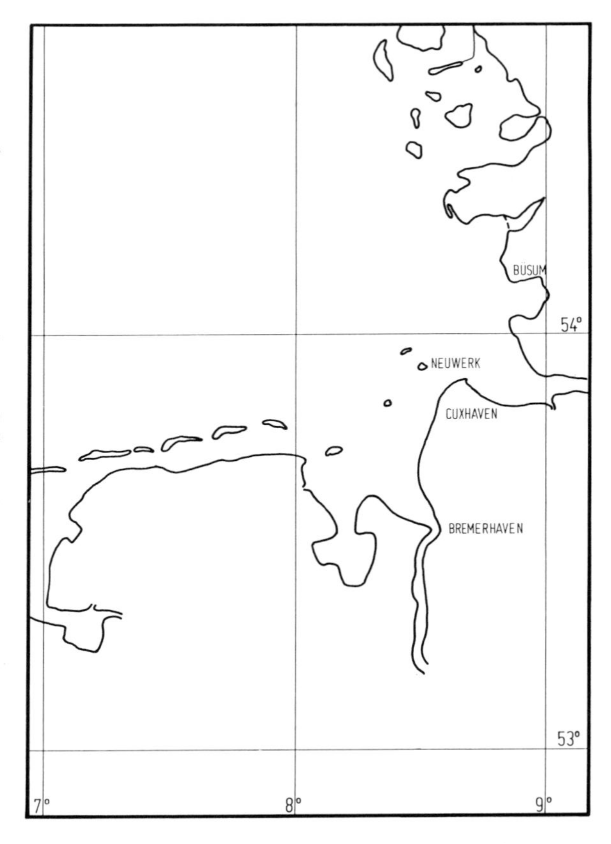

Fig. 1. Map of the German Bight with tidal flats.

4. ACTUAL MEAN SEA-LEVEL HEIGHT

Detailed investigations have been carried out for the southeastern North Sea by analysis of hydrographs from ~200 gauges in estuaries, at the coast and in the German Bight itself. Mean tidal conditions were presented in a detailed manner (SIEFERT & LASSEN, 1985); further investigations are still going on.

Preliminary results of this project from 150 Wadden Sea stations show that the mean sea level (MSL), expressed as the overall mean of the normal tide cycle, varied from 5 cm below German datum (NN) along the Dutch northern coast, 1 cm below NN along the German northwestern coast, 3 cm below NN to 4 cm above NN in Jade, Weser and Elbe estuaries to 5 cm above at the German northern and Danish western coasts. A general value of 4 cm below NN can be given for the seaward border of the Wadden Sea. Details about the local and regional variations will be published at the end of the project activities (LASSEN, 1989).

The absolute MSL in the North Sea or in the oceans are not yet known. It should be the first aim

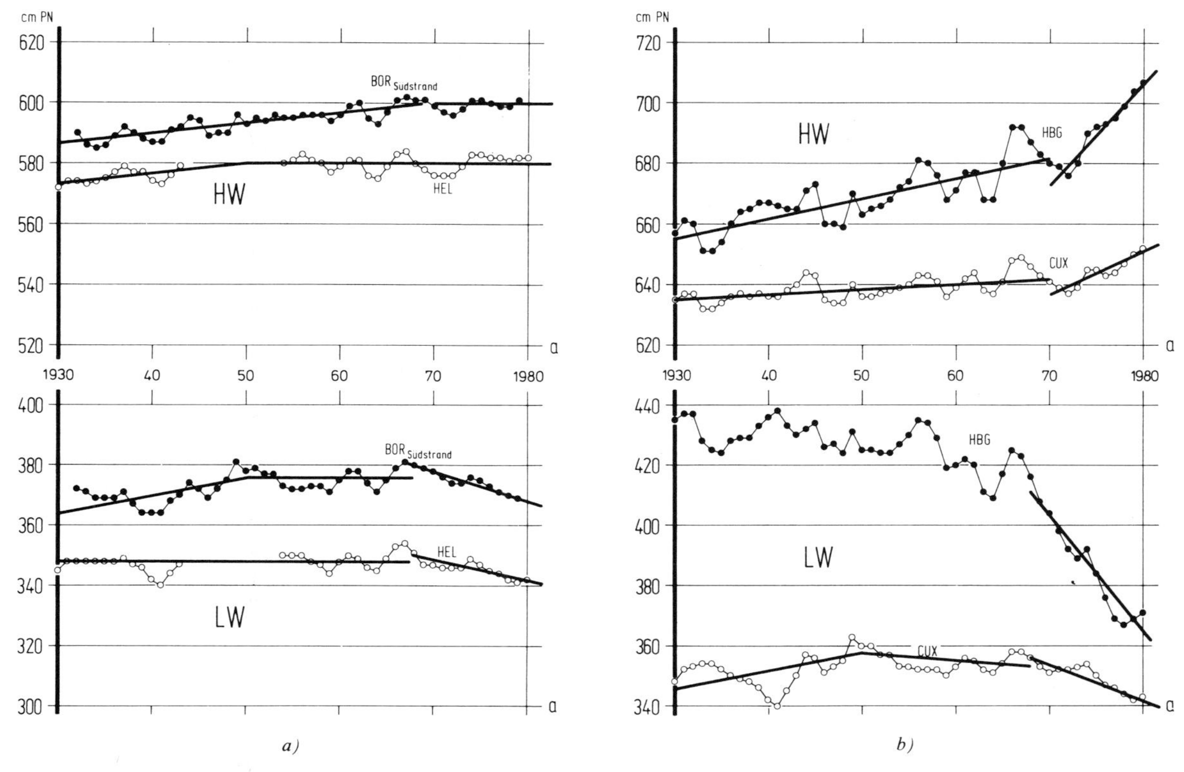

Fig. 2. Mean HW and LW developments (3-y running means) at (a) Helgoland (HEL) and Borkum (BOR), and (b) Cuxhaven (CUX) and Hamburg- St.Pauli (HBG).

to achieve tide records all along the coasts, especially in the southern hemisphere. Only after this first step a rough approximation of a 'worldwide MSL' can be given.

5. RECENT CHANGES IN TIDAL WATER LEVELS

The changes in tidal conditions can be demonstrated by the heights of mean high water (HW) and mean low water (LW). Three typical examples for the German Bight are given in Fig. 2, *viz.* for the islands Borkum (BOR) and Helgoland (HEL) and for Cuxhaven (CUX) at the seaward boundary of the Elbe estuary. At the islands (10 to 50 km in front of the coast), HW has been constant since 1970 (Fig. 2a), while at the coast (Cuxhaven is typical!) HW is rising (Fig. 2b). At the same time LW is falling both at the islands and at the coast. Thus the tidal range increased rapidly during the last few decades.

In the tidal estuaries the HW and LW developments are governed by conditions at sea and in the river. The sea influence in the Elbe is shown in Fig. 2b. Hamburg (HBG) is situated 100 km upstream of Cuxhaven, and qualitatively the developments are the same as in Cuxhaven. Especially the LW data show strong effects of the deepening of the fairway. The same is to be expected in other estuaries.

The number of indications about changes of tidal characteristics in the North Sea is increasing:
- at the German coast tidal ranges are now measured that have never occurred before
- tidal-current velocities in the outer parts of the estuaries (seaward of fairway improvements) increased
- shapes of mean tide curves have changed distinctly.

However, as the factor $k = \frac{HW-MSL}{HW-LW} \cdot 100\%$

is locally constant with very small scattering (BOR: 47.0%; HEL: 48.1%; CUX: 47.0%), the behaviour of MSL at these stations can be used to investigate sea level rise. An analysis of mean tidal curves has been applied to 3 stations over the last 50 to 120 years (the method is given by SIEFERT & LASSEN, 1985). It results in an estimate of the mean long-term MSL rise of ~10 cm per century. This is the same order of magnitude as the value for the Dutch northern coast (14 cm; VAN MALDE, 1986) and the one for the global sea-level rise in the past century (14cm ± 50%; BARNETT, 1984). Not any acceleration of this rise during the last one or two decades is obvious.

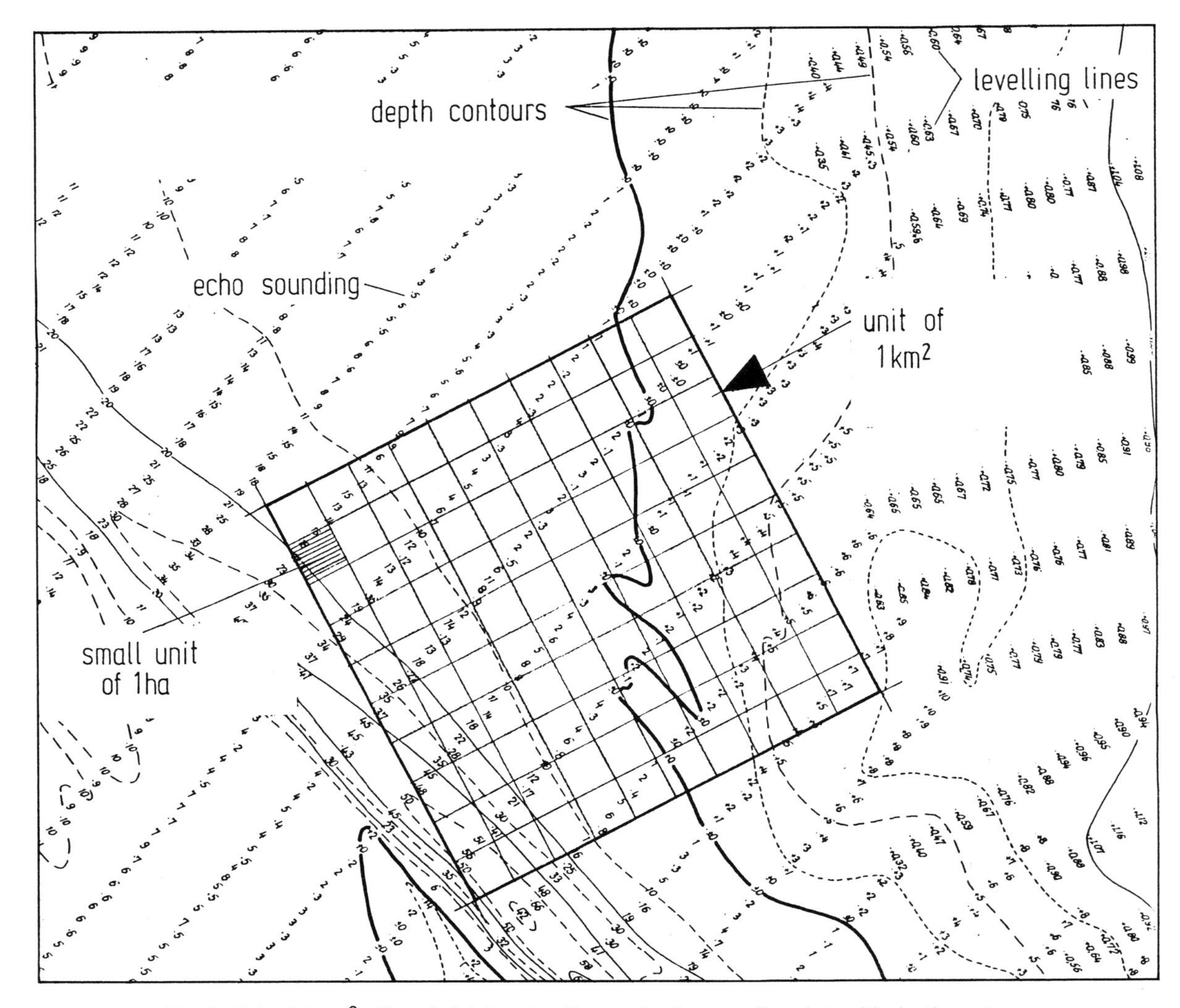

Fig. 3. Unit of 1 km^2 with subdivision; levelling and echo sounding data with depth contours.

6. METHOD OF MORPHOLOGICAL ANALYSIS

Usually the changes in volume of material in tidal-flat areas are evaluated by different maps between two surveys. But analyses like these have to be studied carefully. Extrapolation is not reliable, the amount of redistribution of material within the area remains unknown, and so are the natural variability and the variability due to the methods of levelling, echo sounding and data treatment.

That is why the MORAN project was started (SIEFERT, 1983; 1987). The philosophy of the analysis is the comparison of a number of surveys, each of them with all others, and to calculate the turnover height h_u and the net-balance height h_b as functions of wave climate and current energy during the time between the surveys. For that purpose the coastal area was divided into units of 1 km^2 size. Each of these were subdivided into 100 cells, *i.e.* small units of a size of 1 ha (Fig. 3).

Each comparison of two surveys gives a table such as the one shown in Fig. 4 for the cells and mean values for the 1 km^2-unit, respectively, whereby:
- balance=change of mean height of the unit; h_b (+ or −) in dm within time interval a in years,
- turnover=mean topographical height change (omitting the sign) over all cells within the unit; h_u in dm within the same time interval a.

Theoretical considerations (see SIEFERT, 1987) lead to the representation of the turnover height h_u over the time interval a by a saturation function such as

$$h_u = \bar{h}_u \cdot (1 - e^{-a/a_o})$$

as given in Fig. 5.

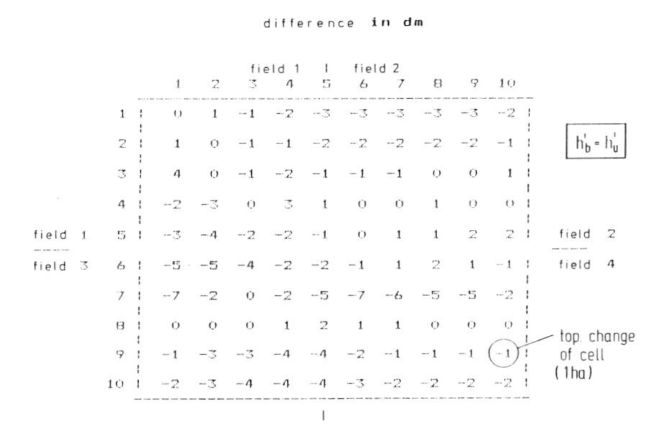

difference in dm

		1	2	3	4	5	6	7	8	9	10	
	1	0	1	-1	-2	-3	-3	-3	-3	-3	-2	$h'_b = h'_u$
	2	1	0	-1	-1	-2	-2	-2	-2	-2	-1	
	3	4	0	-1	-2	-1	-1	-1	0	0	1	
	4	-2	-3	0	3	1	0	0	1	0	0	
field 1	5	-3	-4	-2	-2	-1	0	1	1	2	2	field 2
field 3	6	-5	-5	-4	-2	-2	-1	1	2	1	-1	field 4
	7	-7	-2	0	-2	-5	-7	-6	-5	-5	-2	
	8	0	0	0	1	2	1	1	0	0	0	
	9	-1	-3	-3	-4	-4	-2	-1	-1	-1	-1	top. change of cell (1ha)
	10	-2	-3	-4	-4	-4	-3	-2	-2	-2	-2	

		field 1	2	3	4	1km² sum	mean
sedimentation	(dm)	10	8	3	6	27	0.27
erosion	(dm)	31	25	66	44	166	1.66
balance	(dm)	-21	-17	-63	-38	-139	h_b = -1.39
turnover	(dm)	41	33	69	50	193	h_u = 1.93
points		25	25	25	25	100	

Fig. 4. Height difference in dm for each unit between two surveys within a time interval of '*a*' years (example).

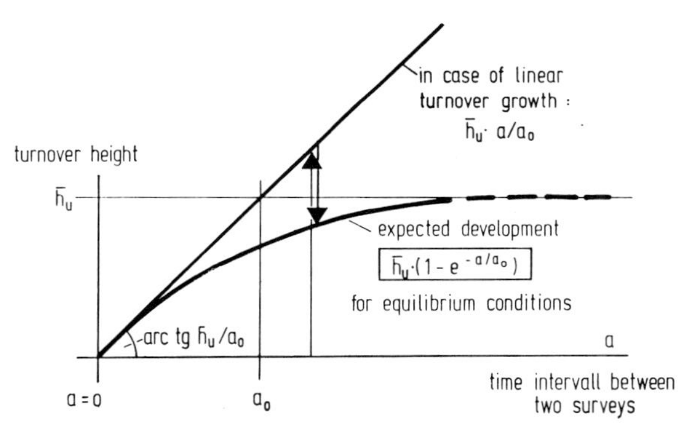

Fig. 5. Saturation function for h_u over *a*.

The morphological character of an area can thus be expressed by two parameters $\bar{h}_u$ and a_0. The value $\bar{h}_u/a_0$ in cm per year represents the mean resedimentation rate; $\bar{h}_u$ is the asymptotic turnover height that turns out as the result of long-term topographical comparisons in equilibrium conditions. As $h_b \leqslant h_u$ by definition, the change of the mean topographical height has to be $h_b(a) \leqslant \bar{h}_u$ and statistically reaches values of $h_b \approx$ (0.1 to 0.4)$\cdot\bar{h}_u$ (SIEFERT, 1987). This leads to the assumption that the mean height of such a partial area remains constant over a period if the turnover height function can be proven to be horizontally asymptotic (Fig. 6).

7. TIDAL-FLAT STABILITY AND SEA-LEVEL RISE

To give an impression of the weakness of derivations from only a limited number of comparisons between

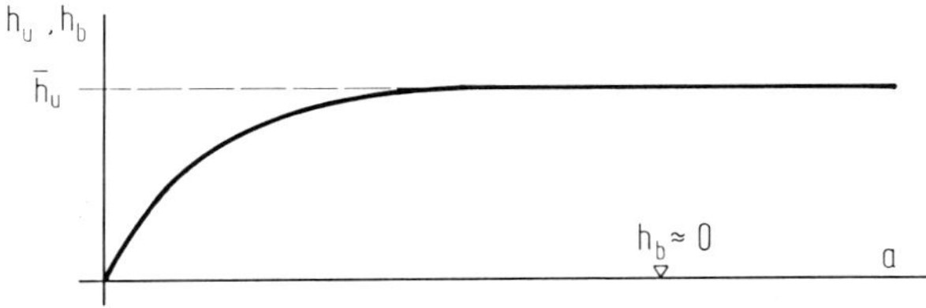

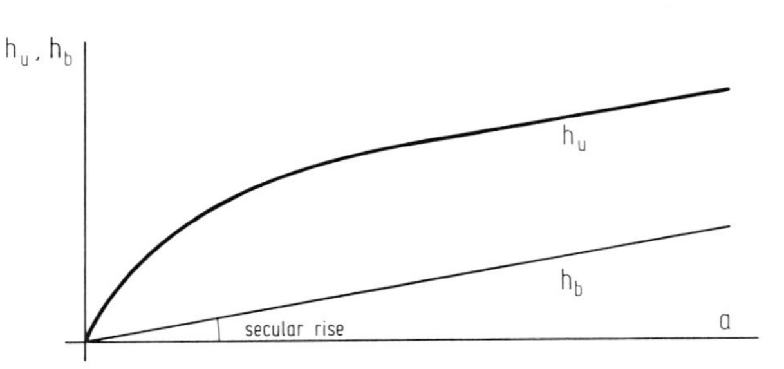

Fig. 6. Saturation function with horizontal asymptote presuming $h_b \approx 0$ for long *a* (top) and presuming secular rise of tidal flat height ($h_b \approx 0$ for long *a*) (bottom).

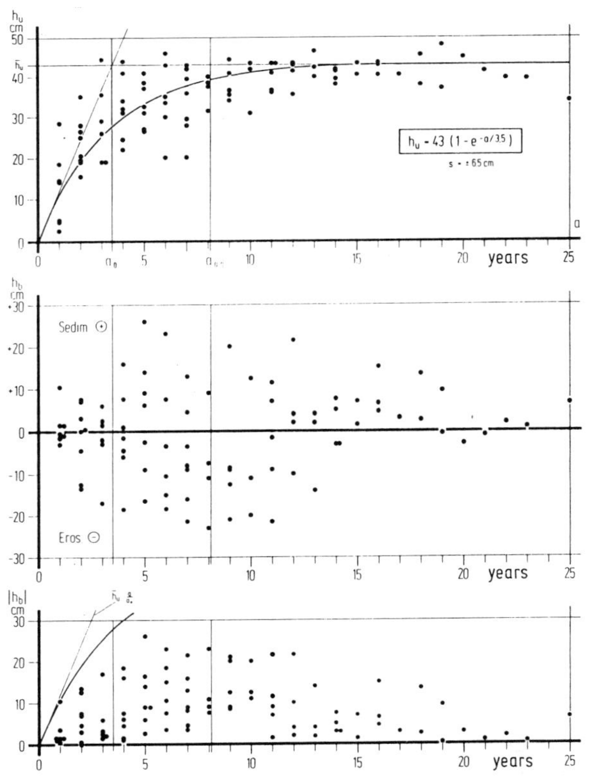

Fig. 7. Turnover and balance heights in a unit of 1-km² size for 91 topographical comparisons; time interval up to 25 years.

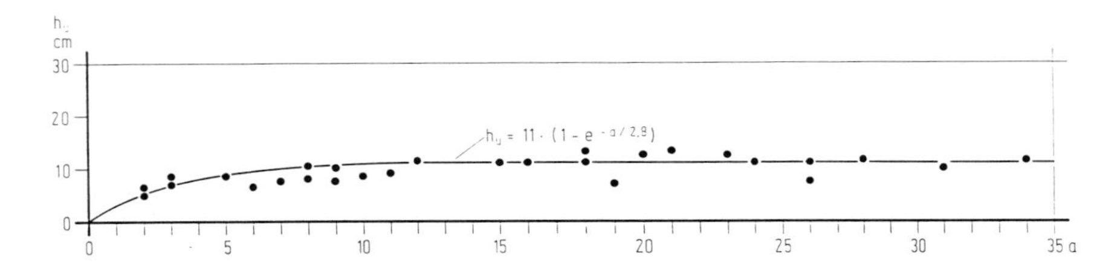

Fig. 8. Horizontal asymptote over 35 years indicates height stability without any long-term trend.

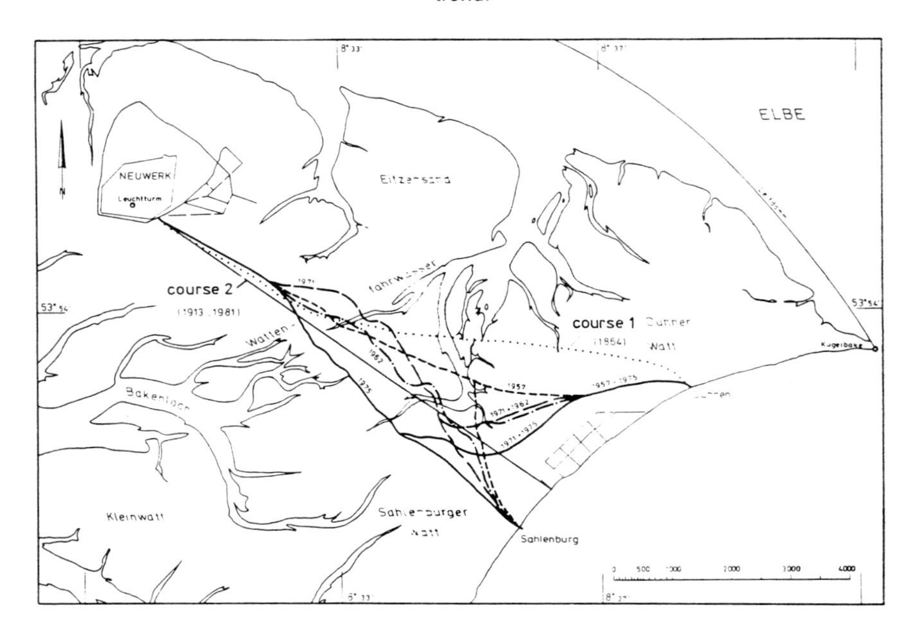

Fig. 9. Location of surveys across the tidal flat from the mainland (Cuxhaven) to an island (Neuwerk).

topographical maps, Fig. 7 gives 91 data points for turnover and balance heights out of 91 comparisons over $a = 1$ to 25 years. While the turnover heights can be described relatively precisely by a saturation function with a confidence interval of about 0.15 . $\bar{h}_u$, the individual balance heights varied between +26 cm and −24 cm. So one of the comparisons over $a = 5$ y gave a sedimentation height of 26 cm, another gave an erosion depth of 17 cm, a third nearly zero. Of course, all of them are correct, but we have to keep in mind that each survey and especially each balance evaluation includes an element of chance.

Analyses show that especially in the higher parts of the tidal flats no secular change in height can be discerned within the last 35 years, while in the same period the MSL remained nearly constant, though the more important mean high water level rose by ~15 cm, the mean storm surge level even more (example on Fig. 8). Fig. 7 already indicates a horizontal asymptote over 25 years.

Of particular interest is the evaluation of surveys of the tidal flat between Cuxhaven and Neuwerk (Fig. 1) from 1864 and 1913 and the comparison with the actual situation. In this case reliable levellings across this tidal flat over a distance of ~10 km (Fig. 9) could be reconstructed from the original data set and be compared with modern data (SIEFERT & LASSEN, 1987). The dependences demanded a $\bar{h}_u$ value about halfway between 20 and 45 cm and an a_0 value of ~4 y for tidal-flat areas like this one. The best-fit saturation function resulted in
$\bar{h}_u = 30.(1 - e^{-a/4})$
with a spread of ±5.4 cm (Fig. 10). So the resedimentation was at the small rate of $\bar{h}_u/a_0$ = 7.5 cm per year and the horizontal asymptote existed over 120 y. Hence, within the last 120 years there was no rise of the mean height of the tidal flats, while the MSL rose by ~15 cm, the mean high water level by at least 20 cm and the mean storm surge level by twice that amount.

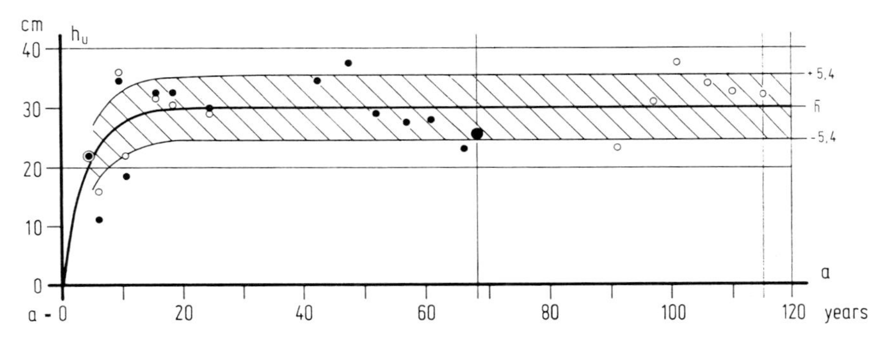

Fig. 10. Turnover height for tidal flat surveys of Fig. 9. The horizontal asymptote can be proven to have existed over 120 years (SIEFERT & LASSEN, 1987).

This result may be confirmed by another evaluation: the comparison of 2 topographical maps of another area of 4500 km^2 with a time interval of $a = 5$ y showed a mean 'sedimentation' of a mere 1 cm, while the turnover height was 70 cm. With a value $h_b/h_u = 0.014$ this 'trend' must be ignored. Any extrapolation is a matter of speculation.

8. TIDAL-GULLY EROSION AND TIDAL-RANGE INCREASE

While the mean sea level has slightly been rising during the last centuries, a remarkable increase in tidal range could be recognized over the last two decades (see chap. 5). The reasons for this development are still unknown. Only in certain areas it can be identified as the result of artificial changes. One famous example is the influence of the dam separating the Zuiderzee from the North Sea in 1932, that caused an increase in tidal range of 0.5 m in Harlingen and of 0.2 m in Den Helder (VAN MALDE, 1986).

The effect of this construction on the morphology of the outer Vliestroom, separaring the islands of Vlieland and Terschelling, may be given as an example (G. Gönnert, pers. comm.): an analysis of maps from 1933 to 1982 (as in Fig. 6) showed turnover heights between 1.5 and 4.0 m in deeper portions, while accretion up to 2.5 m or stability predominated in the shallow areas.

A second example (A. Schüller, pers. comm.) of tidal gully erosion is the system of the Till between Weser and Elbe (about 25 km^2): while the tidal range increased by 5 to 10% to ~3 m, the share of shallow areas remained constant, whereas the areas deeper than 6 m below LWL grew steadily. There are no artificial influences.

9. CONCLUSIONS FOR FUTURE DEVELOPMENT

The analysis of the behaviour of a large number of tidal-flat units over 35 years, as well as the analysis of surveys in two courses across a tidal flat over the last 120 years, indicate that a significant height change parallel to the rising sea level has not taken place. If the future global sea-level rise would proceed more rapidly, the question is whether tidal flats will rise in height with the same rate. Our data suggest that this may not be the case. This poses serious problems, as wave and current climates will change significantly. Moreover it is even possible that the mean tidal-flat height may be reduced - due to increasing turbulence with rising water level, higher wave and current energy (see Fig. 11). In that case the outer parts of the gullies and estuaries will become deeper, with large sedimentation areas in between.

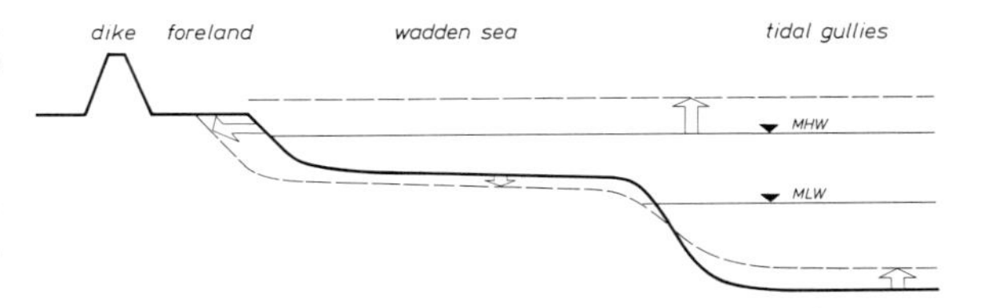

Fig. 11. Possible height development in the Wadden Sea as a consequence of rising sea level and increasing tidal range.

10. REFERENCES

BARNETT, T.P., 1984. The estimation of 'global' sea level change: A problem of uniqueness.—J. Geophys. Res. **89:** 7980-7988.

LASSEN, H., 1989. Oertliche und zeitliche Variationen des Meeresspiegels in der südöstlichen Nordsee.—Die Küste **50** (in prep.).

MALDE, J. VAN, 1986. Relatieve rijzing van gemiddelde zeeniveaus. In: Zee in-zicht. Symp., T.H. Delft: 53-85.

REINECK, H.E., 1978. The tidal flats on the German North Sea coast.—Die Küste **32:** 66-83.

SIEFERT, W., 1983. Morphologische Analysen für das Knechtsand-Gebiet.—Die Küste **38:** 1-57.

——, 1987. Umsatz- und Bilanzanalysen für das Küstenvorfeld der Deutschen Bucht.—Die Küste **45:** 1-57.

SIEFERT, W. & V. BARTHEL, 1980. The German MORAN Project. 17th ICCE, Sydney 1980, ASCE: 443-444.

SIEFERT, W. & H. LASSEN, 1985. Gesamtdarstellung der Wasserstandsverhältnisse im Küstenvorfeld der Deutschen Bucht nach neuen Pegelauswertungen.—Die Küste **42:** 1-77.

——, 1987. Zum säkularen Verhalten der mittleren Watthöhen an ausgewählten Beispielen.—Die Küste **45:** 59-70.

LONG-TERM BEACH AND SHOREFACE CHANGES, NW JUTLAND, DENMARK: EFFECTS OF A CHANGE IN WIND DIRECTION

CHRISTIAN CHRISTIANSEN[1] and DAN BOWMAN[2]

[1] *Department of Earth Sciences University of Aarhus, Ny Munkegade, Bygn. 520, 8000 Aarhus C, Denmark*
[2] *Ben-Gurion University of the Negev Department of Geography Beer Sheva 84105, P.O.B. 653, Israel*

ABSTRACT

Beach and near-shore changes during the 1944–1986 period along a 80 km stretch of coast in NW Denmark were measured on sequential air-photos and by repeated echo sounding. The study area showed an overall sediment deficit with a trend towards a positive sediment budget in the NE part of the area. Dune foot and water line showed similar behaviour throughout the study area with maximum erosion central in bays and maximum accumulation on NW parts of headlands. Apparently, this is a response to a changing wind regime with the predominant wind moving from W towards NW. Areas with accumulation had a decrease in longshore sediment transport under the new wind regime whereas areas with erosion have an increase in longshore sediment transport.

1. INTRODUCTION

In understanding the effects of a raise in sea-level on the coastal environment availability and transport of sediments become keywords. Higher sea-levels do not necessarily cause coastal erosion as suggested by Bruun (1962). In the Holocene history of the Danish coasts many examples are found where a change in sediment budget more than compensated for the sea-level rise (Christiansen *et al.*, 1985; Bowman *et al.*, 1987). Therefore longterm data of subaerial and shoreface changes are essential for coastal planning and management. Beach erosion and progression has been intensively dealt with in literature (Dolan *et al.*, 1979; McGreal, 1977; Shuisky & Schwartz, 1980; Zimmerman & Bokuniewcz, 1978). Studies on shoreline changes have mainly aimed at defining trends and rates. The questions of which beach element, water line, dune foot or coastal scarp, has been mainly affected, and how far they behave similarly has, however, not been frequently addressed. The link between the subaerial beach fluctuations and the bathymetric shoreline changes has only seldom been dealt with.

The aim of this study is to determine the longterm rates of coastal advance or regression along a 80 km coastal segment of NW Jutland, Denmark, from Klitmøller to Blokhus (Fig. 1). The study emphasizes the waterline/cliff morphodynamics and thereby differs from the spatial-temporal evolutionary sequence of Short (1988) in South Australia.

2. STUDY AREA

The study area extends over 80 km of the NW coast of Jutland, from Klitmøller to Blokhus (Fig. 1). The beaches are of median sand size with occasional gravel at Senonian and Danian exposures. The coastal cliffs are topped with dunes. Hinterland is composed of unconsolidated dunes with Danian chalk promontories dividing the study area into 4 compartments. Up to 1974, sand and gravel mining took place at a number of beaches and was since then prohibited, without a significant beach response (Christiansen & Møller, 1980).

Westerly winds, of 650 km fetch, prevail for 60% of the year with a WNW (292°) main storm direction, *i.e.*, almost parallel to the W-E coastal compartments (Fig. 2). The westerly component is clearly reflected south of Hanstholm by the modal dune direction: 98°-118°. During 12 % of the year westerly wind velocities peak to >10 m.s^{-1}. Tide ranges up to 0.3 m and and is insignificant compared to effects of storms from the W-NNW, which raise sea level 0.6 to 2.5 m above MSL. The westerly winds trigger the east-going longshore drift which is also significantly influenced by wave refraction which reduces the breaker- beach angle to 10°–20°. No major reversal of the longshore current occurs. Based on the 1931–1960 wind statistics, longshore transport ranges 0.3x106 m^3.y^{-1} to 0.6x106 m^3.y^{-1} (Hansen, 1968). The longshore current velocities range from 0.9 km.h^{-1} to 7.2 km.h^{-1} (Royal Hydrographic Office, 1967). Max wave height reached 9 m (Harbormaster of Hanstholm, pers. comm.) and storm wave period ranges from 5.9 to 8.8 s (T of 6 to 7 s being most frequent).

J.J. Beukema et al. (eds.), Expected Effects of Climatic Change on Marine Coastal Ecosystems, 113–122.

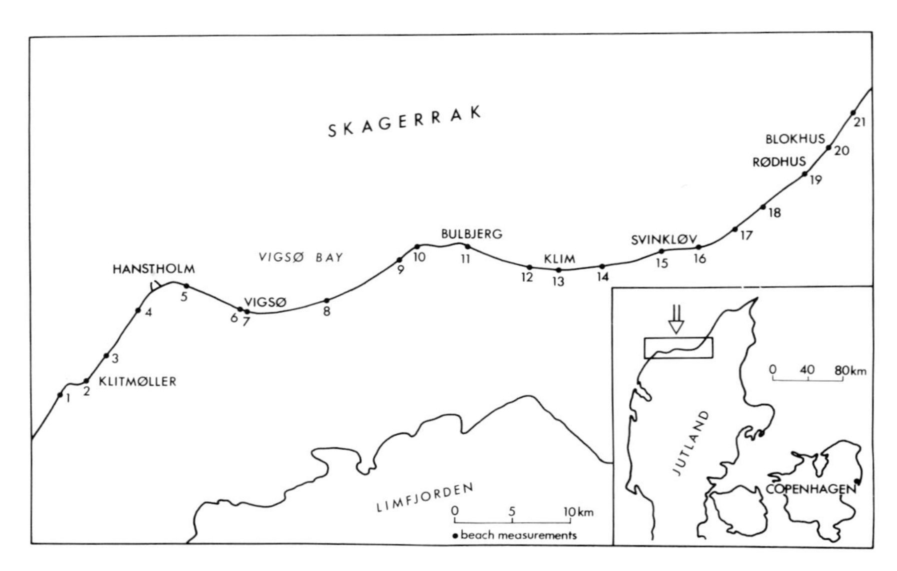

Fig. 1. The study area.

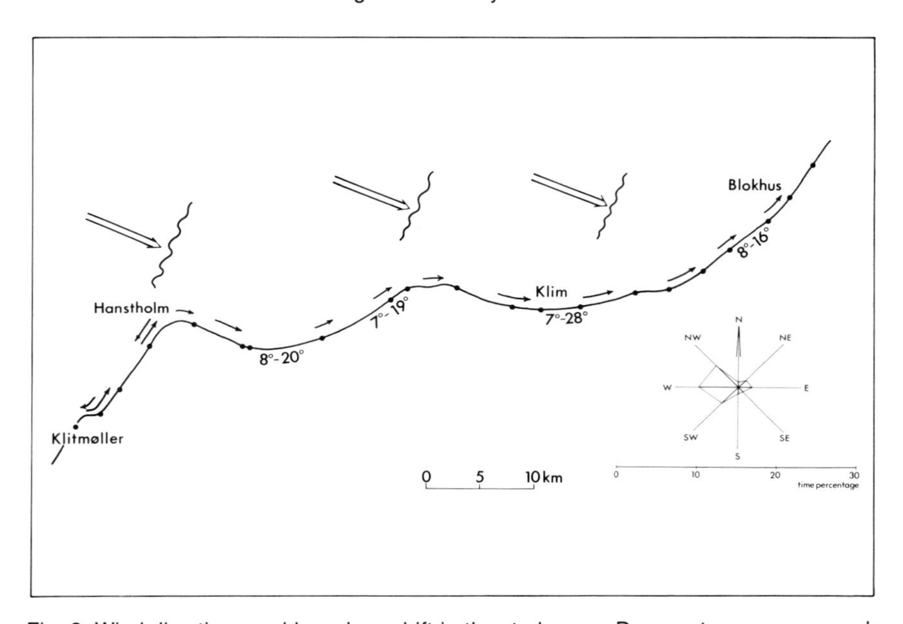

Fig. 2. Wind directions and longshore drift in the study area. Deep water wave approach and that of the breakers after refraction are indicated.

3. METHODOLOGY

Part of the data set consists of sequential air-photos extending over a period of 42 years, starting with German intelligence photos from 1944, through repeated photos from 1954, 1959, 1963, 1967, 1974, 1976, 1981 up to 1986. Such sequential data points however, do not resolve the gradual or catastrophic rates at intermediate time intervals (5 to 8 y) and completely miss their fluctuations.

The distances on the air photos between sharp reference points and the beach were measured with a micrometer to the nearest 0.1 mm, making a resolution limit of 1 to 2.5 m dependent on the scale of the air photos used (1:10,000 to 1:25,000). In order to estimate the accuracy of the air photo study, a further evaluation of the combined inherent errors stemming from repeated photo matching and from the measuring act, as well as from scale variations and the resolution limit, was undertaken. Six field

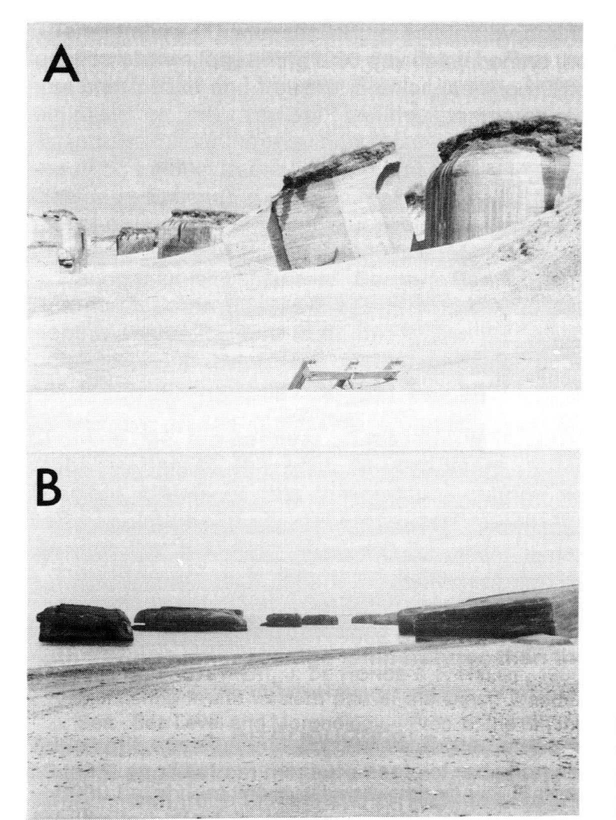

Fig. 3. German bunkers: A. on the beach (Klitmøller). B. In the water (Vigsø).

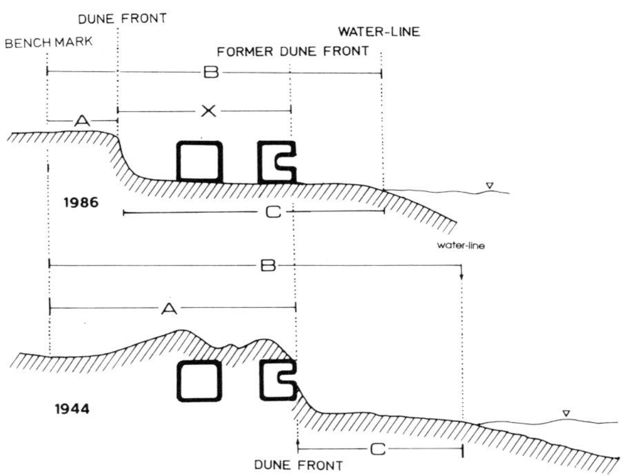

Fig. 4. The use of German bunkers for recording scarp and waterline migration:(A) Distance from bench-mark to dune front;(B) Distance from bench-mark to waterline; and (C) Beach width (B-A). The front German bunkers indicate the dunefront in 1944. X = dune front retreat relative to the bunker measured in air photos and verified in the field.

measurements between German bunkers which served as stable references and the fore dune scarp were conducted in 1987 and compared to the equivalent distances on 1986 air photos. The uncertainty associated with the air photo measurements ranged from 2 to 5.7 m, an unknown part of which may reflect the gap of 1 year between the compared measurements. It was decided to regard all air-photo-measured changes, when <10 m, as insignificant and such coast as stable, thereby filtering out the smaller coastal changes. A study of 42 years which allows for a combined 10 m error accepts a potential annual error of $10/42 \approx 0.2$ m.y^{-1}.

For precisely recording beach location in aerial photos we followed two reference lines: the coastal scarp or dune-front and the beach line, defined as the high waterline. Beach width was recorded as well (Fig. 4). The coastal scarp usually displayed a sharp morphology, independent of seasonal cyclic beach fluctuations and mainly reflect longterm coastal trend. Bluff morphology was of great use for marking erosive trends shown by nearly vertical slopes with upper vegetable cover, whereas accumulative trends were demonstrated by gentlier slopes with a patchy vegetation cover.

In order to quantify and make long-term shoreline trends visible, the German blockhouses along the western Danish coast served as unique, clearly recognizable long-term reference points. The unique advantage of the German bunkers for marking beach changes compared to the traditional demonstration of property damage is their endurance and stability. They do not collapse or disappear although within the reach of the swash or waves (Figs 3 and 4). The total coastal changes during the study period, including all trends, were summed up in each of the 21 stations (Fig. 5A) and regarded as a 'dynamic index' which demonstrates the level of activity unrelated to trend.

Bathymetric data was provided by the Danish Department of Coastal Engineering. Recurrent profiling was conducted in 1970, 1971, 1972, 1975, 1978, 1983, 1985, and 1987. The profiles, of 1 km interval, were measured with a subaerial standard leveling technique combined with echo sounding, performed by (ATLAS instruments) and location was determined by TORAN. Depth reading was performed at 150 m intervals up to 3.2 km offshore, to an average depth of 15 m. Standard deviation of the depth measurement was up to 1975 ~ 20 cm, ~ 10 cm in 1983 and improved to ~ 5 cm in 1985. Location was accurate to 10 m. In order to resolve the major bathymetric modes of variation over the study period the profiles in front of the 21 beach stations were drawn and analyzed. Profiles transects as well as volumetric changes and sediment budget calculations along the entire nearshore and beachface were further conducted. Changes in wind regime during the study

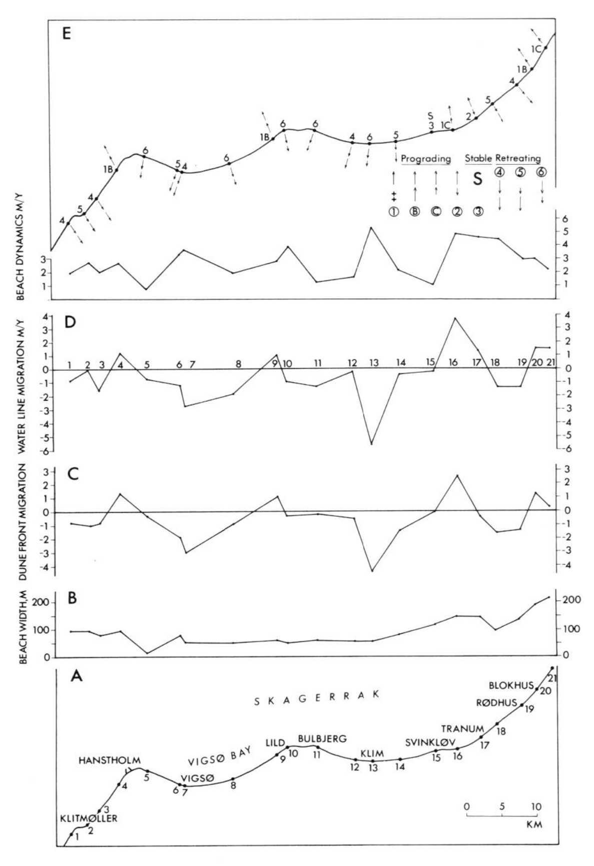

Fig. 5. Long-term beach trends.

period were examined and compared to the bathymetric and subaerial data. For this purpose, two sets of wind statistics were used: 1931–1960 (METEOROLOGICAL INSTITUT, 1971) and 1968–1982 (HANSTHOLM HAVNERAAD, 1983).

4. RESULTS

4.1. LONG-TERM BEACH-CHANGE CATEGORIES

The trends and rates of the waterline and of the dune front migrations, including beach width fluctuations and dynamic index of all stations were cluster analyzed, resulting in an 8-class beach model with the following characteristics (Fig. 6):

- Group 1

Class 1a shows dune front progression which is faster than that of the waterline, resulting in a narrow beach. This type represents the extreme end-member of the model which has not been, however, encountered in the field. Because of its height, much sand supply is needed for a sand dune to prograde shorewards. High-water table makes the beach saturated and further inhibits sand supply. Moreover, narrowing beaches limit sand supply and act as a negative feedback for further dune front progation. The studied Danish beaches, as other beaches of the world, are nowadays in a transgressive, narrow mode. It seems that only a regressive sea and abundant longshore sediment supply onto wide beaches of low water table would fulfil the primary condition for 1a class. At stations 20 and 21 classes, 1b and 1c were encountered with stable and widening beachwidth, respectively (Fig. 5E). The gradual seaward advance of the fore-dune is manifested by a moderate patchy vegetative dune ramp on an accreting beach. Stations 20 and 21 reflect a clear 9° change in coastal direction, thus more parallel to westerly-driven waves, causing a slowdown of the longshore drift and building up the widest beaches in the study area. Station 4 reflects the damming effect of Hanstholm Harbor. Class 1c at Station 16 reflects an almost totally buried groin. Station 9, however remains unexplained.

- Group 2

This is a transitional category, demonstrating contradictory trends (Fig. 5E, Station 17): beach progression with a retreating stable dune front. It is located between progradational beaches (Station 16) and a retreating segment (Station 4 and 5). These beaches are widening fast and show high dynamics.

- Group 3

This is a stable beach (Station 15) shown by a very slow waterline retreat over the study period (5m) and very low (0.9 $m.y^{-1}$) waterline dynamics. The rate of scarp progression somewhat exceeds that of the waterline pointing to the eolian sand transfer.

- Groups 4 and 5

These groups spatially alternate and indicate overall retreating beaches with different waterline and dune-front rates of migration, resulting in retreating beaches of stable width (4) or in a widening beach type because of faster scarp retreat (5). These groups represent a less severe waterline retreat but high scarp (9 $m.y^{-1}$) and waterline (14.8 $m.y^{-1}$) dynamics. They are located as a transitional segment between retreating to prograding beaches (Stations 14 and 18). Their similar beach width (97, 82, 82, 95 m) points to self-regulation by the limit of the run up.

- Group 6

Fast-eroding category with the waterline taking the lead, resulting in narrow beaches and demonstrating the severest coastal retreat and the highest dynamic rates, best represented by Station 13-Klim (Fig. 5E). This station is located downcurrent, beyond the main beach sand storage of Bulbjerg-Thorup, wherefrom

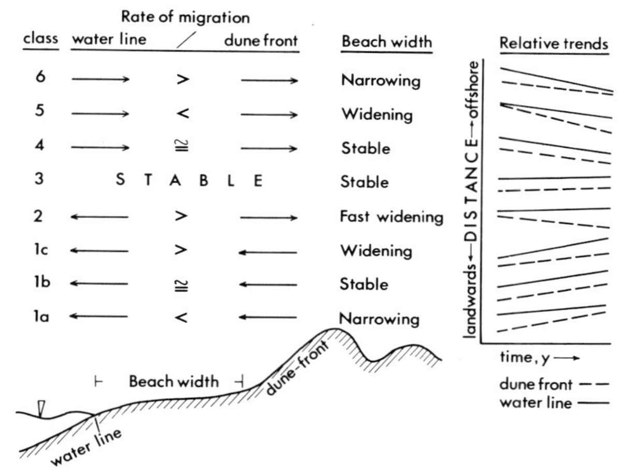

Fig. 6. Categories of waterline and dunefront migrations with trends of the beach width indicated.

bars grow eastwards but die out west to Klim, leaving this site unprotected and its nearshore breached by rips which drain the eastward driven surf water (Fig. 11).

4.2. MAIN COASTAL TRENDS

The retreating beach classes predominate along the study area (Fig. 5E). By their apparent advance towards the sea, the German Bunkers visibly demonstrate this longterm waterline and dunefront retreat (Fig. 7), which reached its extreme at the Bay of Vigsø and Klim. Most beaches are backed by sea cliffs topped with dunes, indicating former dune transgressions (CHRISTIANSEN & BOWMAN, 1986). Progradational beach trends are local but strengthen eastwards, where a smaller angle of wave approach results in decrease of the littoral drift. At Blokhus (Stations 20, 21) the dunefront progrades as sand ramps and climbing modern dunes jeopardize summer houses. This trend strengthens northwards in the study area. Such pattern suggests that the eastward driven sediments, eroded off the beaches make the eastern compartment (Stations 16–21) their main sink.

The trends of the waterline compare favorably with those of the dunefront migrations (Fig. 5C and D; Figs 8 and 9) *i.e.*, retreating and advancing shorelines are backed by the same trend of the dunefront. Beach width is a far less fluctuating parameter and locally independent of the beach trends (Fig. 5B). It gradually increases from Vigsø eastwards where the widest beaches (200 m: Stations 20, 21) serve as an efficient buffer against wave attack on the dune front.

The rates of beach dynamics (Fig. 5E) are neither in phase with regressional nor with progradational trends. Highest dynamics are shown at the sites of severest erosion (Stations 7, 13) as at sites of accumulation (Station 16). The consistency of the dynamic index within the same beach class is low. Along the study area beach dynamics increase eastwards. Downdrift from promotories (Stations 5, 11–12) dynamics is the lowest.

Waterline dynamics ranged along the study area between high (r 4.4 $m.y^{-1}$) to low (s 2.7 $m.y^{-1}$). High scarp dynamics amounted to r 3.3 $m.y^{-1}$ whereas when low it was s 1.1 $m.y^{-1}$ and even reached 0.2 to 0.5 $m.y^{-1}$ The highest mean rate of scarp retreat over the study priod was 4.5 $m.y^{-1}$ and occurred at Klim. Maximum rate of waterline retreat at Klim was 5.5 $m.y^{-1}$. High retreat rates were also encountered at Vigsø with 2.8 $m.y^{-1}$ scarp retreat and 3 $m.y^{-1}$ retreat of the waterline. For some of the intermediate periods waterline and scarp migration rates ranged very high: 6–12 $m.y^{-1}$ with a maximum rate of about 40 m for one event waterline retreat at Klim (CHRISTIANSEN & MØLLER, 1980). The longshore rates of scarp retreat, because of less fluctuations, prove more significant than those of the waterline, which are often insignificant.

Seasonal fluctuations complicate longterm beach measurements and should be taken into account. The minimum long-term detectable trend equals the magnitude of the average short term variation divided by the period of observation (ZIMMERMAN & BOKUNIEWCZ, 1987). Based on 30 $m.y^{-1}$ short term seasonal fluctuations, our maximum rates of retreat are well beyond seasonal variation. Rates below 1 $m.y^{-1}$ are below long-term resolution and may reflect seasonality.

Compared to the most rapid erosional rate along the pacific coast of the United States at Cape Shoalwater, *viz.* 37.8 $m.y^{-1}$ (TERICH & LEVENSELLER, 1986), our extreme rates are low. However, compared to rates of retreat along the Atlantic coast of the U.S.A. and along the Gulf of Mexico shorelines with a few meters per year (EL-ASHRY, 1971; DOLAN *et al.*, 1979) and to cliff retreat in diluvial deposits in Japan of 0.1 to 1.1 $m.y^{-1}$ (HORIKAWA & SUNAMURA, 1967) as well as to longterm retreat rates in the glacial sediments of Northern Ireland sea cliffs of 0.4 $m.y^{-1}$ (MCGREAL, 1977), the long-term retreat at the studied beaches record very high.

4.3. BATHYMETRY

Bathymetric maps based on repeated echo-sounding and the slope gradients up to 1000 m and 200 m offshore (Fig. 10D) indicate that Hanstholm is the only promontory felt offshore at the shoreface. At its leeside, in the Vigsø bay, bathymetry reveals an advanced stage of filling of the western side of the bay. Fig. 11 clearly divides the bay into a western shallow

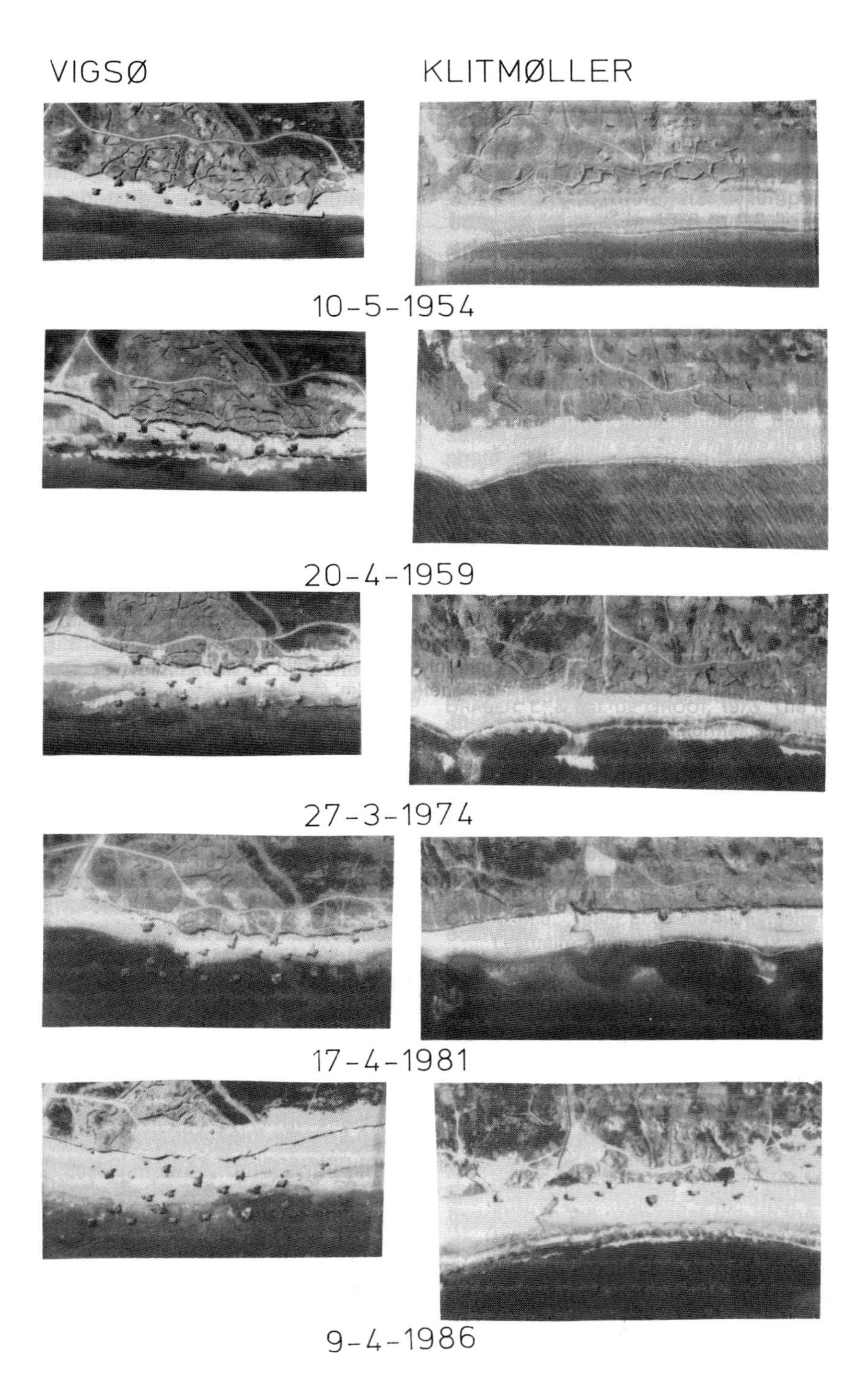

Fig. 7. Sequential air photos demonstrating beach retreat by the apparent advance of German bunkers seawards. (A) Vigsø. (B) Klitmøller. Note significant earlier start of erosion at Vigsø.

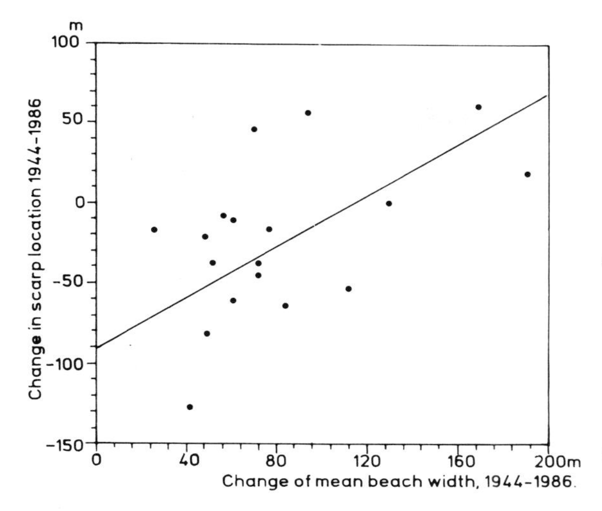

Fig. 8. Correlation of the change in the mean beach width with the change in location of the scarp over the entire study period. (n=20, r=0.525, which is significant at the 95% level.

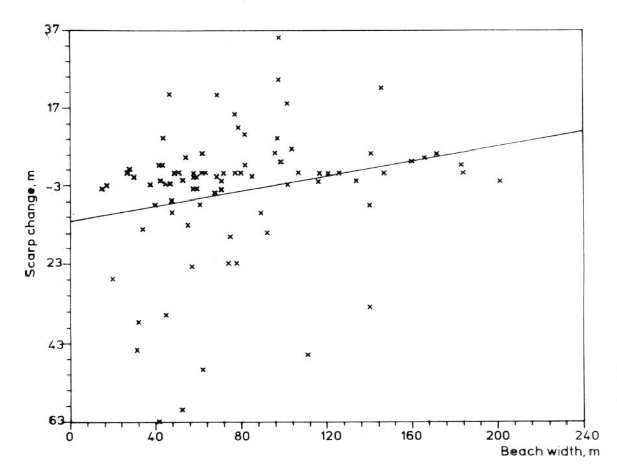

Fig. 9. Correlation of beach width to scarp retreat in the following period. (n=84, r=0.237 which is significant at the 95% level.

part and a deeper and steeper sink at the eastern side. The sedimentary wedge may by refraction to a certain degree focus part of the wave energy towards Vigsø. Following hydrographic maps no. 91 and 93 which range in scale from 130,000 to 1:200,000, the trend of the −20 m isobath is unrelated to the W-E trend of the study area (Fig. 11), suggesting diversion and bypassing of sediments in the offshore from Hanstholm to the NE, thereby partly ending up outside the study area. The fill of Vigsø Bay and the −20 m isobath trace indicate Holocene shoreline simplification and straightening. The sediment trap of Vigsø bay may partly explain the deficiency of sediments further eastwards.

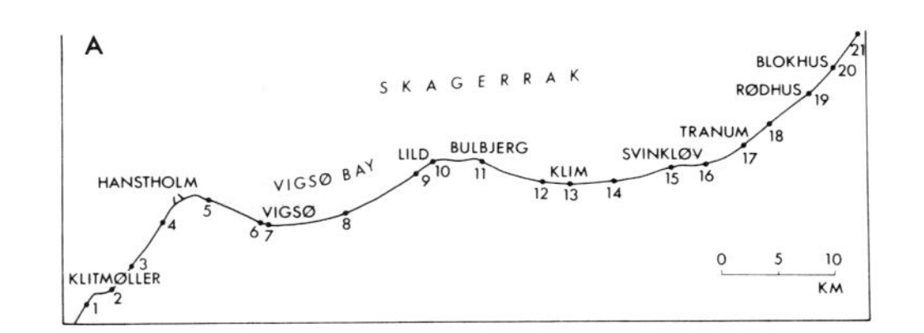

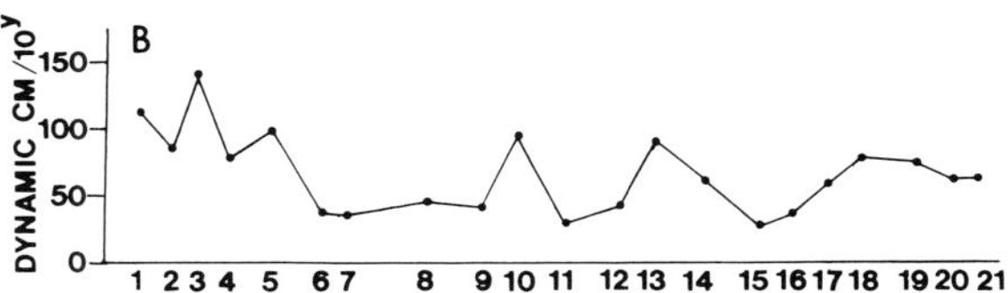

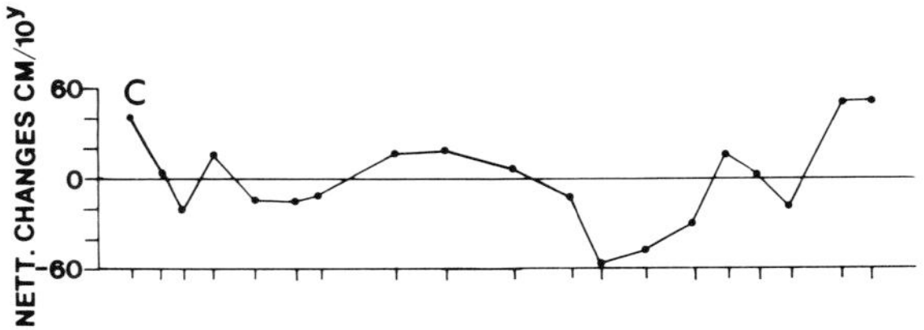

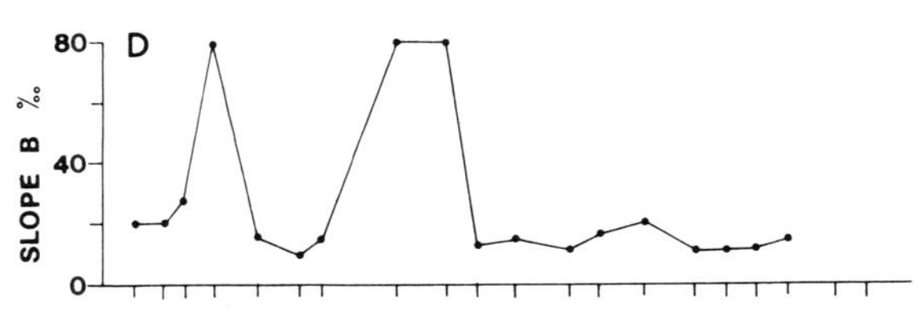

Fig. 10. Longterm bathymetric trends.

The bathymetric dynamic index is the total accumulative depth changes during 10 years in profiles out to 3.2 km from the waterline. Fig. 10B shows highest bathymetric dynamics at the west-facing segment of Klitmøller-Hanstholm with a lowering trend eastwards which is reverse to the trend of the beach dynamics (Fig. 5E). However, beach fluctuations and bathymetric dynamics are usually in phase except at Vigsø which shows higher beach dynamics but is bathymetrically calmer. The bathymetric net change along the study area (Fig. 10C) is well in phase with the water line and dune front migrations (Figs 5D,C), all which suggest that the beaches and the shoreface operate as one cell.

4.4. WAVE CLIMATE

Large-magnitude storm surges, when sea level rises several meters above normal high-water conditions, accomplish most of the geological work and make the most important contribution to the beach budget (VELLINGA, 1982). Long-term storm data may therefore be the key for resolving beach trends. The

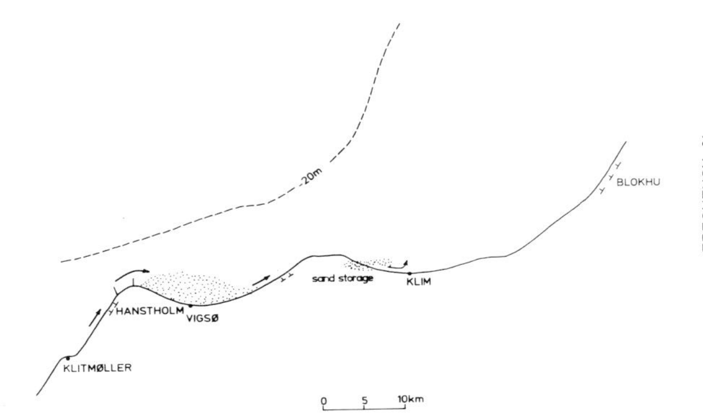

Fig. 11. The Vigsø bay sediment trap, shown (schematically) as a sand wedge (based on bathymetric maps). At Klim the sand supply in form of bars dies out west to the site whereas eastward driven longshore currents drain and breach the inshore (based on air photos). Dips indicate sites of steep beaches (based on bathymetric maps), which reflect sand wedges and favorably agree with prograding beaches (group 1). The 20 m isobath is also delineated.

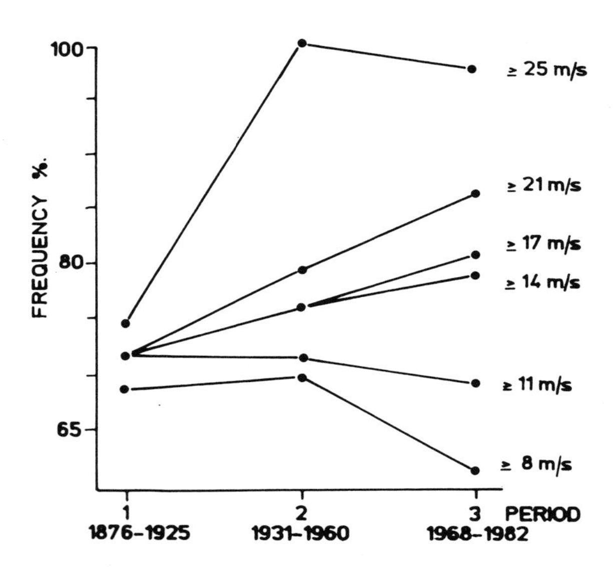

Fig. 12. Frequency of winds with a westerly component and of selected velocities. The trend towards higher velocities is clearly indicated.

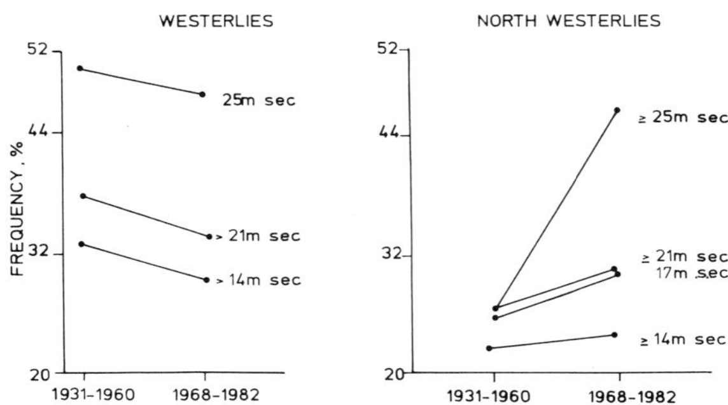

Fig. 13. Changes in wind direction: Increase in the frequency of the north westerlies and decrease of the westerlies is shown during the period of study.

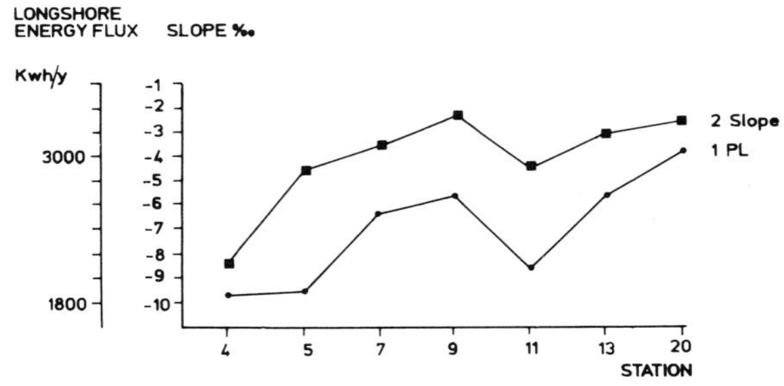

Fig. 14. Change of profiles slope and of longshore power along the study area.

westerly winds of > 2 m.s^{-1} compose 70 to 100% of all winds in the study area. A significant increase in the frequency of the higher-velocity westerlies is shown in Fig. 12. The westerlies of relative low velocity diminished during the period studied. Superimposed are pronounced changes in wind direction (Fig. 13): the frequency of the westerlies fell whereas the north westerlies became more frequent especially those ≥ 25 m.s^{-1}. As a result the north western facing beach segments in the study area became gradually more exposed to the severest storm surges.

The effects of the changes in wind velocity and in direction on the longshore transport were estimated following KOMAR (1976) by expressing the longshore wave energy component for each wind direction as

$$P_L = \sum_{4}^{11} 42.7\, H_m^2 \cdot T_s \cdot f . \sin\alpha . \cos\alpha \;(\mathrm{KWH.y^{-1}})$$

where

H_m = mean wave height
T_s = significant wave period
α = angle between incoming wave crest and shoreline direction
f = frequency of winds at 4–11 Beaufort force.

As no wave data were available for the study area before 1968, wave parameters were estimated according to CERC (1975). The resulting net longshore energy component was converted into net longshore transport by multiplying P_L with 284 (KOMAR, 1976).

Table 1 summarizes the estimations of the longshore drift for 7 stations. The Danish Coastal Engineering Department estimated the longshore

TABLE 1

Changes in longshore transport compared to beach changes.

	Longshore transport in 10^5 $m^3.y^{-1}$							
	St.4	*St.5*	*St.7*	*St.9*	*St.11*	*St.13*	*St.20*	*mean*
1931 - 1960	6.7	3.9	6.4	7.9	4.1	7.2	9.5	6.4
1968 - 1982	5.2	5.2	7.3	7.6	5.9	7.5	8.7	6.7
Change in transport	-	+	+	-	+	+	-	+
Net beach change 1944 - 1986	Acc.	Ero.	Ero.	Acc.	Ero.	Ero.	Acc.	Ero.

transport at station 4 around 1960 to be 6 to 7 x 10^5 $m^3.y^{-1}$ based on actual measurements (HANSEN, 1968) which fits favourably our estimates. All stations with a fall in estimated longshore transport from 1931-1960 to 1968–1982 showed in the present study net coastal propagation whereas stations with a rise in the longshore transport showed net coastal retreat. This is in accordance with theoretical considerations. KOMAR (1976) showed from continuity considerations that if the transport is increasing the shoreline must retreat or erode to supply the higher transport.

Apparently the change in wind direction plays a major role in coastal morphology. Table 1 also suggests that the mean estimated longshore transport for the study area has risen during the study period as a consequence of the changing wind regime. This is in good agreement with the observation of the general coastal retreat. Fig. 14 shows the changes in longshore transport related to the mean slope of the profiles 1 km offshore. There is a general westward trend of flatter profiles concommitent to increase in the longshore power up to Station 20.

5. DISCUSSION

The long-term beach trend, the recurent profiling as well as morphological characteristics suggest an overall sediment deficit. As the study area experiences an average mean sea-level rise of 0.5 $mm.y^{-1}$ (CHRISTIANSEN *et al.*, 1985), this could be interpreted as being in accordance with Bruun's rule (BRUUN, 1962) with coastal erosion as a consequence of sea-level rise. However, a closer inspection of the areal distribution of erosion and accumulation suggests that longshore transport plays a major role in shapening the coastal configuration in the study area. There are maximum erosion central in the bays at Vigsø and at Klim and maximum accumulation on NW parts of the headlands at Hanstholm, Horsebæk and Slette Strand. Apparently, shorelines in each of the coastal segments are approximating themselves to an equilibrium with the new predominant wind direction. The lack of parallisme between shoreface bathymetry and the present shoreline indicates that reshapening of the shoreface is proceeding by filling of sediment traps.

The increase in storminess and change in wind direction in the study area seems to have dictated the erosional trend. J. Nielsen (pers. comm.) has shown that a similar trend exists in the south eastern part of Denmark where longshore sediment transport has reversed totally in the present study period due to the change in wind direction. On a wider scale LAMB & WEISS (1979) have shown a similar meteorological trend for the entire North Sea area.

NAMIAS & HUANG (1972) show that a change in the prevailing wind pattern off southern California was the major factor for the rise in sea-level of 5.6 cm during a decade.

Many authors (TANNER & STAPOR, 1972; WALTON, 1978) have pointed to the potential increase in coastal erosion and the accompanying loss of land and property due to the rising sea-level. The examination of air-photos covering 42 years in the present study has also shown many houses lost to the sea. Less attention has, however, been given to the potential disadvantage of coastal accumulation due to a changing longshore sediment transport.

At Slette Strand (Station 16) coastal progradation has been 159 m in the study period. The progradation has been specially rapid in the last half of the period. The abundance of sediment in the nearshore has resulted in many bars which makes it difficult for the fishermen whose boats are hauled up onto the beach to get their boats out in the sea. Compared to another small fisher community at Torup Strand (Station 12) with a slight erosion the fishers at Slette Strand have 90 days fewer per year at which they are able to get out for fishing. Similarly the fast widening beach has forced the fishermen at Slette Strand to give up their old building containing power-winches and build a new one closer to the sea.

Beach width reflect the phase relationship of the waterline relative to that of the coastal scarp. Although widely accepted that wide beaches are the best shore protection and that relationship between beach width and erosion rate is fairly consistent (CARTER *et al.*, 1986) the present data suggest that the long-term trend of beach width is a misleading indicator of the beach trend, *i.e.* identical beach width may manifest very different coastal trends. Maximum out-of-phase relations which form fast widening beaches, were observed when opposite though not extreme trends dominated (Tranum Station 17). High out-of-phase relations were encountered both when trends were regressional (Klim, Station 13) and progradational (Blokhus N, Station 21). Nevertheles, beach width proved a quite stable parameter along the study area. The spatial relative small variability of the beach width (Fig. 5B) does, however, not indicate equilibrium conditions.

ACKNOWLEDGEMENTS

This study was made possible by a grant (J.Nr. 11–6366) from the Danish Natural Science Research Council.

6. REFERENCES

BOWMAN, D., C. CHRISTIANSEN & H. MARGARITZ, 1987. Late-Holocene coastal evolution of the Hanstholm-Hjardemaal region, NW-Denmark. Morphology, sediments and dating.—Geoskrifter 25, University of Aarhus: 1-36.

BRUUN, P., 1962. Sea level rise as a cause of shore-erosion.—Am. Soc. Civ. Eng. J. Water. Har. Div. **88:** 117-130.

CARTER, C.H., C.B. MONROE & D.E. GUY, 1986, Lake Erie Shore Erosion: The effect of beach width and shore protection structures.—J. Coastal. Res. **2:** 17-23.

CERC, 1975. Shore protection manual Vol. 1, Department of the U.S.A. Army Corps of Engineers, Virginia.

CHRISTIANSEN, C. & D. BOWMAN, 1986. Sea-level changes, coastal dune building and sand drift, North Western Jutland, Denmark.—Geogr. Tidsskrift. **86:** 28-31.

CHRISTIANSEN, C. & J.T. MØLLER, 1980. Beach erosion at Klim, Denmark. A ten year record.—Coast Eng. **3:** 282-296.

CHRISTIANSEN, C., J. T. MØLLER & J. NIELSEN, 1985. Fluctuations in sea-level and associated morphological response: Examples from Denmark.—Eiszeitalter und Gegenwart **35:** 89-108.

DOLAN, R., B. HAYDEN, C. REA & J. HEYWOOD, 1979. Shoreline erosion rates along the middle Atlantic coast of the United States.—Geology, **7:** 602-606.

EL-ASHRY, M.T., 1971. Causes of recent increased erosion along United States shorelines.—Geol. Soc. Amer. Bull. **82:** 2033- 2038.

HANSEN, E., 1968. Vandbygningsvæsenet, 1868-1968. København.

HANSTHOLM HAVNERAAD, 1983. Hanstholm Havn-vind og vandstands-statistik, 1968-1982. Hanstholm.

HORIKAWA, K. & T. SUMAMURA, 1967. A study on erosion of coastal cliffs by using aerial photographs.—Coast. Eng. Japan. **10:** 67- 83.

KOMAR, P.D., 1976. Beach processes and sedimentation.—Prentice Hall, New Jersey: 1-429.

LAMB H.H. & I. WEISS, 1979. On changes of the wind and wave regime of the North Sea and the outlook. Fachliche Mitteilungen, Amt für Wehrgeophysik Nr. 194. 108 pp.

MCGREAL, W.S., 1977. Retreat of cliff coastline in the Kilkeel area of country Down. Queens University of Belfast (unpublished thesis).

METEOROLOGISK INSTITUT, 1933. Danmark Klima. København.

——, 1971. Danmark Klima. I. Vind. Standardnormalen 1931-1960. København.

NAMIAS, J. & J.C.K. HUANG, 1972. Sea level at southern California: a decadal fluctuation.—Science **140:** 979-984.

ROYAL HYDROGRAPHIC OFFICE, 1967. Den danske Lods, **2.** Copenhagen: 1-428.

SHORT, A.D., 1988. The South Australian Coast and Holocene sea-level transgression.—Geogr. Rev. **78:** 119-136.

SHUISKY, J.D. & M.L. SCHWARTZ, 1980. Influence of beaches on development of coastal erosion slopes in the Northwestern part of the Black Sea.—Shore and Beach **48:** 30-34.

TANNER, W.F. & F.W. STAPOR, 1972. Accelerating crisis in beach erosion.—Int. Geogr. **2:** 1020-1021.

TERICH, T. & T. LEVENSELLER, 1986. The severe erosion of Cape Shoalwater, Washington, JF.—Coast. Res. **24:** 465-477.

VELLINGA, P., 1982. Beach and dune erosion during storm surges.—Coast. Eng. **6:** 361-387.

WALTON, T.L., 1978. Coastal erosion-some causes and some consequences.—J. Mar. Tech. Soc. J. **12:** 28-33.

ZIMMERMAN, M.S. & H.J. BOKUNIEWCZ, 1987. Multi-year beach response along the south shore of Long Island New York.—Shore and Beach **55:** 3-8.

CLIMATE CHANGE, SEA LEVEL RISE AND MORPHOLOGICAL DEVELOPMENTS IN THE DUTCH WADDEN SEA, A MARINE WETLAND

ROBBERT MISDORP, FRANK STEYAERT, FRANK HALLIE and JOHN DE RONDE

Ministry of Transport and Public Works, Rijkswaterstaat, Tidal Waters Division, P.O. Box 20907, 2500 EX The Hague, The Netherlands

ABSTRACT

A short review is given of measurements, causes and implications of sea-level rise. Along the Dutch coast not only the mean sea level but also the mean tidal amplitude is changing.

The main part of this paper deals with the expected impacts of sea-level rise on the morphology of the Dutch Wadden Sea. A change in the morphology in this area may be of great importance to its ecology. A hypothesis on the relationship between morphological changes and sea-level rise is presented. In addition, preliminary predictions for the development of the Wadden Sea area are given, as well as some general conclusions and recommendations.

1. INTRODUCTION

This paper gives an indication of possible effects of an acceleration in sea-level rise on the tidal flats of the western Dutch Wadden Sea, which form an essential part of the Wadden ecosystem.

In order to estimate the effects of sea-level rise, the following two steps have been taken:

- Processing and integrating results of field measurements, including tidal water level recordings (DE RONDE, 1983; VAN MALDE, 1989) and systematic bathymetric surveys of tidal basins (GLIM *et al.*, 1987, 1988, 1989)
- Formulation of a hypothesis on future morphological developments (DE RONDE & DE RUYTER, 1987).

The methods of monitoring tidal water levels and monitoring morphology by means of systematic echo-sounding surveys in the western Dutch Wadden Sea have previously been described (MISDORP *et al.*, 1988).

Recommendations for the necessary future research to improve predictions of the effects of expected climatic change on morphological tidal basin developments are formulated.

2. CLIMATIC CHANGE AND SEA LEVEL RISE

Time series of global air temperature, starting in the 18th century, reveal relatively large oscillations (ΔT: 0.6°C) without an increase of the mean during the first century of recordings. This period is followed by a general increase of 0.7°C with smaller variations during the last century (HANSEN & LEBEDEFF, 1987). The melting of mountain glaciers, an indication of general temperature rise, began in the period 1850–1870 (OERLEMANS, 1989)

One of the impacts of the temperature rise is sea-level rise, caused for ~ 50% by thermal expansion of ocean waters, ~ 35% by partially melting of mountain glaciers and of the Greenland Ice Sheet and ~ 15% by melting of part of the West Antarctic Ice Sheet (OERLEMANS, 1989). The important future response of the vast East Antarctic Ice Sheet is most uncertain, influencing the uncertainties of the future sea-level-rise predictions.

The world's longest record of tide gauges, namely of Amsterdam (Fig. 1) indicates that the mean sea level started rising also around the middle of the 19th century, in the pre-industrialized period. The more detailed picture of the records of the Dutch tide-gauge station Harlingen (Fig. 2) shows a continuous rise of mean sea level of ~ 11 cm per century.

The differences in relative mean sea-level rise between the various Dutch tide-gauge stations are rather large (Table 1). This is due to local circumstances: subsidence of Pleistocene subsoil ranges from 0 to 6 cm per century (NOOMEN, 1989) and man-induced influences such as harbor construction and navigation-channel deepening.

The Dutch tide-gauge records do not show any indications of an acceleration in the increase of mean sea level during the last decades. However, an increase in tidal range has been observed, starting during the 1930's to 1950's (Fig. 2: tidal range increase Harlingen: ~ 0.20 m per century). This increase of tidal amplitude has also been observed in the German Bight (SIEFERT, this issue) and in

J.J. Beukema et al. (eds.), Expected Effects of Climatic Change on Marine Coastal Ecosystems, 123–131.

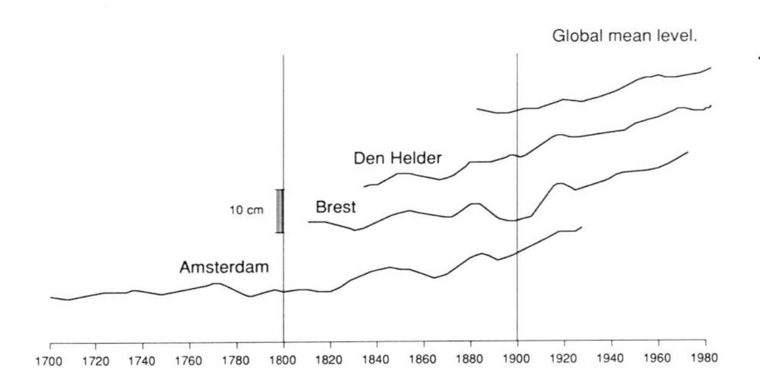

Fig. 1. Time series of filtered mean sea levels of the tide-gauge stations of Amsterdam, Den Helder (the Netherlands), Brest (France) and of global mean sea level (DE RONDE, 1988).

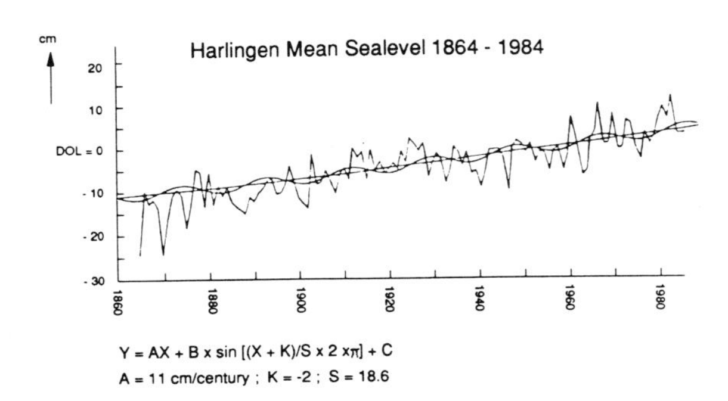

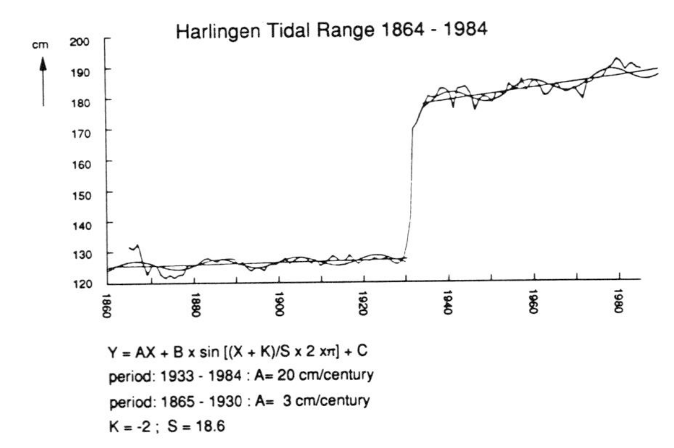

Fig. 2. Time series of mean sea level (top) and of tidal range (bottom) measured at the tide-gauge station Harlingen (1864–1984).

eastern-US tidal records (FÜHRBÖTER, 1989). It has not yet been established whether this increase of amplitude might be related to climatic change or to large scale coastal engineering constructions like damming off firths and estuaries.

The effect of closing off the Dutch Zuyderzee (a former shallow tidal basin) in 1932 by means of a 35 km long Enclosure Dike is illustrated by the sudden increase of the tidal range (Fig. 2: 55 cm) observed

TABLE 1

Trends in sea-level rise and tidal range in the Netherlands during the period 1901 to 1986 (in cm per century).

Tide Gauge Station	Relative Sea Level Rise	Increase Tidal Range
Vlissingen	21	15
Hoek van Holland	20	-
IJmuiden	22	9
Den Helder	14	11 *
Harlingen	11	21 *
Terschelling	10	34 *
Delfzijl	17	-

* period 1933 - 1986

at the surrounding tidal stations. The increase of tidal range and tidal velocities was predicted by the STATE COMMISSION ZUYDERZEE (1926: 99–111). This increase of tidal parameters can be explained by the increase of tidal wave energy in the western part of the Dutch Wadden Sea, which previously dissipated in the former Zuyderzee. The change in tidal volume of the basin of the Texelstroom was negatively influenced by the decrease of surface area of the reduced tidal basin and positively by the sudden increase of tidal range. This resulted in an increase of tidal volume of the Texelstroom tidal basin after the closure of the Zuyderzee (BATTJES, 1961).

Predictions of future sea-level rise are uncertain and, therefore, are made in the form of scenarios. A wide range of scenarios have been published. The impacts of an extreme scenario of 5-m sea-level rise on the coastal protection and the fresh-water management of the Netherlands during the next two centuries was published by DE RONDE & DE RUYTER (1987). Recently a modest, possibly realistic scenario, of global sea-level rise in the order of 0.8 m during the 21th century has been presented (OERLEMANS, 1989).

3. MORPHOLOGICAL DEVELOPMENTS IN THE DUTCH WADDEN SEA

3.1. INTRODUCTION

The origin of the Wadden Sea is related to Holocene sea-level rise accompanied by net sedimentation of sand and mud. The main morphological elements of the Wadden Sea area are:
- the barrier islands, separated from each other by tidal inlets;
- the tidal area with sandy intertidal flats (the area between Mean Low Water=MLW and Mean High Water=MHW) and intersecting tidal channels;
- the supratidal salt marshes adjacent to the coastal area of the mainland and islands, occupying a few percent of the Wadden Sea area.

The southern border of the Wadden Sea is formed

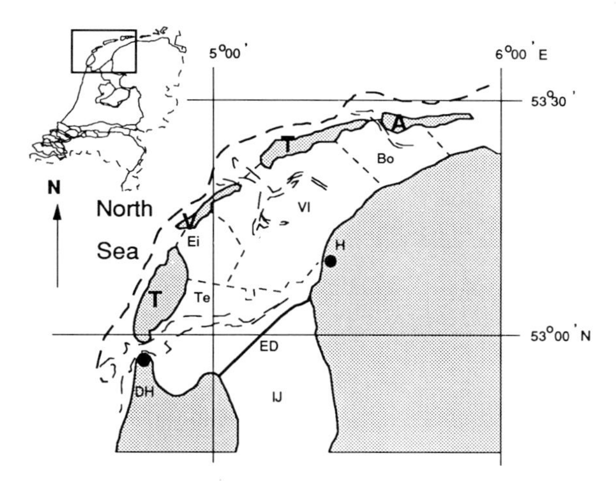

Fig. 3. Map of the western Dutch Wadden Sea investigated, with tidal- basin boundaries (---) and depth contours of −10 m (———). Abbreviations: Te=Texelstroom, Ei=Eierlandse Gat, Vl=Vliestroom and Bo=Borndiep (4 tidal basins); DH=Den Helder, H=Harlingen (2 towns); T=Texel, V=Vlieland, T=Terschelling and A=Ameland (4 Wadden Islands); ED=Enclosure Dike and IJ=IJsselmeer, former Zuyderzee.

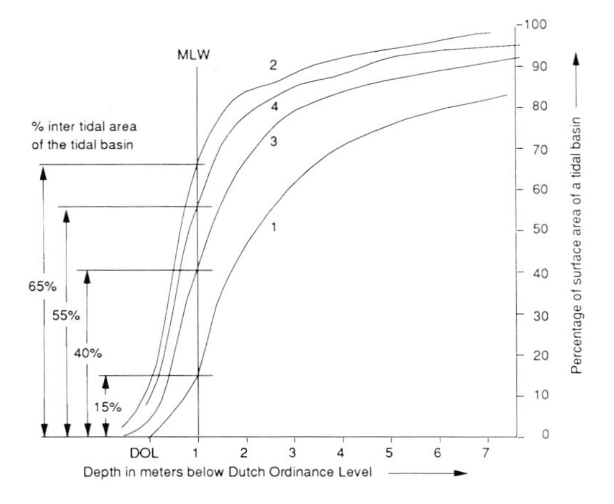

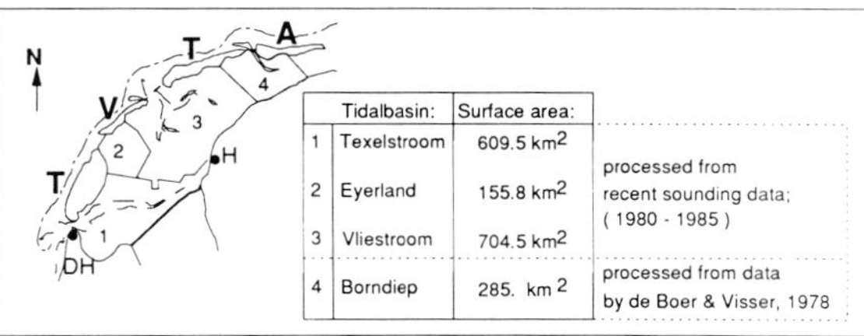

	Tidalbasin:	Surface area:	
1	Texelstroom	609.5 km^2	processed from recent sounding data; (1980 - 1985)
2	Eyerland	155.8 km^2	
3	Vliestroom	704.5 km^2	
4	Borndiep	285. km^2	processed from data by de Boer & Visser, 1978

Fig. 4. The relation (top) between the surface area and the depth of 4 tidal basins (bottom). The surface area is expressed as a percentage of the total area of the tidal basin.

by protective dikes and dams. Most of the Wadden Sea area consists of sandy sediments, while the muddy parts of the intertidal flats and of the salt marshes occupy only a few percent of the total area.

The intertidal flats are abundant with benthic organisms. The density of macrozoobenthos is related to the intertidal level and shows maximum values somewhat below mean sea level (BEUKEMA, 1976). Possible future reduction of the size and/or elevation of these flats would cause changes in functions (*e.g.* fisheries) and natural values of the Wadden Sea ecosystem.

The main question to be answered is: *Will the future sedimentation of the intertidal areas keep pace with the acceleration of sea-level rise?*

The western Dutch Wadden Sea (Fig. 3) consists of the Texelstroom, Eierlandse Gat and the Vliestroom tidal basins occupying ~1500 km^2, which is half of the total Dutch Wadden Sea area. These tidal basins not only vary in extent, but also their geometry is remarkably different: the surface areas present above MLW of the Texelstroom, the Vliestroom and the Eierlandse Gat tidal basins occupy, respectively, ~15%, ~40% and ~65% of their total area (Fig. 4). These differences in percentages of intertidal areas will prove to be important for the future morphological development.

Two coherent tidal parameters mainly determine the long-term morphological development of a tidal basin:

- The cross-sectional area (A_c) of the tidal inlet (Fig. 5a). This is the area, expressed in m^2, of the cross-section of a tidal inlet/- channel, usually calculated below the DOL (= fixed Dutch Ordinance Level).
- The tidal volume (TV), expressed in m^3, of a tidal basin. This is the amount of water entering the tidal basin during flood (=flood volume) + the amount of water which is flushed out of the tidal basin during ebb (=ebb volume).

The tidal volume can be measured in two different ways:

- Tidal discharge measurements in the tidal inlet.
- Determination of the volume of tidal water (=tidal prism; Fig. 5b) present in a tidal basin between the MLW and MHW level based on systematic bathymetric surveys of the tidal basin. The tidal volume is twice the tidal prism.

Simple morphometric relations, derived from hydrographic data exist in tidal basins (POSTMA, 1983). GERRITSEN & DE JONG (1985) drew empirical relations between cross-sectional areas and several tidal parameters like maximum tidal discharges, maximum ebb/flood current velocities, ebb/flood volumes and tidal volumes of the Dutch Wadden Sea. The empirical relations between A_c and TV of several Wadden Sea tidal inlets (GERRITSEN & DE JONG, 1985) and the Dutch Delta area (GERRITSEN & DE JONG, 1983) agree very well with those of the eastern US coast established by BRUUN (1978). Thus the regression line has been proven to be valid in areas with different hydraulic conditions and may therefore be

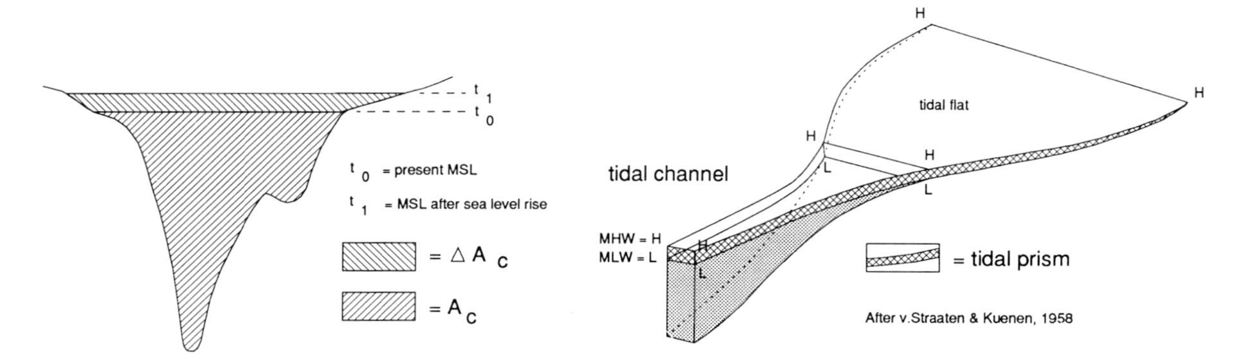

Fig. 5. Schematic presentation of: (left) a cross-sectional area (A_c) of a tidal inlet/channel and (right) a tidal prism (= half Tidal Volume) of a tidal basin (after VAN STRAATEN & KUENEN, 1958).

considered as a line of equilibrium (Figs 8 and 9). The tidal volumes in all of these empirical relations are determined by tidal discharge measurements.

3.2. HYPOTHESIS ON THE IMPACTS OF ACCELERATED SEA-LEVEL RISE ON THE TIDAL FLATS OF THE WADDEN SEA

For the sake of formulating the hypothesis, a theoretical distinction has been made between the period during which the sea level rise (SLR) is accelerated (enlarging A_c and TV), followed by the period of reaction of the morphology to the sea-level rise. In nature, these processes will take place simultaneously.

3.2.1. THE DIRECT EFFECTS OF SEA-LEVEL RISE ON A_c AND TV, DURING THE FIRST PERIOD OF SEA-LEVEL-RISE ACCELERATION

In case of sea-level rise the cross-sectional areas of the tidal inlets will increase with the surface area between the present level of reference and the future MSL (= ΔA_c) of the inlet, as schematically indicated in Fig. 5a.

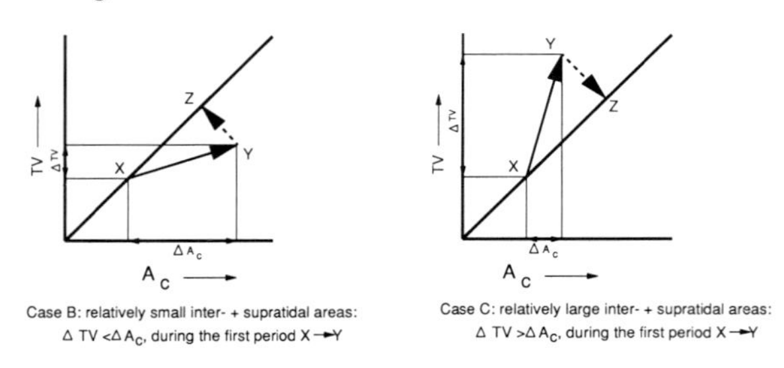

Fig. 6. Schematic presentation of two TV - A_c diagrams:
-(left) Δ TV $<$ ΔA_c, case B: in tidal basins with relatively small inter- and supratidal areas.
-(right) Δ TV $>$ ΔA_c, case C: in tidal basins with relatively large inter- and supratidal areas.

The effect of sea-level rise on the tidal volume depends very much on the geometry within a tidal basin. The tidal volume will not increase in a tidal basin without inter- and supratidal areas (Case A). The increase of tidal volume (= Δ TV) will in first instance occur in tidal basins with inter- and supratidal areas. The larger those areas the larger Δ TV.

The combined effect of increase of A_c and of TV due to a sudden acceleration of sea level rise is illustrated in movement from X on the solid (equilibrium) line to Y, in the following two schematic A_c-TV diagrams (Fig. 6a,b).

3.2.2. THE REACTION OF THE MORPHOLOGY DURING THE ADAPTIVE PERIOD

The impact of an acceleration in sea-level rise on the morphology of tidal basins can be predicted by means of the relation between TV and A_c. The reaction of the morphology is indicated by the movement from Y back to the equilibrium line (to Z) in the two A_c-TV diagrams (Fig. 6).

Three different cases can be distinguished, of which case A, however, concerns tidal basins without intertidal areas and is not present in the Wadden Sea. The two others differ in the extent of their inter- and supratidal areas. This distinction is important for the future development of the Wadden Sea tidal basins under influence of sea level rise (DE RONDE & DE RUYTER, 1987):

- Case B: The tidal volume increases, but the increase is proportionally smaller than the increase of the cross-sectional area of the tidal inlet. This situation is present in the case of tidal basins with relatively small inter- and supratidal areas.

In this case the cross-sectional area of the inlet and channels will become too large in relation to the tidal volume. Consequently we will observe:
- decrease of tidal currents,

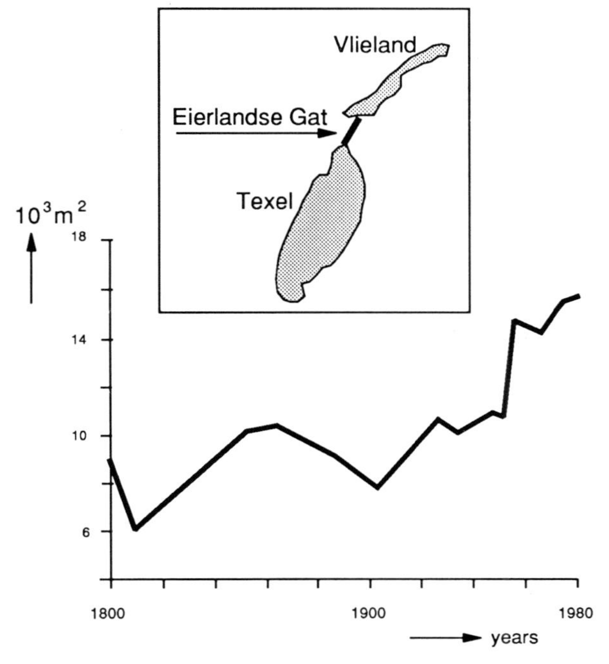

Fig. 7. Development of the cross-sectional area (A_c, in 10^3 m^2) of the tidal inlet of the Eierlandsegat.

- increase of sedimentation of inlet and channels,
- increase of erosion of tidal flats.

Part of the sand necessary for inlet sedimentation will be derived from the increasingly eroding North Sea coast, resulting in a net import of sand from the North Sea into the inlet and channels. Another part of the sand will be derived from the intertidal areas. The constructive tidal currents of the tidal channels decrease, while the destructive wave actions will be the same or might even increase. This will cause a net sand transport directed from the tidal flats to the channels. The effects of constructive tidal currents and destructive wave processes have been measured during extensive morpho-dynamic investigations on the shoals in the Eastern Scheldt tidal basin (KOHSIEK *et al.*, 1986).

The sedimentation in the inlet and channels will thus coincide with erosion of the intertidal flats, which in itself causes an acceleration of the increase of the tidal volume, leading to a new equilibrium.

- Case C: The tidal volume increases proportionally more than the increase of the cross-sectional area of the tidal inlet and channels. This is the case if relatively large intertidal and supratidal areas are present in the tidal basin.

In this case the cross-sectional area of the tidal inlet and channels becomes too small in relation to the tidal volume. Consequently we will observe:

- increase of tidal currents,

TABLE 2

Tidal volumes (TV, in $10^6 m^3$) and cross-sectional areas (A_c, in $10^3 m^2$, at Dutch Ordinance Level) as measured in various years for 3 tidal basins and inlets in the western part of the Dutch Wadden Sea. The 1932-value originates from the situation before the closure of the Zuyderzee. Data from GERRITSEN & DE JONG (1985).

	Texelstroom		Eierlandse Gat		Vliestroom	
	Tidal Volume $\times 10^6 m^3$	Ac $\times 10^3 m^2$	Tidal Volume $\times 10^6 m^3$	Ac $\times 10^3 m^2$	Tidal Volume $\times 10^6 m^3$	Ac $\times 10^3 m^2$
1932 *	1430	63.0				
1933	1843	53.5			2212	58.4
1934			389	10.2		
1949			405	11.0		
1951					2243	62.0
1953	1886	52.1				
1964			413	14.3		
1965	1893	54.9			2320	
1970	1886	55.5				
1972			395	15.2	2305	67.9
1975	1883	56.0				
1976			398	15.6		
1977					2313	69.0
1980	1905	56.5				
1982			405	15.9		
1983					2356	71.5

- increase of erosion in the inlets and channels,
- possible sedimentation of tidal flats.

The eroded sediments will partly be transported towards the outer, ebb-shield delta and partly be directed towards the intertidal areas by the increased tidal currents. The sedimentation of the intertidal areas will slow down the trend of tidal volume increase, which is in agreement with the tendency of reaching a new state of equilibrium.

This sedimentation rate will be relatively small, however, due to the absence of large tidal channels responsible for feeding the tidal flats in this case.

Summarizing:

- The tidal basins with a relatively small inter- and supratidal area will react on an accelerated sea level rise by strong sedimentation of the tidal inlets and channels. The necessary sand will be derived from the increased erosion of the North Sea coastal areas, of the ebb-shield delta and of the intertidal flats. The relatively small intertidal areas will tend to disappear, not only due to sea-level rise itself, but also due to the net sand transport from the flats to the channels.
- The tidal inlets and channels of tidal basins with relatively large inter- and supratidal areas will erode. The drowning effect of sea-level rise on the intertidal areas in those tidal basins will be counteracted by sedimentation of eroded sand from the tidal channels. To which extent this sedimentation can keep pace with the rising sea level depends on: the ratio between tidal channel and intertidal area; the rate of

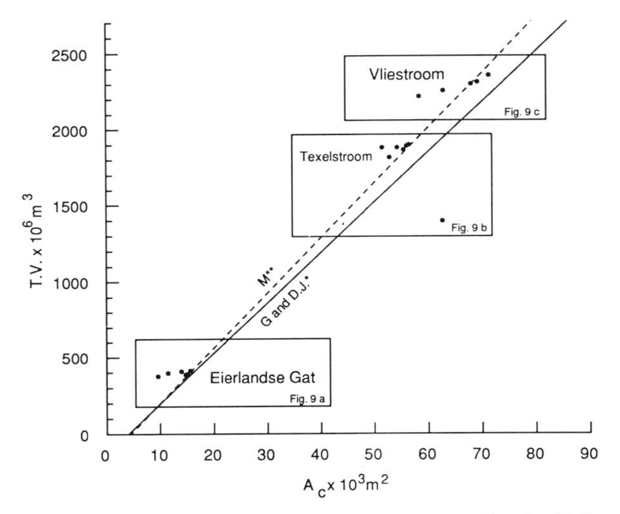

Fig. 8. The TV - A_c relation of the 3 western Dutch tidal basins. TV=Tidal volume of tidal basin; A_c =Cross-sectional area of tidal inlet.
Solid line (G. and D.J.): GERRITSEN & DE JONG, 1985: TV=33198 A_c −127.6 x 10^6 m^3; r=0.981; n=11; for 1969 - 1980.
Dashed line (M.**): Present study: TV=35959 A_c −152.0 x 10^6 m^3; r=0.999; n=9; for 1970 - 1983).

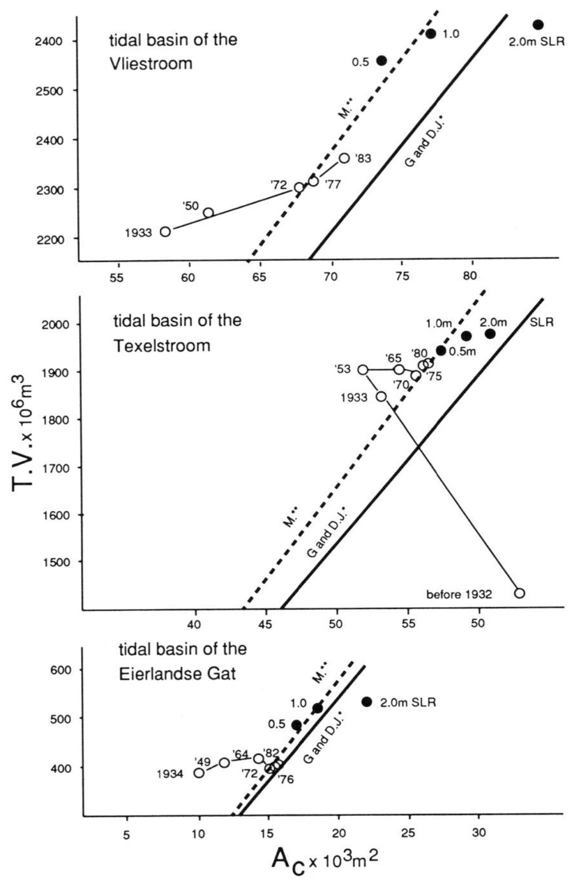

Fig. 9. The TV - A_c relations separately for the 3 western Dutch tidal basins during the last 50 years and for 3 sea-level-rise scenarios: 0.5, 1.0 and 2.0 m. Open points: measured, during the period 1933 - 1983; closed points: calculated for 3 SLR scenarios: 0.5, 1.0 and 2.0 m; based on 1980/3.
Solid lines (G. and D.J.): GERRITSEN & DE JONG, 1985: TV=33198 A_c x 10^6 m^3; r=0.981; n=11; for 1969 - 1980.
Dashed lines (M.**): Present study: TV=35959 A_c - 152.0 x 10^6 m^3; r=0.999; n=9; for 1970 - 1983.

sand-transport increase to the intertidal areas and the rate of sea-level rise. If the sedimentation of the tidal flats cannot keep pace with the rate of sea-level rise, ultimately the morphodynamic processes will lead to the development as described in case B.

3.3. HISTORICAL CHANGES OF TIDAL INLETS AND OF TIDAL VOLUMES OF THREE TIDAL BASINS OF THE WESTERN DUTCH WADDEN SEA

Changes of the cross-sectional areas of the tidal inlets in time reflect changes in tidal parameters such as tidal volumes and morphology of the tidal basins. During the last two centuries the tidal inlets have enlarged. The cross-sectional areas of the tidal inlets of the Texelstroom and Eierlandse Gat were calculated on base of bathymetric maps, starting in 1796 (RIJZEWIJK, 1986).

The cross-sectional areas of the tidal inlets of the Texelstroom and of the Eierlandse Gat (Fig. 7) are increasing by an average of, respectively, about 25% and 35% per century. A tendency of a larger increase might be observed during the 20th century. The A_c of the Vliestroom tidal inlet has also increased by about 30% during 1910 to 1975 (KLOK & SCHALKERS, 1980: fig. 15).

More detailed studies also of other Wadden tidal inlets in combination with changes in tidal basins during the last centuries are needed. Those studies, combined with conceptual model studies, should reveal whether the observed increases of cross-sectional areas of the inlets can be related to a sea-level rise of 15 to 20 cm per century.

The results of systematic echo-sounding surveys of the western Dutch Wadden Sea tidal basins and of the ebb-shield outer deltas from 1933 to 1980 are reported. The tidal basins were surveyed 6 times during the last 50 years. The tidal prisms (= volume of water present between MLW and MHW) of the Texelstroom and Vliestroom tidal basins are computed on base of bathymetric survey data, revised from GLIM *et al.* (1987; 1988). The tidal prisms calculations of the Eierlandse Gat tidal basin are based on GLIM *et al.* (1989). The tidal volumes of the Texelstroom, Eierlandse Gat and Vliestroom tidal basins have in-

	Texelstroom tidal basin/inlet Tidal range 1.61m			Eierlandse Gat tidal basin/inlet Tidal range 1.70m			Vliestroom tidal basin/inlet Tidal range 1.87m		
SLR m:	0.5	1.0	2.0	0.5	1.0	2.0	0.5	1.0	2.0
MHW m:	+ 1.23	+ 1.73	+ 2.73	+ 1.26	+ 1.76	+ 2.76	+ 1.37	+ 1.87	+ 2.87
MLW m:	- 0.38	+ 0.12	+ 1.12	- 0.44	+ 0.06	+ 1.06	- 0.50	0.00	+ 1.00
TV x $10^6 m^3$	1933	1962	1962	486	517	530	2552	2612	2634
A_c x $10^3 m^2$	57.8	59.5	61.3	17.1	18.6	22.1	74.0	77.6	85.1
T.V. %	1.5 %	3.0 %	3.0 %	20 %	28 %	31 %	8.3 %	11 %	12 %
A_c %	2.3 %	5.3 %	8.5 %	7.5 %	17 %	39 %	3.5 %	8.5 %	19 %
case :	B	B	B	C	C	B	C	C	B

TABLE 3
Expected effects of a rise in sea level (SLR, in m) of 0.5 to 2.0 m on tidal volume (TV, in 10^6m^3) and cross-sectional area (A_c, in 10^3m^2) of 3 tidal basins and their inlets in relation to the 1980/83 situation.

creased with, respectively, 3.4%, 4.1% and 6.5% during the period 1933 - 1980/83 (Table 2).

The Wadden Sea computations of GERRITSEN & DE JONG (1985) demonstrate that the cross-sectional areas of the tidal inlets are linearly related to the tidal volumes (= the solid line in Fig. 8: TV = 33198 A_c - 127.6 x 10^6 m^3, r = 0.981, n=11). The TV values were determined by tidal-discharge measurements mainly executed during the 1970's. Such a relation has been proven to be valid in different geographical areas with varying hydraulic conditions. The regression line may, therefore, be considered as a line of equilibrium.

Plotting the A_c and TV values (based on tidal prisms) of the 3 tidal basins (1933 - 1980/83) in Fig. 9 shows three important matters:
- The regular increase of TV and A_c in time.
- The influence of the closure of the Zuyderzee (1932) during the first 3 to 4 decades.
- The tendency of reaching an equilibrium during the last 2 decades.

The influence of the closure of the Zuyderzee is represented by the fact that the plotted values of 1933 to 1970 are well above the equilibrium line of GERRITSEN & DE JONG (1985). This indicates that tidal-volume values are relatively too large. The increase of TV upon closure is mainly caused by the sudden increase of tidal range (Fig. 2).

The plotted values of A_c and TV computed for the period 1970 - 1983 of all 3 tidal basins follow an equilibrium (dashed) line (Figs 8 and 9) with the following characteristics:

$$TV = 35959\ A_c - 152.0 \times 10^6\ m^3\ (r = 0.999,\ n=9)$$

The here presented relation (dashed line) deviates from the one of GERRITSEN & DE JONG (1985) by at most 6 to 8% of the TV values for the Texel- and Vliestroom tidal basin. This deviation is well within the accuracy limits of tidal-volume measurements by means of tidal-discharge measurements and by means of bathymetric surveys.

3.4. INDICATIONS OF POSSIBLE FUTURE CHANGES OF THE THREE WESTERN DUTCH TIDAL BASINS IN RELATION TO SEA-LEVEL RISE

In the following, an estimation will be made of the impacts of an acceleration in sea-level rise on the intertidal-flats development of the Texelstroom, Eierlandse Gat and Vliestroom tidal basins. To gain insight into the development of the A_c and TV values, these values are computed for 3 sea-level-rise scenarios: 0.5, 1.0 and 2.0 m as compared to 1980/83 (Table 3). The values are plotted in Figs 9a, b and c.

The following assumptions have been made for the first stage of accelerated sea-level rise:
- Tidal range will be constant during this period.
- The geometry of the tidal basins remains the same: neither sedimentation nor erosion will occur during the first stage.

The higher parts of the salt marshes present in the Vliestroom tidal basin are excluded.

The relative increases of A_c (= ΔA_c) and TV (= Δ TV) are expressed in percentages with regard to 1980/83 (Table 3). The relative increase of both A_c and TV is smallest in the Texelstroom tidal basin, followed by the Vliestroom and largest in the tidal basin of Eierlandse Gat. This order coincides well with the order of percentages of surface areas present above MLW in the tidal basins being, respectively, ~15%, ~40% and ~65% (Fig. 4).

From Fig. 9 the following conclusions concerning possible future developments can be drawn, assuming that the dams and dikes on the border of the

Wadden Sea remain intact:
- In the Texelstroom tidal basin, ΔA_c is larger than Δ TV in all 3 sea-level-rise scenarios. This indicates that the development of the intertidal flats as described in case B might be followed: strong reduction and likely disappearance of the tidal flats, even with a sea-level rise of only 0.5 m. This might also be true in case of a 2.0-m sea-level rise in the tidal basins of Eierlandse Gat and Vliestroom.
- The development of the tidal flats in the Eierlandse Gat and Vliestroom tidal basins during a sea level rise of 0.5 and 1.0 m might be as described in case C. Whether the tidal flats can keep pace with the sea-level rise depends on the sedimentation rate, mainly determined by the availability of sufficient amounts of sand to be transported from the eroding channels to the flats. The relatively small extent of the Eierlandse Gat and Vliestroom tidal channels implies a relatively long period of adaptation.

3.5. GENERAL CONCLUSIONS

a) The cross-sectional areas of the tidal inlets of the western Dutch Wadden Sea have increased during the last 2 centuries.

b) Within 3 to 4 decades after the construction of the Enclosure Dike, the relation between the cross-sectional areas of the inlets and the tidal volume of the 3 western Dutch tidal basins was restored.

c) The size of the tidal-flat area within a tidal basin is indicative for future developments of the tidal flats and channels in case of a sea-level rise.

d) Future morphological development of the intertidal areas of the Wadden Sea (assuming that the present shape and position of its outline remains intact) during the next century:
- The tidal flats of tidal basins like Texelstroom, with relatively small intertidal areas, will tend to disappear even under influence of a small (0.5 m) sea-level rise.
- Whether the future sedimentation of flats in tidal basins with extended intertidal areas, like in the Eierlandse Gat and Vliestroom basins, can keep pace with a 0.5- and 1.0-m sea-level rise is yet uncertain.
- In case of a much larger sea-level rise ($\sim$2 m) a disappearance of tidal flats of the tidal basins, such as Eierlandse Gat and Vliestroom, is envisaged.
- The development of the tidal flats in eastern Dutch tidal basins — which are comparable with the Eierlands Gat basin — although not yet investigated, might follow the same pattern. Regarding the developments of the Eems-Dollard estuary, nothing can yet be stated since the hydraulic conditions are rather different and the investigations are still in progress.

e) A possible future reduction of tidal-flat areas will have consequences for the Wadden Sea ecosystem, its functional uses (*e.g.* fisheries) and its natural values (*e.g.* birds).

4. RECOMMENDATIONS

The following recommendations are made in order to confirm and extend the here presented hypotheses on the development of the Wadden Sea in relation to sea-level rise:

1. Processing of more historical bathymetric survey data of the Wadden Sea.

2. Extension of morphodynamic studies of tidal inlets and basins to the eastern Dutch Wadden Sea, including the supratidal salt-marsh areas.

3. Extension of conceptual model studies regarding the morphometric relations between tidal and morphological parameters. An important question to answer is: How long will it take to reach future states of equilibrium in relation to different sets of sea-level-rise scenarios?

ACKNOWLEDGEMENTS

The authors are grateful for assistance given by their colleagues, especially those working in Rijkswaterstaat, Division Noord-Holland, Hoorn.

5. REFERENCES

BATTJES, J.A., 1961. Study of the Texelstroom tidal inlet (in Dutch).—Report 62.4, Rijkswaterstaat, Hoorn, the Netherlands.

BEUKEMA, J.J., 1976. Biomass and species richness of the macrobenthic animals living on the tidal flats of the Dutch Wadden Sea.—Neth. J. Sea Res. **10:** 236-261.

BRUUN, P., 1978. Stability of tidal inlets, theory and engineering. Elsevier Scientific Publish. Comp. Amsterdam-Oxford-New York: 1-506.

DIJKEMA, K.S., 1983. Geomorphology of the Wadden Sea area. In: W.J. WOLFF. Ecology of the Wadden Sea. Vol. 1, Balkema, Rotterdam: 1/1-1/135.

FÜHRBÖTER, A., 1989. Changes of the tidal water levels at the German N Sea Coast.—Proc. 6th Int. Wadden Sea Symp. Sylt.—Helgoländer Meeresf. **43** (3/4): in press.

GERRITSEN, F. & H. DE JONG,, 1983. Stability of the tidal inlet of the Westerschelde. (In Dutch.)—Rep. nr. WWKZ-83.V008. Rijkswaterstaat, Vlissingen, the Netherlands: 1-38.

——, 1985. Stability parameters of tidal inlets the Dutch Wadden Sea. (In Dutch).—Note WWKZ-84.V0,16, Rijkswaterstaat, the Hague, the Netherlands: 1-53.

GLIM, G.W., G. KOOL, M.F. LIESHOUT & M. DE BOER, 1987. Erosion and sedimentation in the tidal basin of Texelstroom 1933-1982. (In Dutch.)—Rep. nr. ANWX87.,H201, Rijkswaterstaat, Division Noord-Holland, Hoorn, the Netherlands: 1-33.

——, 1989. Erosion and sedimentation in the tidal basin of Eierlandse Gat 1934-1982. (In Dutch.)—Rep. nr. ANWX-89.H202. Rijkswaterstaat, Division Noord-Holland, Hoorn, the Netherlands: 1-31.

GLIM, G.W., N. DE GRAAF, G. KOOL, M.F. LIESHOUT, & M. DE BOER, 1988. Erosion and sedimentation in the tidal basin of Vliestroom 1933-1983. (In Dutch.)—Rep.nr. ANWX 88.H204, Rijkswaterstaat, Division Noord-Holland, Hoorn, the Netherlands: 1-32.

HANSEN, J. & S. LEBEDEFF, 1987. Global trends of measured surface air temperature.—J. Geophys. Res. **92** (11):

KLOK, B. & K.M. SCHALKERS, 1980. 'Changes in the Wadden Sea related to closure of the Zuyderzee.' (In Dutch.)—Rep.nr WWKZ 78.H238, Rijkswaterstaat, Division Watermanagement and Watermovement, The Hague, the Netherlands: 1-13.

KOHSIEK, L.H.M., H.J. BUIST, P. BLOKS, R. MISDORP, J.H. VAN DER BERG & J. VISSER, 1986. Sedimentary processes on a sandy shoal in a mesotidal estuary (Eastern Scheldt). In: Tide-influenced sedimentary environments and facies, Reidel Publ. Comp: 201-214.

MALDE, J. VAN, 1990. Relative rise of mean sea level in the Netherlands in recent times. Proc. of European Workshop on International Bioclimatic and Land Use Changes, Noordwijkerhout, the Netherlands. Blackwell, Oxford: in press.

MISDORP, R., F. STEYAERT, J. DE RONDE & F. HALLIE, 1989. Monitoring in the western part of the Dutch Wadden Sea - Sea Level and Morphology.—Proc. of the 6th Int. Wadden Sea Symp. Sylt.—Helgoländer Meeresf. **43** (3/4): in press.

NOOMEN, P., 1989. Lecture on subsidence rate in the Netherlands, KIVI, April 1989, Delft. Symposium on Sea Level Rise and Subsidence: 'Which Scenario Will be Reality?'

OERLEMANS, J., 1989. A projection of future sea level.—Inst. of Meteor. and Oceanography, University of Utrecht, Utrecht, the Netherlands: 1-28.

POSTMA, H., 1983. Hydrography of the Wadden Sea: Movements and properties of water and particulate matter. In: W.J. WOLFF. Ecology of the Wadden Sea, Vol. 1, Balkema, Rotterdam: 2/1-2/75.

RONDE, J.G. DE, 1983. Changes of relative mean sea level and of mean tidal amplitude along the Dutch coast.—Proc. NATO Advanced Research Workshop, Utrecht, June 1982.

——, 1988. Past and future sea level rise in the Netherlands. Proc. Workshop on Sea Level Rise and Coastal Processes, Florida.

RONDE, J.G. DE & W.P.M. DE RUYTER, 1987. Die Auswirkungen eines verstärkten Meeresspiegelanstiegs auf die Niederlande.—Die Küste, H. **45:** 123-163.

RIJZEWIJK, L.C., 1986. Bathymetric maps of the tidal inlets of the western Dutch Wadden Sea 1796-1985.—Rep.nr. ANWX 86.H208, Rijkswaterstaat, Division Noord-Holland, Hoorn, the Netherlands: 1-96.

STATE COMMISSION ZUYDERZEE, 1926. Report on consequences of closure of Zuyderzee 1918 - 1926 (in Dutch). the Hague, the Netherlands: 1-345.

STRAATEN, L.M.J.U. VAN & PH.H. KUENEN, 1958. Tidal action as a cause of clay accumulation.—J. Sedimentary Petrol. **28:** 406-413.

SEA-LEVEL RISE AND COASTAL SEDIMENTATION IN CENTRAL NOORD-HOLLAND (THE NETHERLANDS) AROUND 5000 BP: A CASE STUDY OF CHANGES IN SEDIMENTATION DYNAMICS AND SEDIMENT DISTRIBUTION PATTERNS

W.E. WESTERHOFF and P. CLEVERINGA

Geological Survey of The Netherlands, P.O. Box 157, 2000 AD Haarlem, The Netherlands

ABSTRACT

The coastal development of the central part of the present province of Noord-Holland during the Atlantic and Subboreal led to a change in sedimentary environments, which ranged from lagoonal, via a tidal-flat-like depositional system, to a coast where prograding barriers predominated. The sedimentation pattern in the coastal plain is determined not only by rising of the sea-level but also by the pre-existing relief, the supply of sediments, and the hydrodynamic setting. The marked transition from an open (with W-E-running tidal channels) to a closed (by barriers) coast at the end of the Atlantic is discussed.

1. INTRODUCTION

The postglacial amelioration of the climate led to a strong sea-level rise (SLR). On the coastal plain of The Netherlands the Holocene sedimentation was governed by this changing sea-level, which exceeded 2 m per 100 y before 7000 BP, slowed down to about 0.5 m per 100 y around 5000 BP, and averaged 0.15 m per 100 y between 5000 and 2000 BP (JELGERSMA, 1961; VAN DER PLASSCHE, 1982). The sedimentation is also controlled by the configuration of the tidal basins and the position of the rivers (ZAGWIJN, 1986).

The present paper deals with the change of the coastal configuration around 5000 BP, which throws light on the main sedimentation patterns and the source of the sediments deposited on the coastal plain. A brief outline of the relief of the Pleistocene surface underlying the western part of the coastal plain is followed by a description of the Early-Atlantic to Subboreal development as revealed by the geological mapping of the central part of the province of Noord-Holland (WESTERHOFF *et al.*, 1987), illustrated here by paleogeographical maps and schematic cross-sections. The impact of the SLR and the sedimentation dynamics during the Atlantic and Subboreal in the coastal plain are briefly discussed.

2. THE PLEISTOCENE RELIEF

The Holocene deposits of the coastal plain in the western part of The Netherlands consists of a wedge-shaped, more than 25 m thick sedimentary sequence of lagoonal, tidal-flat, estuarine, and swamp deposits protected from the sea by a dune-barrier belt with a width of 8 to 10 km. The plain itself is more than 50 km wide.

This Holocene sequence was deposited on a gently westward-dipping Pleistocene surface incised by two valley systems (Rhine/Meuse and central part of Noord-Holland), and is bounded by the Pleistocene highs of the island of Texel in the north (Texel spur) and the province of Zeeland in the south. Thus, the coastal plain of the western Netherlands can be subdivided from north to south (Fig. 1) as follows:

1) Northern Noord-Holland with a relatively high-lying Pleistocene surface outcropping near Texel.

2) Central Noord-Holland (indicated in Fig. 1), with the Pleistocene surface situated 30 to 35 m below Dutch Ordnance Datum (NAP). The remains of a former drainage system have been found. Since the Saale-Glaciation, drainage of the northern and eastern parts of The Netherlands, including the IJsselmeer region, has taken place via this system. Fluviatile sedimentation is believed to have played a minor role starting in the early Weichselien, when the branch of the Rhine flowing through the IJssel valley was abandoned (VAN DER MEENE & ZAGWIJN, 1978). The alluvial plain belonging to this drainage system is widely covered by aeolian deposits of Weichselian age.

3) The cental part of the coastal plain, where the Pleistocene deposits show a smooth surface with a westward dip of 0 to 15 m below NAP.

4) The Rhine/Meuse valley, where the top of the Pleistocene deposits is situated at a depth of about 25 m below NAP. Unlike the situation in Noord-Holland, sedimentation in this valley system continued from the Weichselien into the Holocene.

5) The southwestern part of the coastal plain, where the Pleistocene surface inclines to almost the NAP level.

J.J. Beukema et al. (eds.), Expected Effects of Climatic Change on Marine Coastal Ecosystems, 133–138.

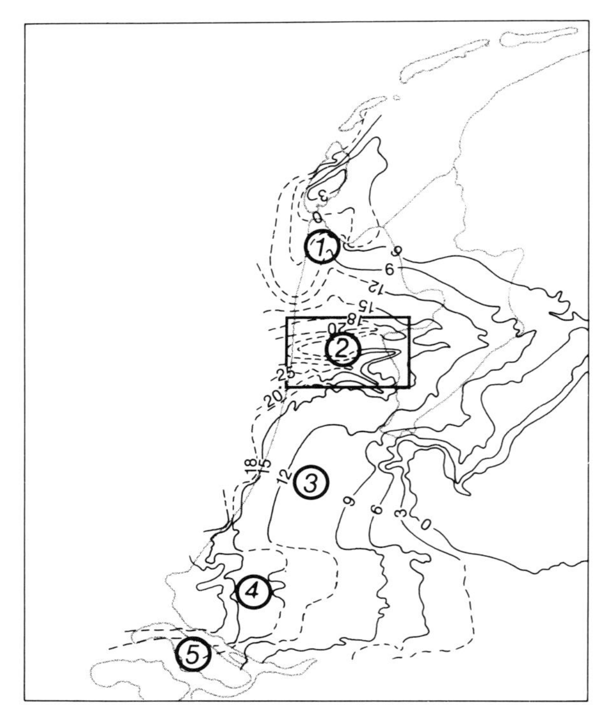

Fig. 1. Depth contours (in m below NAP) of the Pleistocene surface in the western part of The Netherlands (after ZAGWIJN, 1986). The central part of the province of Noord-Holland is indicated (numbers refer to areas cited in the text).

3. COASTAL SEDIMENTATION IN CENTRAL NOORD-HOLLAND

In association with SLR the groundwater tables in the coastal plain tended to rise, which led to the formation of peat in front of a zone where marine clastic sedimentation occurred (DE VRIES, 1977). Under the influence of the SLR, this zone of peat swamps gradually migrated inland (Fig. 2, 7500 BP).

In our study area peat growth started around 7500 BP, although due to the undulating topography of the Pleistocene surface, the onset may have occurred earlier in some places (JELGERSMA, 1961). Because this peat forms the base of the Holocene sequence in the sedimentary record, it is designated Basal or Lower Peat (DOPPERT, 1957; PONS *et al.*, 1963).

The rapidly rising sea-level submerged the peat, and a lagoonal environment was gradually established. In this lagoon a layer of heavy clay, 1 to 2 m thick with laminations of fine sand and silt, was deposited. Locally, the clay is humic or peaty and often contains *Phragmites* rootlets. Signs of burrowing by marine fauna are quite common. Diatom analysis of the clay showed a fresh to brackish depositional environment with a vegetation, shallow water, and minor tidal movements (VOS & DE WOLF, 1988). In the upward direction the diatom assemblages recorded in the clay layer show an increased salinity of the water, which is attributed to intensification of tidal influences. The clay often contains remains of the gastropod *Hydrobia* (see also VAN STRAATEN, 1957), and is of therefore known as *Hydrobia* clay or the Velsen Layer.

Together with the Basal Peat, the Velsen Layer (*Hydrobia* clay) forms the initial tract of the Holocene marine transgression. In the study area this tract started around 7500 BP, transgraded eastward over the Pleistocene surface, and reached its maximum extension around 5500 BP along the eastern fringe of the IJsselmeer region.

Shell-lags on top of the lagoonal sediments, and sandy deposits with mega-cross lamination, parallel lamination, and strong bioturbation indicate that the lagoonal environment was replaced by a tidal-flat-like depositional system. In the western part of the area the transition between the lagoonal clays and the tidal-flat system is characterized by an erosive contact sometimes visible as a shell-lag concentration. In the eastern part the transition between the two systems is more gradual. The sedimentary sequences revealed by undisturbed bore samples range from deep subtidal (inlets and channels), via shallow subtidal, to inter- and supratidal (shoals and tidal-flats). These interpretations are based on analysis of sedimentary structures, and were confirmed by diatom analysis (VOS & DE WOLF, 1988).

During the Early- to Mid-Atlantic, the relatively low-lying area of central Noord-Holland was invaded by the sea, which led to the formation of a large tidal basin promoting a tidal prism penetrating far eastward. The intertidal zone, which is the area between the mean low water (MLW) and the mean high water (MHW) levels, enclosed a large area and provided conditions for a high tidal discharge, which led to the development of W-E-running tidal channels.

The rapid SLR and the tidal movements were accompanied by considerable W-E transport of sediments. During this phase a severe erosion of the Pleistocene surface and the Early-Atlantic deposits took place. Sandy shoals developed between the inlets (Fig. 2, 5300 BP), and large amounts of sand were deposited in ebb and flood tidal deltas. Sedimentation of sandy material was concentrated around the inlets and along the banks of the tidal channels, and a deficit of sediment supply occurred in the interchannel areas and along the margins of the tidal basin, where lagoonal conditions prevailed (JELGERSMA *et al.*, 1985; BEETS, 1987). In the marginal zone of the tidal basin, brackish conditions persisted. The sedimentation pattern during this phase of the coastal evolution is represented schematically in a cross-section shown in Fig. 3.

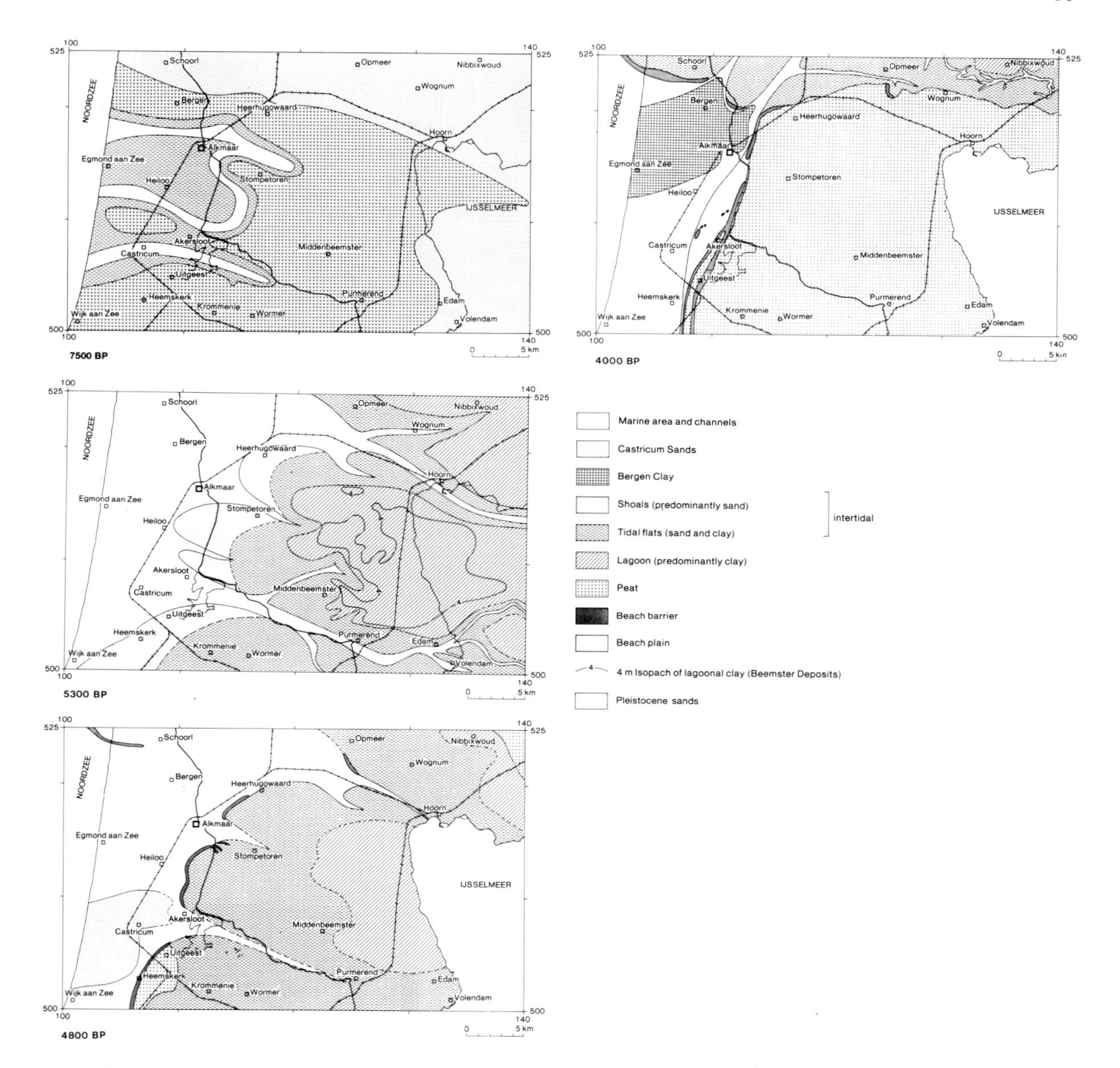

Fig. 2. Paleogeographical maps of the central part of the province of Noord-Holland for the period between 7500 and 4000 BP.

From the geological maps (WESTERHOFF *et al.*, 1987), paleo-environmental data, and radiocarbon dating it is known that the bulk of sedimentation during the Late-Atlantic (6000-5000 BP) was concentrated in the western part of the area (Fig. 3), *i.e.* around the inlets and along the tidal channels.

Up to the present, the coastal development of Noord-Holland has been determined predominantly by the combined effects of a rapid SLR and the low-lying topography of the Pleistocene surface.

At the end of the Atlantic, around or shortly after 5000 BP, when the SLR slowed down, the tidal flats and marshes adjacent to the tidal channels as well as the sandy shoals between the inlets became well expanded and silted up to almost the MHW level. In the interchannel areas, lagoonal clay was deposited. Sand and silt laminae in these clay deposits represent bottom currents, probably induced by a higher discharge rate in the channels and perhaps occurring after stormy weather conditions.

The sedimentation belt along the channels reached almost MHW levels and acted as a kind of levee. The penetrating tidal prism was ultimately restricted to the channel area itself. This led to a strong reduc-

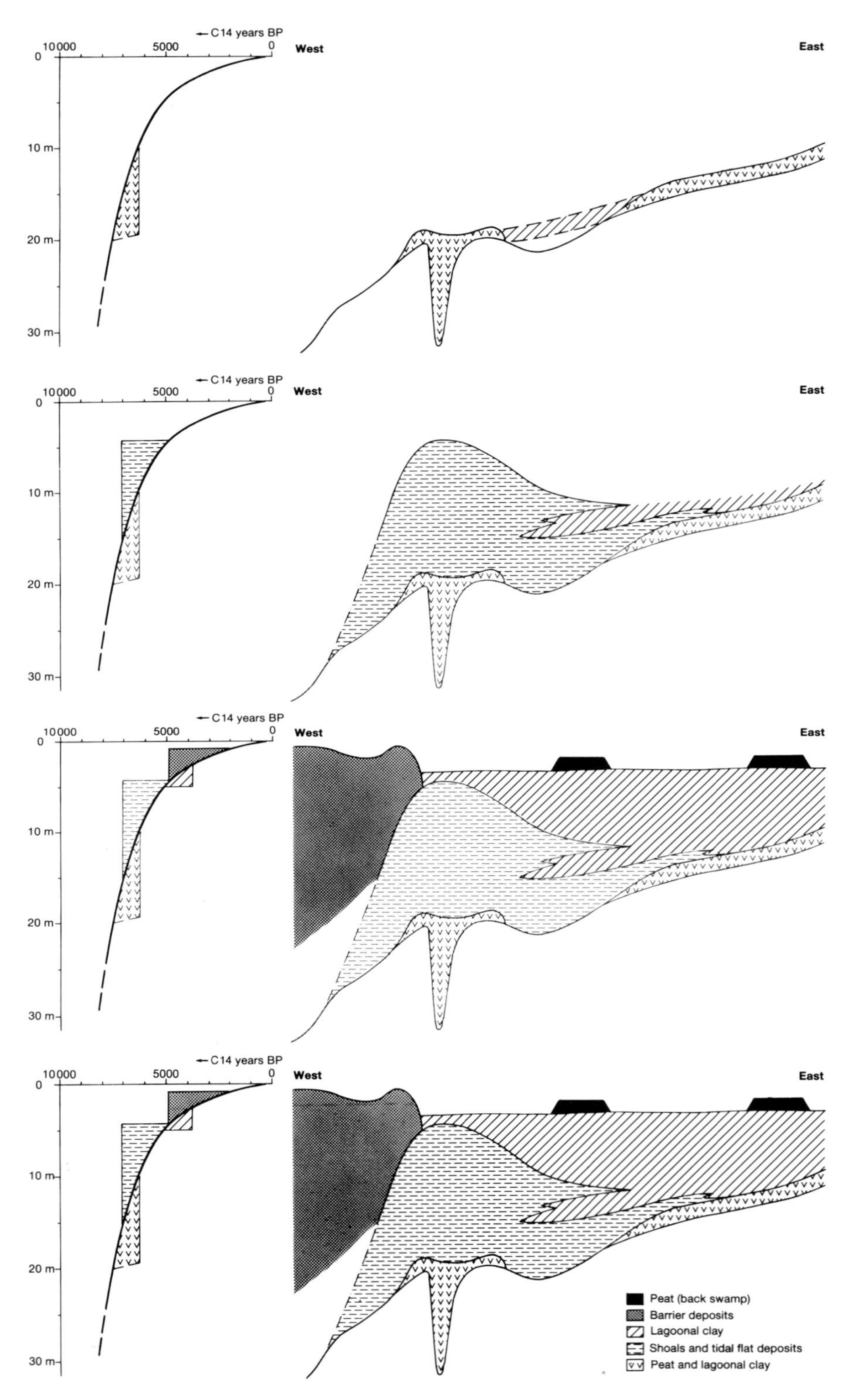

Fig. 3. Schematic cross-sections and development of the sedimentary sequence in the central part of the province of Noord-Holland.

tion of the volume of the tidal discharge, and this reduction in turn affected the size of the channels (VAN DEN BERG, 1986) as well as the sedimentation rate in them. In time this process accelerated and eventually most of the channels were filled. On the shoals between the inlets, barriers started to develop (Fig. 2, 4800 BP). Along the coast, the barriers south and west of Alkmaar prograded westward between 4500-2300 BP (BEETS *et al.*, 1981; ROEP, 1984; WESTERHOFF *et al.*, 1987). This progadation ceased around 2300 BP.

Under the influence of longshore transport, these barriers migrated northward and closed the tidal channels. In the study area only one inlet, called the Bergen Inlet (JELGERSMA *et al.*, 1970; BEETS *et al.*, 1981; DE MULDER & BOSCH, 1982; JELGERSMA, 1983), was still open to about 3500-3200 BP. Through this inlet sediments could enter the back-barrier area. In the interchannel areas formed during the previous phase and now changed into a back-barrier lagoon, an aggradational phase characterized by sedimentation of clay started. The dark-black colour (FeS) of the clays indicates sedimentation in a reduced environment. No benthic diatoms are found in these lagoonal clays. The absence of bottom-living diatoms might be due to the presence of a large amount of suspended material which reduced irradiance too much (VOS & DE WOLF, 1988). Toward the top of these clay deposits, which are 8 to 10 m thick, the fresh water influence increases and gives the sediments a typical blueish-grey color. The filling in of the back-barrier lagoonal area took place between 5000 and 4000 BP. Sedimentation ended around 4000 BP and was followed by extensive peat growth (Fig. 2, 4000 BP).

4. DISCUSSION AND FINAL REMARKS

The most remarkable feature of the coastal evolution of the central part of Noord-Holland is the transition from an open coastal configuration, which existed prior to 5000 BP, to a closed coast where barrier formation predominated (Fig. 2). It seems likely that the slackening of the SLR that took place in the same period enhanced this development. But, as already mentioned, an important role was also played by the amount and rate of sediment supply within the coastal area.

Immense amounts of sand were stored in the ebb-deltas belonging to the inlets of the open coast. During the progradation these sands were reworked and redeposited in the barriers by wave- dominated processes. The supply of sediments may have originated, from sources outside the study area. One important source lay in the northernmost part of the coastal area (Texel spur). Erosion of these sediments, followed by longshore transport, contributed to the sedimentation in the northern part of Noord-Holland, especially in the tidal basin behind the Bergen Inlet. The mineralogy and the characteristic absence of lime of the coarse-grained, quartz-rich spits bordering the northern part of this inlet confirm this conclusion. A supply of sediment by longshore transport from a source to the south may have contributed to the barrier progradation.

The impact of the Holocene SLR on the sedimentation sequences of the coastal plain in the central part of Noord-Holland can be divided into three major phases, as follows (Fig. 4):

1. Formation of a landward transgressive tract reflected in the sedimentary record by peat (Basal Peat) and lagoonal clays (Velsen Layer) deposited in a fresh to brackish environment with minor tidal influences. This phase started before 8000 BP and reached its maximum inland extension around 5000 BP.
2. An aggradational phase, preceded by an important erosion of the Pleistocene surface, during the open-coast period, roughly between 7000 and 5000 BP, reflected by the development of large W-E-running tidal channels separated by sandy shoals and ebb and flood tidal deltas.
3. A phase of westward progradation of barriers over a distance of at least 10 km between 5000 and 2300 BP, as a result of which the open coast became a closed coast. The sand destined for barrier formation originated mainly from remolding of the ebb-deltas formed in the preceding phase. In the back-barrier

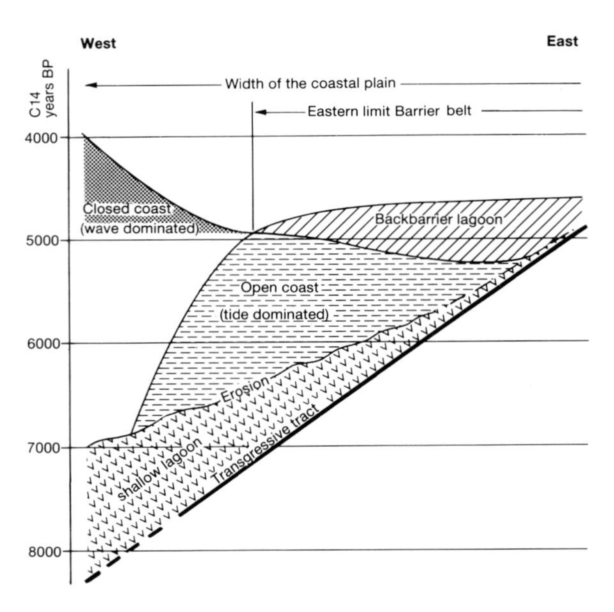

Fig. 4. Schematic representation of the main phases of Holocene sedimentation in the coastal plain of the central part of Noord-Holland.

lagoon, sedimentation of clay continued, and peat growth started.

The pattern of the sea-level changes throughout the Holocene certainly seems to be the most plausible mechanism to explain the coastal setting of the western part of The Netherlands. The sedimentary development of the coastal plain is thought to be also influenced by the morphology of the Pleistocene surface, the wave climate, the tidal regime, and the availability of sand.

The Holocene sedimentary record is the result of a long-term accretionary development. Single events such as storm surges and floods undoubtedly occurred during this development, but they were rarely registered. In our opinion the study of sedimentation processes and patterns in the geological record, on a coast formed during a period with a rising sea-level, offers a basis for a better understanding of the impact of a future rise in sea-level on the development of the coast in The Netherlands.

ACKNOWLEDGEMENTS

We are indebted to the director of the Geological Survey of the Netherlands for permission to publish. The stimulating discussions with D.J. Beets on the genesis of the coast of The Netherlands are greatly appreciated. We also wish to thank D.J. Beets, E. Oele, W. de Gans, S. Jelgersma and J. de Jong for critical comments and valuable suggestions on the manuscript, and R. Metten for preparing the illustrations. The English text was read by Mrs. I. Seeger. The field research was carried out by District West, a department of the Geological Survey of The Netherlands.

5. REFERENCES

BEETS, D.J., 1987. Rapport Kustgenese I.—Rijks Geol. Dienst, Haarlem: 1-12.

BEETS, D.J., TH.B. ROEP & J. DE JONG, 1981. Sedimentary sequences of the subrecent North Sea coast of the Western Netherlands near Alkmaar.—J. Sediment. Spec. Publ. I.A.S. **5:** 133-145.

BEETS, D.J., S. JELGERSMA & W.E. WESTERHOFF, 1985. Holocene coastal development in the Western Netherlands; indications for major changes in tidal range in time. In: Abstracts, Symposium on Modern and Ancient Clastic Tidal Deposits. Utrecht: 35-36.

BERG, VAN DEN J.H., 1986. Aspects of sediment and morphodynamics of subtidal deposits of the Oosterschelde (the Netherlands). Rijkswaterstaat Comm. **43:** 1-128.

DOPPERT, J.W.CHR., 1957. The lower Peat. In: L.M.J.U. VAN STRAATEN & J.D. de Jong. The excavation at Velsen. A detailed study of Upper Pleistocene and Holocene stratigraphy.—Verh. Kon. Ned. Geol. Mijnbouwk. Gen., Geol. Serie **17:** 154-157.

JELGERSMA, S., 1961. Holocene sea-level changes in The Netherlands.—Meded. Geol. Sticht. Serie C, VI, **7:** 1-100.

———, 1983. The Bergen Inlet, transgressive and regressive Holocene shoreline deposits in northwestern Netherlands.—Geol. Mijnb. **62:** 471-486.

JELGERSMA, S., J. DE JONG, W.H. ZAGWIJN & J.F. VAN REGTEREN-ALTENA, 1970. The coastal dunes of the western Netherlands; geology, vegetational history and archeology.—Meded. Rijks Geol. Dienst, Nwe serie, **21:** 93-167.

JELGERSMA, S., D.J. BEETS & R. SCHUTTENHELM, 1985. Een geologische kijk op de kust, kustontwikkeling in Nederland.—PT/Civiele Techniek **(40)**11: 3-8.

MEENE, E.A. VAN DE & W.H. ZAGWIJN, 1978. Die Rheinlaufe im Deutsch-Niederlandischen Grensgebiet seit der Saale-Kaltzeit. Uberblick neuer geologischer und pollenanalytischer Untersuchungen.—Fortschr. Geol. Rheinld. Westf. **28:** 345-359.

MULDER, E.F.J. DE & J.H.A. BOSCH, 1982. Holocene stratigraphy, radio-carbon datings and paleogeography of central and northern North-Holland (The Netherlands).—Meded. Rijks Geol. Dienst, Haarlem **36-3:** 111-160.

PLASSCHE, O. VAN DER, 1982. Sea-level changes and water-level movements in The Netherlands during the holocene.—Meded. Rijks Geol. Dienst, Haarlem **36-1:** 1-93.

PONS, L.J., S. JELGERSMA, A.J. WIGGERS & J.D. DE JONG, 1963. Evolution of the Netherlands coastal area during the Holocene.—Verh. Kon. Ned. Geol. Mijnbouwk. Gen., Geol. Serie **21-2:** 197-208.

ROEP, TH.B., 1984. Progradation, erosion and changing coastal gradient in the coastal barrier deposits of the western Netherlands.—Geol. Mijnb. **63:** 249-258.

STRAATEN, L.M.J.U. VAN, 1957. The Holocene deposits. In: L.M.J.U. VAN STRAATEN & J.D. DE JONG. The excavation at Velsen. A detailed study of Upper-Pleistocene and Holocene stratigraphy.—Verh. Kon. Ned. Geol. Mijnbouwk. Gen., Geol. Serie **17:** 93- 218

VOS, P.C. & H. DE WOLF, 1988. Paleo-ecologisch diatomeeënonderzoek in de Noordzee en provincie Noord-Holland in het kader van het kustgeneseprojekt. Taakgroep 5000 RGD, Afd. Paleobotanie Kenozoicum, Diatomeeën.—Rapportnr. 500: 1-144.

VRIES, J.J., 1977. The stream network in the Netherlands as a groundwater discharge phenomenon.—Geol. Mijnb. **56** (2): 103-122.

WESTERHOFF, W.E., W. DE GANS & E.F.J. DE MULDER, 1987. Toelichting bij Geologische Kaart van Nederland 1 : 50.000. Blad Alkmaar (19O+W).—Rijks Geol. Dienst, Haarlem: 1-227.

ZAGWIJN, W.H., 1986. Nederland in het Holoceen-Geologie van Nederland, Deel I.—Rijks Geol. Dienst, Haarlem.

ECOLOGICAL IMPACT OF SEA LEVEL RISE ON COASTAL ECOSYSTEMS OF MONT-SAINT-MICHEL BAY (FRANCE)

J.C. LEFEUVRE

Muséum National d'Histoire Naturelle, Université de Rennes I, Laboratoire d'Evolution des Systèmes Naturels et Modifiés, 36, Rue Geoffroy Saint Hilaire, F - 75005 Paris, France

ABSTRACT

Over the next few centuries a warming up of the lower atmosphere of several degrees is expected as a result of the greenhouse effect. A rise in temperature would cause a rise in sea level, so the transition area between land and ocean is particularly at risk. This paper puts forth a number of hypotheses concerning the future of a well-studied littoral system: Mont-Saint-Michel Bay. These hypotheses are based on the past history and knowledge of current organization, structure and functioning of this system, considered by sedimentologists as one of the best models of sedimentation in the world.

The different steps of the bay's sedimentation, linked to sea-level changes, are well-known. Since 7500 BP, the sea rose in a serie of successive oscillations sometimes leading to sedimentation of the intertidal area and sometimes of the supratidal area. In the historical period, the patterns of sedimentation have been greatly influenced by human activities, especially reclamation, damming and channelling. They mainly concern the immediate surroundings of the Mont-Saint-Michel. The sedimentary balance is presently largely positive in this part of the bay. The annual average input is estimated at 2 cm per year and the average extension of the salt-marsh area is 30 ha per year.

As the estimation of the average increase in sea level over the next century has a large margin of uncertainty, two scenarios of the future of this bay can be foreseen. One follows current trends (20 cm per century), the other considers the possibility of a maximal rise in sea level (1 to 2 m, or even 3.5 m per century). In the hypothesis of weak and constant rise in sea level, only the general distribution of organisms could be subject to important changes, partly due to a large extension of salt marshes and modifications in the courses of the rivers in the estuarine part of the Bay. In the hypothesis of a significant rise in sea level, salt marshes and polders risk undergoing important fragmentation. Such a transformation would forbid sheep breeding and could cause some of the ducks and geese to abandon this wintering site. Nevertheless, the marine environment could benefit from this structural change which would enhance production of salt-marsh - tidal-flat interfaces. Because of the difficulties in prediction, it is urgent to establish on a national, European and world scale a coherent network of permanent observatories such as Mont-Saint-Michel Bay to develop long-term research on the evolution of environments, variations in their ecological functioning and consequent economic and social changes as an aid in decision-making.

1. INTRODUCTION

Over the course of geological time, Europe's shoreline has changed considerably due to great variations in sea level, particularly during the Quaternary. Such sea level fluctuations are related to climatic changes caused by alternating glacial and interglacial epochs, with an average interval of about 100 000 years between glaciations. Over the next few centuries a warming up of the lower atmosphere by several degrees is expected as a result of the greenhouse effect. Compared to the slow rate of former variations, this could create a major and relatively rapid disruption of the entire globe. The transition area between land and ocean is particularly at risk. A rise in temperature would cause a rise in sea level under the joint effects of three phenomena: thermal expansion of oceanic mass, melting of continental glaciers and melting of ice caps (notably in the arctic).

There is uncertainty about estimations of temperature and sea level increases over the next few centuries. This should lead to prudence in making predictions of the ecological consequences which might result from such phenomena. Caution is also conditioned by another important fact. Responses to changes in the atmosphere's composition will induce

J.J. Beukema et al. (eds.), Expected Effects of Climatic Change on Marine Coastal Ecosystems, 139–153.

movement of climatic zones toward the poles. This would lead to changes in wind, temperature, and rain conditions, difficult to predict locally. Aside from the impact of human activity (polderisation, drainage etc.) a large part of changes which are likely to affect littoral ecological systems will result from the union of three phenomena: rise in sea level, increase in CO_2 (causing, among other things, different C_3 and C_4 plant responses) and local climatic changes.

We can, nevertheless, put forth a number of hypotheses concerning the future of littoral systems, based on the past history and current knowledge of organization, structure and functioning in a few well known systems. The importance of such an exercise, while fascinating from an intellectual standpoint, is to provide a basis for long term monitoring of certain selected sites. This enables validation of hypotheses which are generally a part of large international programs like 'Global Change'.

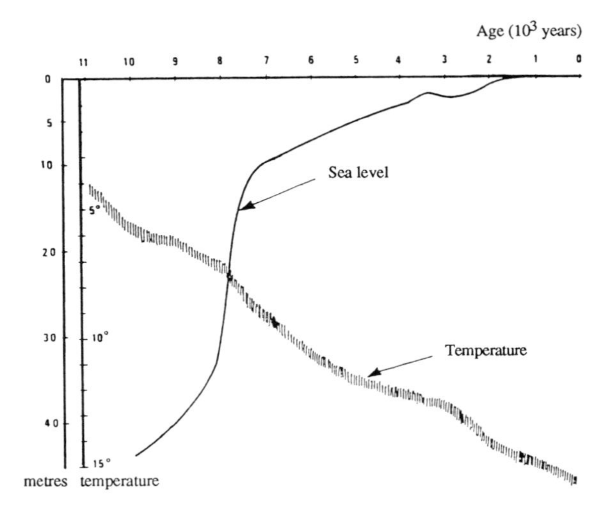

Fig. 1. Sea level and temperature change during the Holocene (in LABORATOIRE CENTRAL D'HYDRAULIQUE DE FRANCE, 1977).

2. MONT-SAINT-MICHEL BAY, YESTERDAY AND TODAY

2.1. CHOICE OF SITE

Mont-Saint-Michel Bay is considered by sedimentologists as one of the most beautiful models of current sedimentation in the world (LARSONNEUR, 1982). The bay is located at the south end of the gulf of Normandy and Brittany, in the angle formed by Brittany and Cotentin (lat. 48°40′N, long. 1°40′W). It is characterized by its exceptional tides (15 m at spring tide). It is also the site of intense filling-in which threatens the island character of Mont-Saint-Michel, the marvellous architectural monument registered, along with the bay, as a World Heritage Site in 1979.

These features were at the origin the bay's two biggest development projects:

- a tidal power station calling for closing of the bay by joining the two points which limit it toward the English channel (BONNEFILLE, 1976);
- a group of works aimed at reestablishing a maritime character to the surroundings of Mont-Saint-Michel. This includes both destruction of old seawalls and construction of particular dams called 'flushing' dams, allowing for evacuation of sediments accumulating at the end of the bay.

These two projects, of which the first was abandoned, were the source of much research on hydrology, sedimentology, botany and ecology. Different types of environments from marine to terrestrial were prospected, and particular emphasis was put on the functioning of this interrelated complex of ecosystems. Knowledge acquired in the Mont-Saint-Michel Bay not only allows for a certain number of hypotheses to be made concerning the ecological consequences of a rise in sea level, but also serves as reference for long-term monitoring of the changes, and for verifying such hypotheses.

2.2. USE OF PAST SEDIMENTARY HISTORY AS A BASIS FOR PREDICTION

The granulometric distribution of superficial sediments, as well as the date of their deposit, influence the distribution of plant and animal species and their assemblage in communities. Analysis of the conditions of sediment deposits and granulometric variations is made according to past sea levels. This provides information for the prediction of the ecological consequences of the current marine transgression and its future evolution in relation to the warming up of the atmosphere.

In less than 20 000 years the sea level has risen considerably. As BROUNS (1988) points out, at the end of the last ice age (Weichselian), 18 000 years ago the sea level was approximately 140 metres below what it is now. The over all warming up of the atmosphere brought about by melting of the vast ice caps which covered Northern Europe, as well as the thermal expansion of the ocean mass.

This provoked a rapid rise in sea level (Fig. 1). Ten thousand years ago, it was still at −50 metres. Between then and 7500 BP, the sea level went from −50 to −10 m (LABORATOIRE CENTRAL D'HYDRAULIQUE DE FRANCE, 1982). The flandrian transgression slowed considerably afterwards, with the rise in waters estimated at an average of 1.5 mm per year. Mont-Saint-Michel's sedimentary history, which is linked to this transgression, was reconstructed by MORZADEC-KERFOURN (1974, 1975, 1977, 1985), DELIBRIAS &

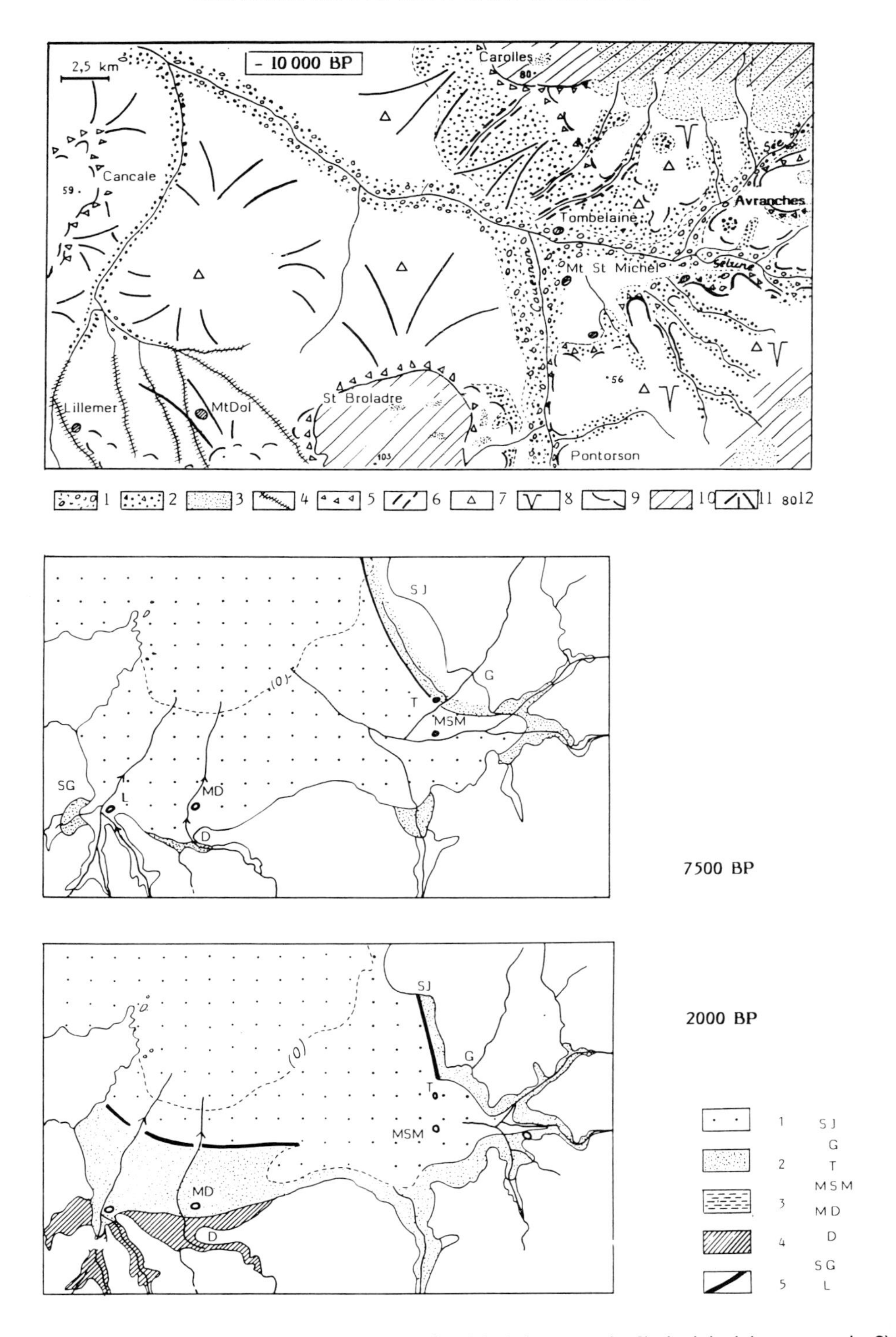

Fig. 2-(top). The Mont-Saint-Michel Bay at the end of the Late Glacial. 1) river gravels; 2) pleniglacial cover sands; 3) Loess; 4) late glacial loams; 5) head; 6) late glacial dunes; 7) aeolisation; 8) sand-wedges; 9) break in the slope; 10) plateaux more than 75 m high 11) periglacial glacis; 12) altitude (metres).

2-(bottom). Two stages of flandrian transgression in the Mont-Saint-Michel Bay. 1) marine sand and silty sands; 2) salt marshes; 3) tidal marsh: silt, clay, peat, and sandy peat on the cover sands (podzolic soil becoming peaty with the rising of flandrian sea); 4) peat bog; 5) coastal barrier (in LAUTRIDOU & MORZADEC, 1982).

TABLE 1

Chronology of variations in sea level and sedimentation in the Mont-Saint-Michel Bay since 9500 B.P.

Subdivision of the Holocene (Flandrian)	Starting dates of climatic periods	Dates	Variation in sea level	Dominant sedimentary deposit types	Notes
Boreal	9500 BP	8000 BP	Fast transgression		The sea reaches the end of the bay in the west
Atlantic	7800 BP	7500 BP	Continuation of the transgression	*Cardium* marine sands	Maximum Flandrian transgression in the west
		6500 BP	Transgression begins to slow	Silt	Development of peat bogs in the north-eastern valleys Extension of *Hydrobia* schorre toward the north, in the west
		6000 BP			The sea once more invades the Bay. Sea level 6 to 7 m below present level
Subboreal	5700 BP	5850 to 5400 BP	Regressive phase	Peat	Marshes of *Cyperaceae* toward Mont-Dol. Development of schorres in the west and in the east
		5000 to 4400 BP	Important transgressive phase	Silt	Sea level 3 m below current level
Subboreal-Subatlantic	3600 BP	3900 to 3450 BP	Transgression slows or stops	Peat	Large development of peat-bogs and schorres
		3300 to 3000 BP	Transgressive phase	Silt	Formation of a littoral strand in the west and development of peat-bogs
Subatlantic	3000 BP	3000 BP	Regressive phase	Peat	
		2300 BP			Formation of a second strand near Mont-Dol with a lacustrine marsh between the two
		2000 BP	Renewal of the transgression	Silt	Schorre sediments spread and reach the altitude of todays highest seas
		1400 BP	Pulsation. Transgression		

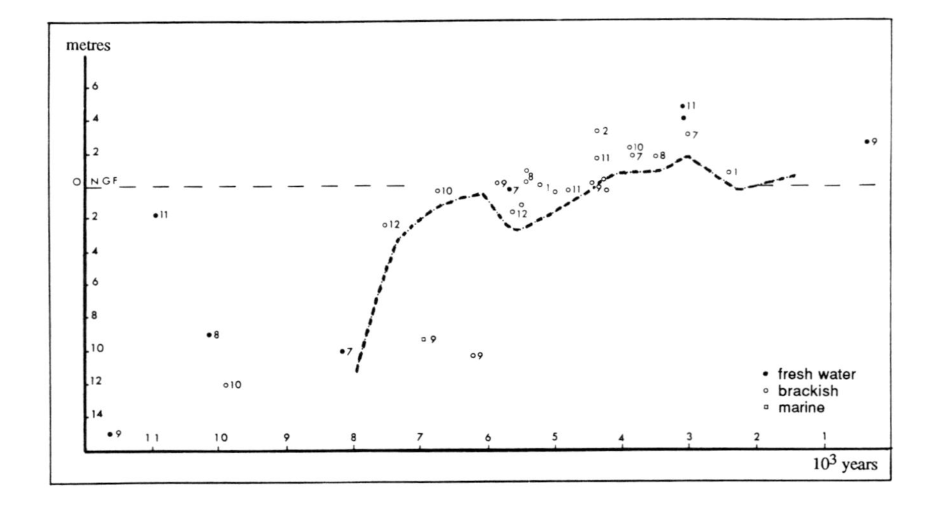

Fig. 3. Oscillation of level of the highest tides at the end of the bay during the Flandrian. Heavy compaction of sediments prevents drawing a curve of mean sea level. Nevertheless, three regressive patterns can be seen in the formation of peaty layers (in DELIBRIAS & MORZADEC, 1975).

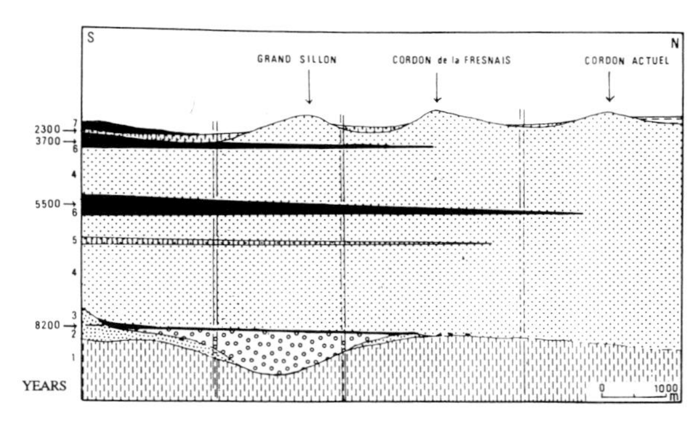

Fig. 4. Sedimentation due to variation in sea level at the end of the bay, during the Flandrian. Embankment of the Dol-de-Bretagne marsh (in MORZADEC-KERFOURN, 1985.) 1) Insular shelf; 2) Limon and tardiweichselian sand; 3) Boreal clay; 4) Sand and silt; 5) Hydrobia silt; 6) Peaty clay; 7) Silt and recent peat.

MORZADEC-KERFOURN (1975), LAUTRIDOU & MORZADEC (1982) (Fig. 2). More recent and current patterns of sedimentation, notably influenced by human activities, are also well known, thanks to the work of LARSONNEUR (1973, 1980, 1982), L'HOMER (1974, 1981), J. DOULCIER (1977), P. DOULCIER (1977), DOULCIER *et al.* (1978), CALINE (1981), LE RHUN (1982), MIGNIOT (1982) as well as the LABORATOIRE CENTRAL D'HYDRAULIQUE DE FRANCE (1971, 1977, 1979, 1982).

2.2.1. END OF THE QUATERNARY

The different steps of the bay's sedimentation history are summarized in Table 1. The sea reaches the end of Mont-Saint-Michel Bay starting in the Boreal. Maximum extension of the transgression could be delimited thanks to *Cardium* sand deposits. The sea rose in a series of successive oscillations (Fig. 3), sometimes leading to sedimentation of the intertidal area and sometimes of the supratidal area (MORZADEC-KERFOURN, 1975; DELIBRIAS & MORZADEC-KERFOURN, 1975; MORZADEC-KERFOURN, 1985). Any transgression is shown by silt deposits, while any regression can be seen in peaty formations (Fig. 4).

Beginning at the end of the Atlantic, the rate of transgression having slowed, we go from sandy sedimentation to a finer silt sedimentation (MORZADEC-KERFOURN, 1985). Filling-in of the bay began, and salt marshes of *Chenopodiaceae* developed. Then the marine influence stopped on the areas furthest inland. The recession of the sea is seen in the first peat deposits between the end of the Atlantic and the beginning of the Subboreal. The *Chenopodiaceae* marshes were then replaced by freshwater marshes of *Gramineae* and *Cyperaceae*. Throughout the history of the bay, two other important peaty formations, one at the end of the Subboreal, the other during the Subatlantic, mark the existence of such phases of regression.

The marine silt covering the peaty formations represents on the other hand an acceleration of the general transgression, which once more allowed for the establishment and spread of *Chenopodiaceae* marshes. Three major transgressive phases are distinguished and can be compared with the phases Calais III, Dunkirk 0 and Dunkirk II, defined in the Netherlands (MORZADEC-KERFOURN, 1985). The third and last Dunkirkian transgression took place at the beginning of the Middle Ages with the temporary submersion of emerged land.

The thickness of the sediment deposit since 7500 BP is on the average 15 metres. This represents a material accumulation of nearly 10^9 m^3. The average annual input has been 1.5 x 10^6 m^3 for all 500 km^2 of the bay, with fine particles selectively transported toward land where we find today's marshes and polders.

2.2.2. HISTORICAL PERIOD

Current patterns of sedimentation are greatly influenced by human activity (Table 2). The first works carried out in the Mont-Saint-Michel Bay date to the 11th century. Localised in the western part of the bay, the works allowed consolidation of the offshore bar, isolating the Dol marshes. In the 13th century the works continued and resulted in the construction of a 40 km seawall, Duchess Anne's, which was completed at the beginning of the 14th century. It is only in the 18th and especially in the 19th centuries that other big works were undertaken in the bay. Essentially they concerned the immediate surroundings of the Mont and consisted of fixing the flow channels of certain rivers. The Couesnon, for example, shifted its course freely from one side of the Mont to the other

TABLE 2

Chronology of major development works in the Mont-Saint-Michel Bay throughout history: dyking, polderization, channeling of rivers.

MAIN DATES	TYPE OF WORK
XIth Century	Consolidation of the offshore bar isolating the Dol marshes
XIIIth Century	The Duchess Anne Sea Wall
1769	Quinette de la Hogue concession (1,000 ha to the south of the Mont) Sea walls destroyed from 1815 to 1857
1856	Concession of land between dry land, the Chapelle Saint-Anne, the Mont-Saint-Michel and Roche-Torin to the Compagnie Mosselmann
1858	Canalisation of 5,600 m of the Couesnon
1858-1934	Construction of 50 km of sea walls in the west, isolating 2,400 ha of polders (change in the line of the outer seawall in 1914 in order to keep Mont-Saint-Michel isolated)
1859	Construction begun on Torin seawall (4,900 m built, out of 6,300 m planned)
1879-1884	Diversion of the rivers situated in the east (Guintre and Ardevon)
1878-1879	Construction of the causeway linking the Mont-Saint-Michel to dry land
1966-1969	An estuarine dam on the Couesnon, at Caserne, 2 km from the Mont, in order to protect 125 ha of agricultural land

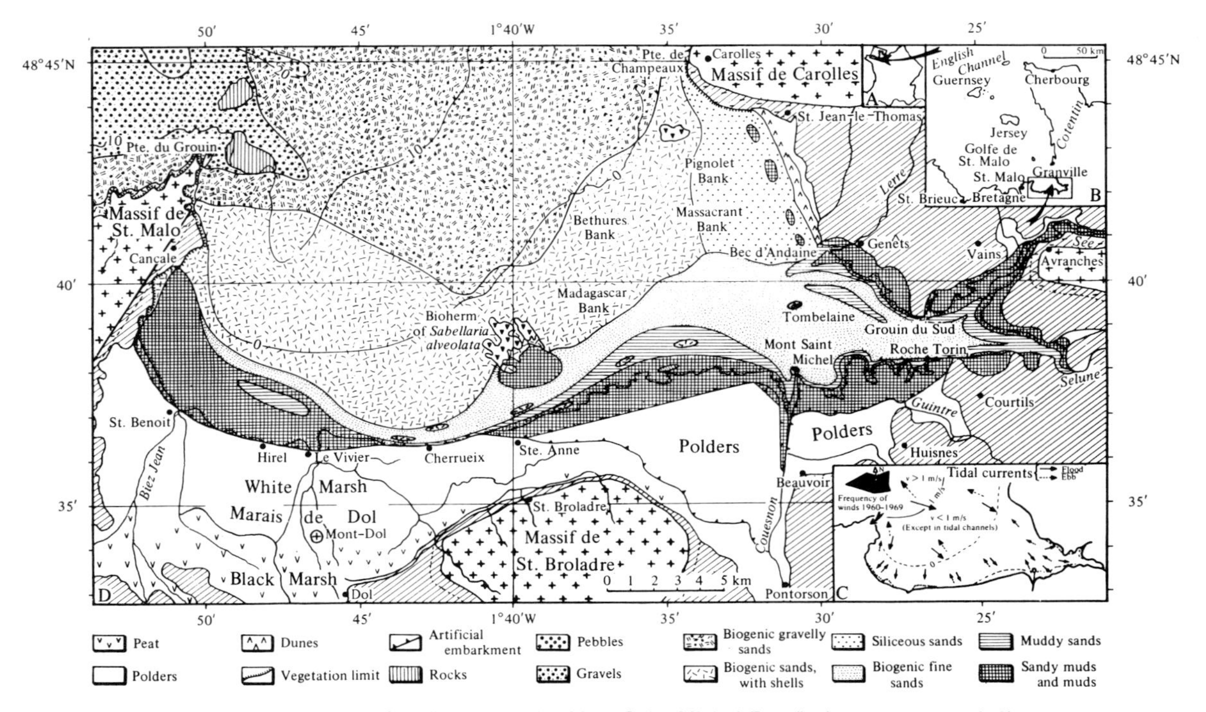

Fig. 5. Distribution of sediments in the Mont-Saint-Michel Bay (in LARSONNEUR, 1973).

before channeling. The works also included reclamation of 3500 ha of polders (3100 in the west). We estimate that between 1850 and 1950, nearly 5000 ha of tidal flats were either polderized or transformed into salt marshes.

2.3. CURRENT STATE OF THE SITE

According to natural sedimentation processes and space transformations influenced by human activities, the bay appears as a heterogeneous whole, formed by a terrestrial element and a marine element (Fig. 5).

In the terrestrial part we distinguish:

- the Dol marsh, formed by two very well differentiated sectors, the white marsh (silt substratum) of about 10 000 ha and the black marsh (peaty substratum) of about 1000 ha;
- polders (3500 ha);
- marshes peripheral to the lower valleys of the Couesnon (600 ha) and of the Sée-Sélune association (250 ha).

The marine part, or the bay in the strict sense of the term, opens widely onto the English Channel. It is characterized by a large tidal flat (20 000 ha) and France's largest area of salt marshes (in French: herbus or prés salés; 4000 ha). On the median axis, the distance between the entrance and the inside of the bay is nearly 30 km.

The assymmetrical shape and the orientation of the bay as related to dynamic agents (tidal currents, dominant north-westerly winds, swells) determine two distinct areas at the tidal flats (CALINE *et al.*, 1985):

- In the west, sheltered from dominant swells and alternating currents between Cancale and the Hermelles massif, a weak and regular incline occurs. Here, the influence of the estuary is either absent, or minimal and very localized. No morphological or sedimentary break enables distinction between the lower and the higher tidal flats, both consisting of fine sands. The western part is the center of an accumulation of mud of variable thickness (10 to 60 cm), saturated with water and rich in organic matter. This part of the bay is characterized by great morphological and sedimentary uniformity (CALINE, 1982).
- In the east, a succession of morphological traits has developed related to the tidal channels. This pre-estuary and estuary zone extends from the outer zone with a tidal bore and megaripples up to the mouths of the three main rivers, the Sée, the Sélune and the Couesnon.

The transition between these two areas takes place in the Cherrueix sector. The effects of swells can be seen clearly. They appear in the formation and the migration of numerous sand and shelly banks which come together in the intertidal area. The tidal flats are dominated by scouring sands and sand particularly rich in carbonate. Soft mud, whose thickness and distribution vary from tide to tide, are preferentially deposited near obstacles such as *bouchots* (a *bouchot* is a line of 110 stakes, 2.5 m high, driven into the ground over a length of 100 m, around them are wound ropes carrying spat. They serve as supports for growing mussels). This transition zone also differs from the rest of the bay by the presence, near the lower water mark, of a reef of *Sabellaria alveolata*, the Hermelles Bank, on which depends the bay's most voluminous sand bank (GUILLAUMONT & HAMON, 1987).

These two big morphological-sedimentary areas are currently the center of two different evolutions. DURANT (1978) emphasizes the important regression of salt marshes in the non estuarine end of the gulf since 1947 to 1961, between which dates oysters and especially mussel farming began (250 km of *bouchots*). From then on, the progression of grasslands slowed considerably. It seems as if the fine sedimentary matter available in this section of the bay was preferentially fixing in the middle and lower part of the intertidal zone in the *bouchot* area (NIKODIK & SORNIN, 1981). Muddy deposits are estimated at 2 million m^3 (LABORATOIRE CENTRAL D'HYDRAULIQUE DE FRANCE, 1979). The sedimentary balance for this part of the bay is positive, showing a rise in the sea bed from 60 to 70 cm since 1829 (up to 1.5 m in the old oyster beds).

In the estuarine bay, the process is altogether different. The accretions, of which the southern shores have been the center since the second half of the 19th century, brought about a new balance in this part of the bay (CALINE *et al.*, 1982) with, as its main consequences:

- the displacement of the axis of the Sée and the Sélune channels toward the northern banks, and the correlating periodic recession of the coast between Saint Jean-le-Thomas and the Bec d'Andaine by erosion of silt and dune sands along the littoral (L'HOMER, 1974);
- the filling-in of the Sélune and especially the Sée valleys and the downstream displacement of all the benthic population from the inner zones of the estuarine part of the bay;
- an extension of the 'Cherrueix flat' type facies toward the east, in place of the former mouth of the Couesnon;
- in the outer estuarine zone, the upstream development and displacement of sand bars which more and more tend to congest the tidal flats between the channels.

According to MIGNIOT (1982), the major works of polderization, channeling, dam construction, shifting of river courses and localized banking on the southern shores on both sides of Mont-Saint-Michel, favoured the deposit of more than 100 million m^3 of silt and sand between 1858 (at the beginning of polderization) and 1934. Again we find in this sector, of only 50 km^2, an annual deposition rate of 1.3 million m^3, comparable to that which is noted for the entire bay (511 km^2) over the last few thousand years.

All of the studies carried out in the estuarine zone conclude that there are very large sedimentary movements, sometimes reaching several decimetres per month. Erosion, active at the level of the channels, does not compensate for the input, and the sedimentary balance is largely positive. We estimate the annual average input at a thickness of 2 cm. Knowing that the erosion is far from equally distributed, the local sedimentation rate can be very high. On many flats it can reach 5 to 10 cm per month, or about 1 m per year (COMPAIN *et al.*, 1982).

Since the beginning of the century, the sea bed to the west of the Mont has risen nearly 3 m. The area covered by grasses (*herbus*) between the water marks + 6 m and + 8 m N.G.F. (*i.e. nivellement général de la France*, situated in the Bay at an average of + 7.36 m above the lowest spring tide level or zero on marine charts) went from 1000 ha to 1500 ha between 1947 and 1978. The average extension was 30 ha per year. This value has tended to increase since 1970 (MIGNIOT, 1982) (Fig. 6).

3. MONT-SAINT-MICHEL BAY, TOMORROW

The estimation of the average increase in sea level over the next century has a large margin of uncertainty. Different authors consider estimates to vary between 0.5 to 3.5 m. The evolution of the Bay will be quite different depending on which of the hypotheses we choose. Two scenarios are analyzed, the one follows current trends, the other considers the possibility of a maximal rise in sea level.

3.1. FIRST SCENARIO: CURRENT TRENDS CONTINUE

3.1.1. SEDIMENTARY EVOLUTION

In Mont-Saint-Michel Bay, the sedimentary prism has grown at an average annual rate of 1.5 million m^3 since about 8000 years ago. This sediment deposit is correlated with an increase in the average sea level of approximately 0.2 cm per year (20 cm per century).

On the bay scale, the filling appears to occur at a constant scale but the localization of deposits is modified through time (COMPAIN *et al.*, 1982). Since last century, the main part of sedimentation, under the effect of development works, is concentrated in the estuarine zone. We can therefore consider, with LARSONNEUR (1982), that the current configuration of the bay represents only a stage of the evolution since the last glaciation, marked by the accumulation of a

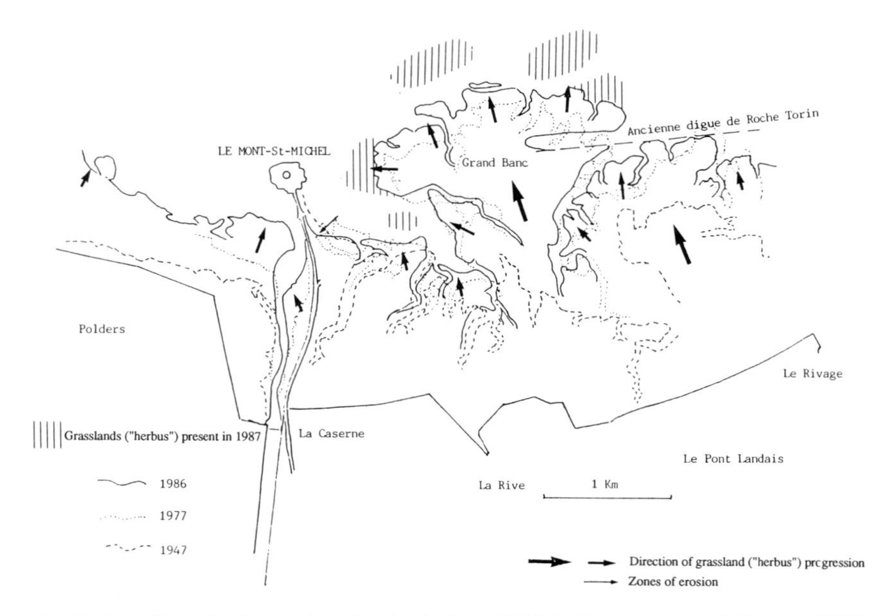

Fig. 6. Evolution of salt marshes (*herbus*) since 1947 (in LARSONNEUR & BARTH, 1988).

sedimentary prism against the land. It is certainly difficult at times to evaluate locally the reality of a rise in sea level. This is especially due to vertical movements of the globe's surface, *i.e.* subsidence, raising (PIRAZZOLI, 1986). Continued filling-in of the bay, however, seems to indicate that we are still in a phase of transgression. This opinion is widely held: ROBIN (1986) estimated the rise in the average sea level at around 10–15 cm in the last 100 years. These figures are very close to those evaluated for a period of 8000 years. They authorize a first hypothesis that over the next century the process of sedimentation could continue to follow current patterns.

According to such a hypothesis, we could reasonably predict the following changes for the next 100 years:
- in the western part near the *bouchots* and oyster beds, a 40 cm (0.4 cm per year) rise in the sea bed;
- in the estuarine part, a sediment deposit of 130 million tons (1.3 million tons per year), provoking an average rise of 2 m (2 cm per year) and an increase of 3000 ha of salt marshes (30 ha per year).

In the absence of new developments, polderization or construction of 'flushing' dams intended to prevent sedimentation surrounding Mont-Saint-Michel, this extension of marshes, which practically doubles the current area, will considerably change the bay's estuarine physiognomy. Indeed, reinforcement of sediment deposits in the west between Sainte-Anne and Mont-Saint-Michel and in the east between the rivers Couesnon and the Sélune, as well as in the area between the Sée and the Sélune, should contribute to carrying the estuarine zone toward the west and the north. The islet of Tombelaine might be caught in the sedimentation zone because of the new south-north orientation of the Sélune (based on studies of the Laboratoire Central d'Hydraulique de France which modelled the probable changes in bottom topography over a period of ten years). The confluence of the three rivers currently situated in the bay and characterized by important shifts, could be fixed for ever to the longitude of Tombelaine by progressive settling of the sediments deposited.

3.1.2. EVOLUTION OF ECOLOGICAL SYSTEMS

The majority of the intertidal zone is currently dominated by the *Macoma balthica* community. The fine sands of the middle levels make an ideal biotope for these animals and *Macoma* densities are among the highest observed in the intertidal biotopes of north-western Europe (GUILLAUMONT & HAMON, 1987).

Aside from the *Macoma*-type population, three other facies are distributed in the tidal zone (Fig. 7):
- very fine level sands with *Corophium volutator*, situated at the bay's periphery, bordering the salt marshes in the supratidal area, corresponding to high tidal flats above +2 m N.G.F.;
- mid level fine to medium grained sands with *Haustoridae* (pre-estuarine zone) or *Arenicola marina*;
- medium grained and silty sands with *Lanice*. This facies, situated 3 m below zero N.G.F. covers about 100 ha north-east of Hermelles Bank.

All of these communities should persist, with local redistribution of the different facies. Modifications in biomass are difficult to estimate. The disappearance

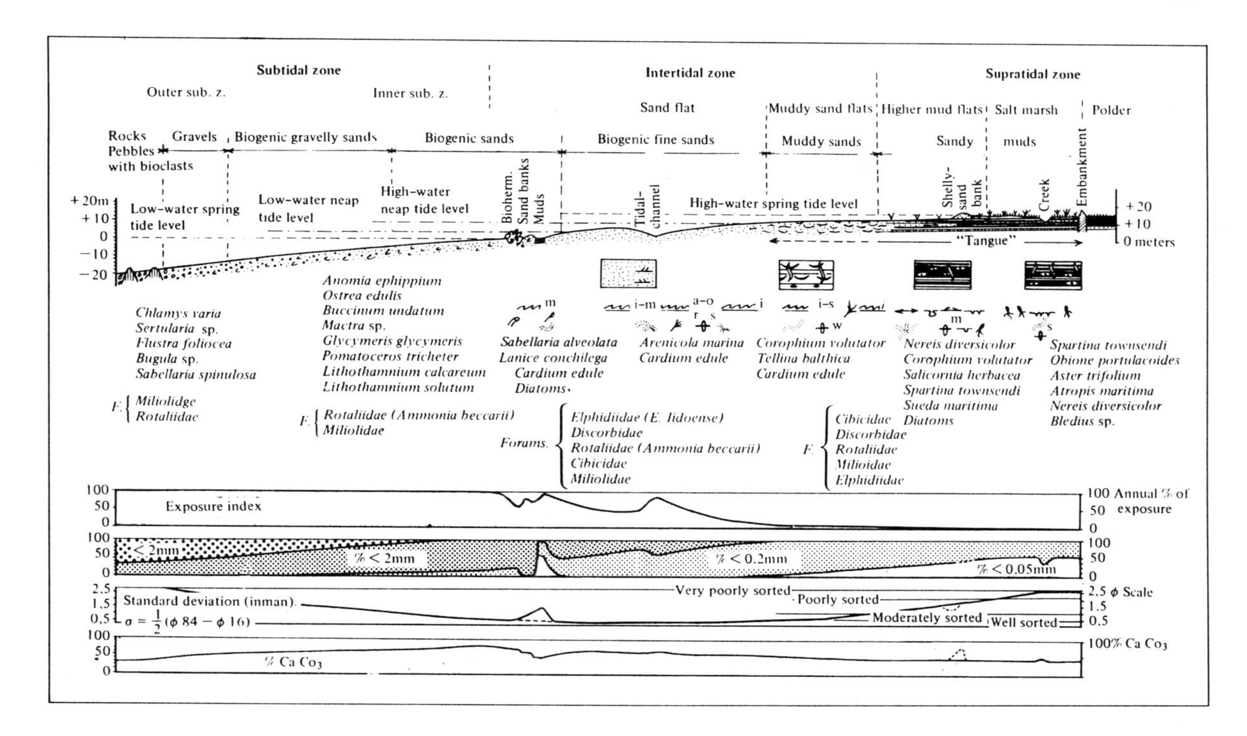

Fig. 7. Typical sedimentary sequence and animal distribution (in LARSONNEUR, 1973).

of 3000 ha of tidal flat in its eastern part would have only limited effect on the general functioning of the bay. This zone, undoubtedly because of sedimentary instability currently provoked by shifting of the rivers Sée and Sélune indeed presents a low number of benthic macrofauna species and is characterized by a low biomass (often below 1 g dry weight per m^2, GUILLON & LEGENDRE, 1981).

Mont-Saint-Michel Bay, because of the diversity of the habitats exploited by waterbirds, proves to be an extremely important site for shore birds, ducks, geese, gulls and terns (LEFEUVRE & LANDRÉ, 1974; BORET, 1981; MAHEO & BORET, 1983; SCHRICKE, 1983).

The transformation of the bay's estuarine zone could have great effects on the distribution of waterbirds in the bay. Their numbers, however, would be relatively unaffected.

The variations in plant communities in relation to the age of recent sediment deposits are well known (GÉHU, 1979; GUILLON, 1980, 1984a, b; GUILLON & LEGENDRE, 1981; LEVASSEUR, 1987). Plants occur in an ordinated sequence of communities. This zonation depends on three principal factors (GUILLON, 1984a, b): the tidal and submersion frequency (Table 3), physico-chemical characteristics of soils in formation (granulometry, variations in humidity, salinity...), and sediment dynamics.

According to GUILLON (1984a, b) pioneer vegetation of the high tidal flats includes:

- an open vegetation of annual saltworts, the dominant species being *Salicornia dolichostachya*, a tetraploid species,
- patches of *Spartina anglica (townsendi)*, a recently established species, noted for the first time in the bay in 1930.

The vegetation of the salt marsh is dominated by *Pucinellia maritima*. We can distinguish, starting at the lower levels:

- open *Puccinellia maritima* cover with tufts of *Arthrocnemum perenne*;
- *P. maritima* cover with a summer dominance of *Suaeda maritima*;
- low and dense *P. maritima* cover: this is the typical *pré salé* (salt meadow) of the bay. Grazed by sheep, it covers vast surfaces on both sides of the Mont (about 2000 ha for all of the bay).

TABLE 3

Distribution of main plant species according to tidal characteristics.

MAIN PLANT SPECIES	TIDAL COEFFICIENT	ANNUAL FREQUENCY OF COEFFICIENTS	UPPER LIMIT OF THE FLOODED ZONE	
Salicornia tetraploides	80	65 %	High slikke	
Puccinellia	80 - 95	23 %	Low schorre	
Halimione	95 - 105	8 %	Mid schorre	variable cover
Festuca rubra var. *littoralis*	105 - 113	3 %	High schorre	according to meteo-
Agropyron	113	1 %	Upper limit of the tide	rological conditions

We also distinguish in these marshes: a semi-woody dwarf formation of *Halimione portulacoides*, a grassland formation dominated by *Festuca rubra* var. *littoralis* which, depending on the degree of grazing, can be a short grassland, where tall grasses are cut yearly; a dense grassland of tall grasses, *Agropyron pungens*; a short, grazed, *Agrostis stolonifera* var. *salina* dominated grassland, and depending on the amount of fresh water arriving by flow or infiltration, an association of *Juncus gerardii* (favoured by grazing) and *Juncus maritimus*. Then, at the extreme of the marine/freshwater sequence, restricted marshy zones of *Scirpus maritimus* and reedbeds of *Phragmites australis* occur.

Current salt-marsh features and especially their degree of extension, lead us to distinguish a western and an eastern region in terms of future evolution.

- Western region

Here marsh extension is limited. This area is often characterized by the presence of bioclastic flats of more or less perennial accumulation. These flats are responsible for the establishment of a *Spartina anglica* type high tidal flat vegetation, and are marginally invaded by a lower marsh vegetation (*Pucinellia maritima, Halimione portulacoides*). Because of the slight upward slope of the intertidal zone sea bed, they may increase in size and become fixed. As is now the case for limited areas, they could then enable the establishment of a mixohaline flora (reedbeds of *Phragmites australis, Scirpus maritimus, Juncus gerardii, Carex extensa, Glaux maritima*...) colonizing the lowest parts which receive meteoric or resurgent fresh water.

- Eastern region

The retrospective analysis of the evolution of salt marshes in the estuarine part of the bay enables us to distinguish two groups:

- Marshes developing either continuously or by jumps, according to a unidirectional process in time.
- Marshes submitted to random extension then regression, directly related to river positions.

This distinction allows us to evaluate the degree of stability of these places from the point of view of plant succession. In particular this is possible for those which, brought to term, are at the origin of developing mesophile grassland formations, especially for *Festuca rubra* var. *littoralis*. This is a indication of land formation. Salt marshes of uncertain future development are for the most part localized in the northern part of the estuarine zone. The others currently localized in the southern part, are likely to know greater expansion.

As LEVASSEUR (1987) points out, once established, a salt marsh changes not only on the seaward side, but also on the landward side. In the hypothesis of continued expansion, the area of *Spartina anglica, Salicornia dolichostachya* and *Puccinellia maritima* (primary community) pioneer zone might be a constant while the area becoming more terrestrial (*Festuca rubra* var. *littoralis* and *Agropyron pungens*) should be in continual progression.

The relative importance of extensive formations, which could develop over the next century in the spaces conquered from the sea, depends also on the evolution of livestock densities.

In our current state of knowledge, this different distribution of vegetation types and especially their physiognomy, could act directly on the differential distribution of the marsh' three dominant groups of arthropods: carabids, arachnids and detrivorous amphipod crustaceans of the genus *Orchestia* (FOUILLET, 1986).

Waterfowl can benefit from the maintenance or abandonment of livestock on the marshes. SCHRICKE (1983) has shown that if the populations of mallards (*Anas platyrhynchos*) use the freshwater marshes in the oldest sedimentary part of the bay essentially to feed, the same is not true for brent geese (*Branta bernicla*) and wigeon (*Anas penelope*). The latter frequent the salt marshes and particularly the secondary groups of *Puccinellia* maintained by grazing. This is the grass they prefer in the Mont-Saint-Michel Bay (up to 80% of their diet).

The withdrawal of sheep from a hunting reserve has shown that in 3 years, a *Halimione* group replaces a *Puccinellia* group; so this zone abundantly frequented by *Anas penelope* becomes deserted. Inversely, the extension and non-grazing of *Festuca rubra* var. *littoralis* and *Agropyron* vegetation can favourably influence the evolution of a relict population of quail, *Coturnix coturnix*, localized late in the year in this type of cut grassland (end of July-beginning of August).

It is more difficult to decide on the evolution of the bay's functioning in relation to its annual production of salt marsh due to an increase in the area of this habitat. The works of GUILLON (1980) and DANAIS (1985) allow estimation of the annual production by plant community (Table 4).

Current research has up to now, concentrated on organic matter exchanges (organic carbon particles and floating organic matter) at spring tides in the fall.

TABLE 4

Primary production of salt marsh vegetation types. After GUILLON (1980) and DANAIS (1985).

PLANT COMMUNITY	ANNUAL PRODUCTION (in g.m^2)
Salicornia	10
Spartina and *Salicornia*	50 to 250
Puccinellia (not grazed)	1300
Puccinellia (grazed)	115
Puccinellia and *Halimione*	400 to 850
Halimione	1000 to 1500
Festuca and *Agropyron*	475 to 1000
Agropyron	800 to 1000

It shows that the marshes of mainly *Halimione* with the highest annual production rate, would be entirely importers of organic matter, while grazed marshes of *Puccinellia* have either an even balance or are slightly exporters.

By modifying plant cover and its recycling, grazing therefore constitutes a very important factor for the balance of organic matter exchange. So, it is difficult to make predictions for over 100 years.

But our considerations cannot be based on only rise in sea level. The doubling of CO_2 might favour the extension and productivity of a C_4 plant like *Spartina anglica*. The localization of this species in pioneer zones favours exchanges with seawater. The establishment of vast marshes of *S. anglica* could lead to a completely different balance possibly enriching the marine environment in carbon, as is the case on the east coast of the United States (ODUM, 1980).

Normally, secondary production, dominated by deposit and suspension-feeders, whether commercially exploitable or not (BRÉGEON, 1977; RETIÈRE, 1979) should not be much affected by such modifications. No more, incidentally, than the sole (*Solea solea*), plaice (*Pleuronectes platessa*), thornback (*Raja clavata*), whiting (*Merlangus merlangus*), bib (*Trisopterus luscus*), shrimp (*Crangon vulgaris*) and cuttlefish (*Sepia officinalis*) nurseries which characterize the bay (BEILLOIS *et al.*, 1979; LEGENDRE, 1984).

3.2. SECOND SCENARIO: A GREAT INCREASE IN SEA LEVEL

The scenario above corresponds to a minimal rise in sea level, as predicted by a number of authors, while others predict a higher rise (1 to 2 metres, or even 3.5 m) over the coming century (HOFFMAN *et al.*, 1983; HOFFMAN, 1984; ROBIN, 1986). Such an elevation, 5 to 6 times the previously considered scenario, could lead to a very different evolution, especially in the estuarine part of the bay.

The only examples at our disposal for evaluating the consequences of such a great rise in waters are either:

- zones of relatively important subsidence, like the Venice lagoon; until 1930 the relative rise in sea level was about 13 cm per century (similar to Trieste). Since then it has increased to 50 cm per century and certain predictions count on closer to 2.2 m between 1930 and 2030, because of the exponential character of this increase (LEFEUVRE & PARIS, 1982).
- bays and estuaries modified by developments which artificially increase the duration of salt marsh flooding; this is the case for the estuary of Rance (RETIÈRE, 1979). A tidal power station closed off this estuary and provoked submersion of salt marshes with weakly saline water for 2 years. It is now creating an increase in the general water level. Moreover, the voluntary retention of water during a cycle of two tides, in order to meet the needs of the tidal power plant, provokes flooding of marshes for twice as long as normal (RETIÈRE, 1979).

In both situations, especially for slightly sloped, more or less flat salt marshes, the microtopography determines a mosaic structure with alternating areas of stagnation and slightly elevated areas. The stagnation of salt water favours the regression of plants from the highest part and the reinstallation of pioneer plants such as saltwort. The disappearance of plants in certain areas leads to breaking down of the marshes into small islets within a complex network of deep and silted drainage canals. Most of these islets have a small basin structure which, in relation to the length and timing of tidal flooding, favours plant microzonation. In most complex cases, this includes a peripheral belt of *Halimione* (zone of slow, even filtration), a zone of *Puccinellia* and a zone of *Salicornia*; the center of the islet can be entirely bare.

In case of a great rise in sea level, we can predict a similar evolution in the Mont-Saint-Michel Bay. The fragmentation of salt marshes could lead to an important decrease in their area. Their futures would differ according to their features:

a) For salt marshes which evolution is constant, there could be:

- a decrease in or even disappearance of zones of *Agropyron* and *Festuca*, which might possibly be replaced by *Halimione* (this would entail the disappearance of the relict population of *Coturnix coturnix*).
- a drastic reduction of *Puccinellia* plant cover. Marsh fragmentation is incompatible with raising sheep, already responsible for the extension of secondary *Puccinellietum* (GUILLON, 1980, 1984).

b) Unstable marshes, submitted to unpredictable sedimentation rates and erosion (LEVASSEUR, 1987), would progressively cover more area than the marshes whose evolution is constant.

The regression of the area of marshes would have marked effects on the fauna, notably on wigeon and brent geese. It could even cause some of the ducks and geese to abandon this wintering site. The decrease in feeding habitats is not the only reason. Local variation of the climate, in particular the milder winter temperatures which would accompany a rise in sea level, could considerably accentuate this trend. Indeed, SCHRICKE (1983) and MAHÉO & BORET (1983) have shown that while the number of mallards remains relatively stable from year to year the numbers of other ducks and geese depend essentially on Europe's climate. Few birds are present during

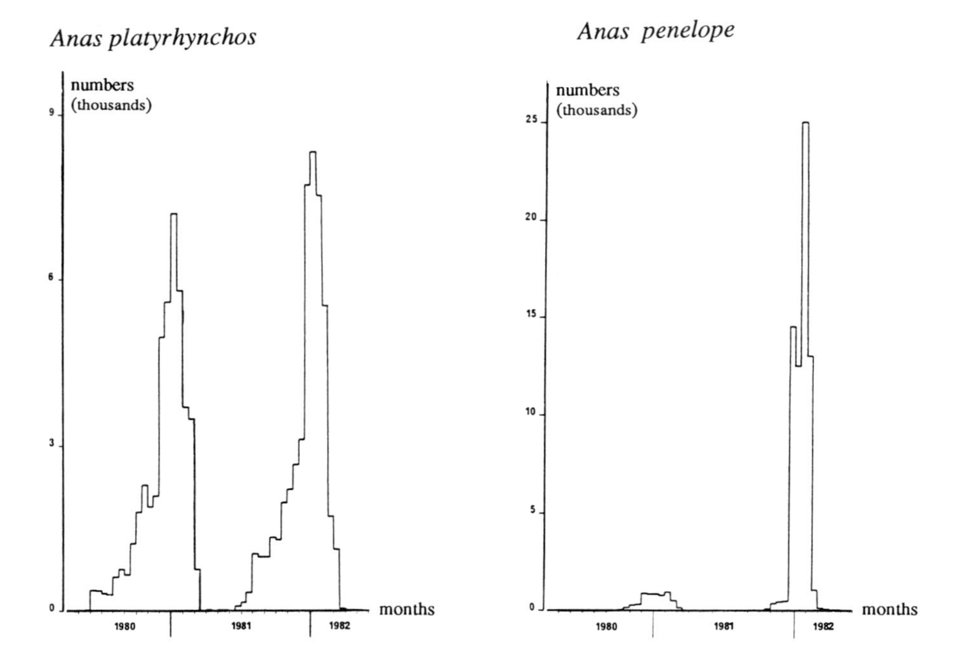

Fig. 8. Comparative evolution in numbers of mallards (*Anas platyrhynchos*) and wigeon (*Anas penelope*) during two different wintering periods: 1980–1981: cold winter; 1981–1982: 'mild' winter (in SCHRICKE, 1983).

mild winters, but during cold periods in Northern Europe, the number of wintering wigeon increases from 50 (1978) to 20 000 (1979), or from 850 (1981) to 25 000 (1982) (Fig. 8). This example allows us to emphasize the difficulty of dissociating ecological effects of interrelated factors (increase in CO_2, average temperature, sea level). A scenario's value is related to the simultaneous analysis of these factors.

This overall approach, also requiring the consideration of parameters such as competition between C_4 plants and C_3 plants, is all the more necessary as a general rise in temperature might considerably modify plant associations in the salt marsh. The Mont-Saint-Michel Bay is indeed included in the biogeographical province of the Central Atlantic, but is situated at the edge of the Southern Atlantic province (DIJKEMA *et al.*, 1984). Warming up could enable establishment of the latter's characteristic flora and fauna.

The combination of these negative environmental effects could be accompanied by positive effects, among which we should note:

- a possible rise in phytobenthic productivity due to an increase in the range of contact between marsh and the upper tidal flats. This would result from marsh fragmentation. LEGENDRE (1985) has shown that the essential part of primary productivity due to diatoms is localized in this zone;
- growth in suspension and deposit feeder populations and, concomitantly, of fish nurseries resulting from an increase in food resources;
- an increase in wetland bird populations. The bay, with an average of 30 000 to 60 000 individuals belonging to 13 species, is already one of the major French wintering sites, especially for the knot *Calidris canutus* (5000 on the average), dunlin, *Calidris alpina* (30 000) and oystercatcher, *Haematopus ostralegus* (15 000) (LEFEUVRE & LANDRÉ, 1974; BORET, 1981). It is also the only French wintering site for ruff, *Philomachus pugnax*.
- an increase in shelduck populations, a species related to the marsh-flat interface.

Finally, in the case of an important rise in the average sea level, the rise in the level of the underground salt water behind the seawalls and in the polders could lead to a change in exploitation of soils: abandonment of crops and a return to grazing. This change would be highly favourable for white-fronted geese since the bay was France's principal wintering area for this species before the turning over of grasslands to maize farming (LEFEUVRE & LANDRÉ, 1974; AUBERT *et al.*, 1975). These areas, newly reconquered by the grasslands could even partly compensate for the loss of *Puccinellia* zones, and thus maintain the brent geese and wigeon populations, if halophytic plants were to increase. On the other hand, a warmer climate might lead to a northward shift of the wintering quarters of these species.

4. CONCLUSIONS

After a century the two scenarios pictured would have very different outcomes.

- In the hypothesis of weak and constant elevation in sea level, only the general distribution of organisms could be subject to important changes. This is especially true in the estuarine part of the bay, whose overall functioning should remain relatively stable;
- In the hypothesis of a significant rise in sea level, salt marshes and polders risk undergoing great transformations or regressive changes. As for the marine environment it could benefit from structural changes which would heighten production, such as an increase in salt marsh-tidal flat interfaces and mud deposits.

These predictions based on acquired knowledge should not conceal the margin of uncertainty which remains. An ecologist cannot allow himself to neglect the importance of factors other than those whose impact has been analyzed. It would be a mistake to believe that examination of the factor 'rise in sea level' could alone decide with certainty the state of Mont-Saint-Michel Bay in 2090. Such predictive models belong in a context of the Global Change project and lead us at the same time to wonder about the consequences of a rise in average temperature, variations in CO_2, and an increase in U.V. rays related to the destruction of the ozone layer. Local variations in pH ('acid' deposits or rains carrying NH_4OH) dependant on human industrial and agricultural activities must also be taken into account.

Furthermore, we should note the necessity for refining current analyses: the preceding scenarios can be entirely modified if we consider other patterns in the evolution of the greenhouse effect. This will be the case, in particular, if the phenomenon, instead of being constant, is irregular, with brusque accelerations and latent periods. Such variations could result, for example, from legal measures reducing discharge of specific gas or products, or from the appearance of new compounds with unknown effects.

Moreover, the lack of knowledge concerning the evolution of local climatic conditions deserves attention. The manner in which these conditions will change will be the determinant. An organism's response can be quite different depending on whether the contrast between seasons grows (cold winters, hot summers) or if, on the other hand, annual thermal amplitude decreases. It can also depend on variations in rainfall (increase, fractioned distribution or not, etc...).

Changes in human activity brought on by transformation of the environment should also be considered (extension or regression of salt marshes, salt intrusion).

Finally, a rise in sea level in Mont-Saint-Michel Bay would not have only ecological consequences, but it would also have important socio-economic and socio-cultural effects. It could challenge for example, the usefulness of major development works destined to keep Mont-Saint-Michel's insular character.

Indeed, the first scenario, the progressive surrounding of Mont-Saint-Michel by salt marshes seems inevitable, in spite of levelling the Torin seawall (1983), transforming the Caserne dam into a 'flushing' dam, and building two more of these reservoirs in the east. This is because the discharge channels of the three reservoirs would sink into the sediment instead of discharging it as intended. Only continued removal of about 1 million m^3 per year of silt (sediment containing 40 to 60% $CaCO_3$, of good agronomic quality) could stabilize the filling-in phenomenon.

In the second scenario with a 3.5 m rise in sea level, if the amplitude of tides remains constant, 65% of annual tides (coefficient 60, maximum sea level at water mark +4,65 m N.G.F.) would reach the 8.15-m watermark and would therefore be the same as the highest current spring tides. These, coefficient 115, currently reach the 8-m water mark. All of the salt marshes are covered by the tide and Mont-Saint-Michel is surrounded by water. At spring tide, the 11.5-m watermark would be covered. Under this hypothesis, not only would the development project destined to keeping Mont-Saint-Michel's maritime character be unneccessary, but protective measures should be taken to prevent flooding in the lower parts of the three rivers.

It therefore is urgent, because of the difficulties in prediction mentioned above, to establish on a national, European, and world scale a network of well identified observation posts (LEFEUVRE, 1985). These should be capable of permanently describing the evolution of environments and variations in their ecological functioning, as well as the study of economic and social changes as an aid to decision-making.

ACKNOWLEDGEMENTS

The author wishes to thank M.T. Morzadec-Kerfourn, W.J. Wolff, C. Legendre, J. le Duchat d'Aubigny, A. Abbas and G. Barnaud for helpful discussions and critical reviews on previous drafts of this paper; P. Duncan and L. Toussaint for the English translation.

5. REFERENCES

AUBERT, B., N. LANDRE & J.C. LEFEUVRE, 1975. Aménagement et mise en valeur des richesses naturelles de la Baie du Mont-Saint-Michel. Doc. Ministère de la Qualité de la Vie, Inspection Générale de l'Environnement, régions de Bretagne et de Basse-Normandie. Ed. Association Normande d'Economie Rurale Appliquée (A.N.E.R.A.) et Société d'Etudes pour la Protection de la Nature en Bretagne (S.E.P.N.B.): 282 pp.

BEILLOIS, P., Y. DESAUNAY, D. DOREL & M. LEMOINE, 1979. Nurseries littorales de la Baie du Mont-Saint-Michel et du Cotentin Est. Rapport Institut Scientifique et Technique des Pêches Maritimes: 115 pp.

BONNEFILLE, M., 1976. Les réalisations d'Electricité de France concernant l'énergie marémotrice.—Houille blanche **2:** 87-149.

BORET, P., 1981. Recherches sur les oiseaux d'eau séjournant dans la Baie du Mont-Saint-Michel. Rapport Office National de la Chasse, Université de Rennes (C.R.E.B.S.): 55 pp.

BREGEON, L., 1977. Richesses et productions marines de la baie du Mont-Saint-Michel. La mytiliculture. - Science et pêches, Bull. Inst. Pêches marit. **267:** 1-29.

BROUNS, J.J.W.M., 1988. The impact of sea level rise on the dutch coastal ecosystems. Report Neth. Inst. Sea Res., Res. Inst. Nature Management: 101 pp.

CALINE, B., 1981. Le secteur occidental de la Baie du Mont-Saint-Michel: morphologie, sédimentologie et cartographie de l'estran. Thèse de 3ème cycle, Université de Paris-sud, centre Orsay: 308 pp.

——, 1982. Le secteur occidental de la Baie du Mont-Saint-Michel: morphologie, sédimentologie et cartographie de l'estran.—Documents du BRGM n° 42, Orléans: 250 pp.

CALINE, B., A. L'HOMER & C. LARSONNEUR, 1982. Conclusion. In: C. LARSONNEUR & A. L'HOMER. La Baie du Mont-Saint-Michel. Voyage d'études. 15-18 septembre 1982. Rapport Association des Sédimentologistes Français: 73-74.

CALINE, B., C. LARSONNEUR & A. L'HOMER, 1985. La Baie du Mont-Saint-Michel: principaux environnements sédimentaires. Livre jub. G. Lucas: 37-51.

COMPAIN, P., C. LARSONNEUR, B. SIMON & P. WALKER, 1982. Dynamique sédimentaire sur le littoral de Saint-Jean-le-Thomas au Bec d'Andaine. In: C. LARSONNEUR & A. L'HOMER. La Baie du Mont-Saint-Michel. Voyage d'études. 15-18 septembre 1982. Rapport Association des Sédimentologistes Francais: 61-63.

DANAIS, M., 1985. Production primaire du schorre et transports de matière organique flottante en Baie du Mont-Saint-Michel. Rapport Fonctionnement des Systèmes Ecologiques en Baie du Mont-Saint-Michel. C.E.E. Environnement et I.R.I.E.C.: 76 pp.

DANAIS, M. & C. LEGENDRE, 1986. La Baie du Mont-Saint-Michel. Premier bilan de fonctionnement. Rapport Fonctionnement des Systèmes Ecologiques de la Baie du Mont-Saint-Michel, C.E.E. Environnement et I.R.I.E.C.: 34 pp.

DELIBRIAS, G. & M.T. MORZADEC-KERFOURN, 1975. Evolution du marais de Dol-de-Bretagne au Flandrien.—Bull. de l'Association francaise pour l'étude du quaternaire, 1975 (2): 59-70.

DIJKEMA, K.S., W.G. BEEFTINK, J.P. DOODY, J.M. GEHU, B. HEYDEMANN & S. RIVAS MARTINEZ, 1984. La végétation halophile en Europe (prés salés).—Rapport Comité Européen pour la Sauvegarde de la Nature et des Ressources Naturelles, Conseil de l'Europe, Strasbourg: 179 pp.

DOULCIER, J., 1977. Réflexion sur le travail d'un modèle très réduit. Résultats. Espoirs de résultats. Thèse Docteur-Ingénieur, Univ. Paris XI, Orsay, dactyl: 420 pp.

DOULCIER, P., 1977. Le Mont-Saint-Michel. Le problème du caractère maritime du site. Thèse Docteur-Ingénieur, Univ. Paris XI, Orsay, dactyl: 218 pp.

DOULCIER, J., P. GEFFRE, C. MIGNIOT, P. PRESCHEZ, G. SIMON & P. VIGUIER, 1978. Le Mont-Saint-Michel entre terre et mer.—Monuments historiques **3:** 33-44.

DURANT, M.A., 1978. Études des contraintes et des potentialités du milieu en vue de l'aménagement de l'herbu de la baie du Mont-Saint-Michel.—Mém. Maîtrise, Univ. Rennes, dactyl: 43 pp.

FOUILLET, P., 1986. Evolution des peuplements d'Arthropodes des schorres de la baie du Mont-Saint-Michel. Influence du pâturage ovin et conséquences de son abandon. Thèse doct. 3ème cycle, Ecologie, Université de Rennes I: 330 pp.

GEHU, J.M., 1979. Étude phytocoenotique analytique et globale de l'ensemble des vases et prés salés et saumâtres de la façade atlantique française. Rapport de synthèse. Ministère de l'Environnnement et du Cadre de Vie, Mission des Études et de la Recherche. Faculté de Pharmacie, Université de Lille II et Station de Phytosociologie, Bailleul: 514 pp.

GUILLAUMONT, B. & D. HAMON, 1987. Golfe Normano-Breton. Carte morphobiosédimentaire de la zone intertidale. Côte ouest Cotentin et Baie du Mont-Saint-Michel. Notice explicative.—IFREMER, DERO-86. 27-EL: 50 pp.

GUILLON, L.M., 1980. Les moutons de prés salés en Baie du Mont-Saint-Michel. Rapport Université de Rennes: 121 pp.

——, 1984a. Les schorres de la Baie du Mont-Saint-Michel. Unités de végétation et facteurs du milieu. Rapport Fonctionnement des Systèmes Ecologiques de la Baie du Mont-Saint-Michel. Ministère de l'Environnement. Muséum National d'Histoire Naturelle et Université de Rennes I. Laboratoire d'Evolution des Systèmes Naturels et Modifiés. Ecole Pratique des Hautes Études. Laboratoire de Géomorphologie: 78 pp.

——, 1984b. Carte de végétation et notice explicative des schorres de la Baie du Mont-Saint-Michel. Rapport Fonctionnement des Systèmes Ecologiques de la Baie du Mont-Saint-Michel. Ministère de l'Environnement. Muséum National d'Histoire Naturelle et Université de Rennes I. Laboratoire d'Evolution des Systèmes Naturels et Modifiés. Ecole Pratique des Hautes Études. Laboratoire de Géomorphologie: 8 pp.

GUILLON, L.M. & C. LEGENDRE, 1981. Baie du Mont-Saint-Michel. Maintien du caractère maritime aux abords du Mont. Étude d'impact de l'arasement de la digue de Roche-Torin. Hydraulique. Sédimentologie. Faune et flore. Rapport Ministére de l'Environnement, D.R.A.E., D.D.E. de la Manche, Muséum National d'Histoire Naturelle: 73 pp.

HOFFMAN, J.S., 1984. Estimates of future sea level rise. In: M.C. BARTH & J.G. TITUS. The greenhouse effect and sea level rise. Van Nostrand Reinhold, New York.

HOFFMAN, J.S., D. KEYES & H.G. TITUS, 1983. Projecting future sea level rise: methodology, estimates to the year 2100, and research needs. Washington D.C.: Office of Policy and Resource Management, U.S. EPA, 2nd edition.

LABORATOIRE CENTRAL D'HYDRAULIQUE DE FRANCE, 1971, 1977, 1988. La baie du Mont-Saint-Michel: études

sédimentologiques et hydrologiques. Rapports inédits, Maisons-Alfort.
——, 1979. Étude de l'envasement des parcs ostréicoles de Cancale. Rapport inédit, Maison-Alfort.
——, 1982. Catalogue sédimentologique des côtes françaises. T. 3. De la Baie de Seine au Mont-Saint-Michel. Ministère des Transports, rapport inédit.
LARSONNEUR, C., 1973. Tidal deposits, Mont-Saint-Michel Bay, France. In: R. GINSBURG. Springer-Verlag, New York: 21-30.
——, 1980. La Baie du Mont-Saint-Michel: un modèle de sédimentation en zone tempérée. In: A. KLINGEBIEL & C. LARSONNEUR. Modèle de sédimentation littorale actuelle en zone tempérée. La façade maritime française de l'Atlantique à la Manche. Bull. de l'Inst. Géol. Bas. d'Aquit., Bordeaux, n° 27, 113-164. Excursion 130A du 26ème Congr. Géol. Inter., Paris, 1980: 23-32.
——, 1982. La Baie du Mont-Saint-Michel: un modèle de sédimentation en zone tempérée. In: C. LARSONNEUR & A. L'HOMER. La Baie du Mont-Saint-Michel. Voyage d'étude. 15-18 septembre 1982. Rapport Association des Sédimentologistes Français: 8-77.
LARSONNEUR, C. & P. BARTH, 1988. Extraction de tangues en Baie du Mont-Saint-Michel. Rapport Université de Caen, Centre Régional d'Etudes Côtieres, Lab. de Géologie Marine, D.R.A.E. de Basse- Normandie: 31 pp.
LAUTRIDOU, J.P. & M.T. MORZADEC, 1982. L'évolution pléistocène à flandrienne de la baie. In: C. LARSONNEUR & A. L'HOMER. La Baie du Mont-Saint-Michel. Voyage d'étude. 15-18 septembre 1982. Rapport Association des Sédimentologistes Francais: 13-21.
LE RHUN, J., 1982. Etude physique de la Baie du Mont-Saint-Michel. Thèse de 3ème cycle, Université de Paris I, Panthéon Sorbonne: 227 pp.
LEFEUVRE, J.C., 1985. Des observatoires des changements écologiques, économiques et sociaux en zone rurale. In: Actes du Colloque 'Recherches sur l'environnement rural. Bilan et perspectives'. Rapport C.N.R.S.-P.I.R.E.N.: 1-4.
LEFEUVRE, J.C. & N. LANDRE, 1974. Inventaire des richesses naturelles de la Baie du Mont-Saint-Michel. Doc. Ministère de la Protection de la Nature et de l'Environnement et Société pour l'Etude et la Protection de la Nature en Bretagne. Ed. Bureau d'Etudes S.E.P.N.B.: 32 pp.
LEFEUVRE, J.C. & Y. PARIS, 1982. Aménagement intégré du littoral européen. Cas particulier de la lagune de Venise. Rapport Commission des Communautés européennes, Comprensorio dei communi della laguna e dell' entroterra di Venezia: 109 pp + annexes.
LEGENDRE, C., 1984. La pêche artisanale sur le domaine intertidal de la Baie du Mont-Saint-Michel. Rapport Fonctionnement des Systèmes Ecologiques de la Baie du Mont-Saint-Michel. Ministère de l'Environnement. Muséum National d'Histoire Naturelle et Université de Rennes I. Laboratoire d'Evolution des Systèmes Naturels et Modifiés: 121 pp.
——, 1985. Aspects qualitatif et quantitatif de la microflore en Baie du Mont-Saint-Michel. Rapport C.E.E. Environnement, I.R.I.E.C.: 61 pp.
LEVASSEUR, J., 1987. Végétation phanérogamique des marais maritimes. Vol. 4. Estran et zones humides. Etude régionale intégrée du Golfe Normano-Breton. C.C.E., IFREMER. DERO-EL: 110-182.
L'HOMER, A., 1974. Liaison morphologie-sédimentation dans la zone intertidale de la baie du Mont-Saint-Michel. In: 2° R.A.S.T., Pont-à-Mousson.
——, 1981. Bilan de sédimentation-érosion en baie du Mont-Saint-Michel depuis 1857. In: Séminaire 'La gestion régionale des sédiments'. Documents du BRGM, n° 30: 245-252.
MAHEO R. & P. BORET, 1983. Recherches sur les oiseaux d'eau séjournant dans la Baie du Mont-Saint-Michel. Rapport Office National de la chasse et C.R.E.B.S., Université de Rennes I: 45 pp.
MIGNIOT, C., 1982. Le problème de l'insularité du Mont. In: C. LARSONNEUR & A. L'HOMER. La Baie du Mont-Saint-Michel. Voyage d'étude. 15-18 septembre 1982. Rapport Association des Sédimentologistes Français: 68-73.
MORZADEC-KERFOURN, M.T., 1974. Variations de la ligne de rivage armoricaine au quaternaire. Analyses polliniques de dépôts organiques littoraux.—Mém. Soc. géol. minéral. Bretagne **17:** 1-208.
——, 1975. Evolution paléogéographique du Marais de Dol de Bretagne (Ille et Vilaine) durant le Flandrien.—Bull. Soc. Géol. minéral. Bretagne, série C, **7:** 49-51.
——, 1977. La baie du Mont-Saint-Michel et le Marais de Dol. In: S. DURAND. Bretagne, guides géologiques régionaux. Ed. Masson: 28-30.
——, 1985. Variations du niveau marin à l'Holocène en Bretagne (France).—Eiszeitalter u. Gegenwart, Hanovre **35:** 15-22.
NIKODIC, J. & J.M. SORNIN, 1982. Modifications apportées à la dynamique sédimentaire par les aménagements. Rôle de la biodéposition. In: C. LARSONNEUR & A. L'HOMER. La baie du Mont-Saint-Michel. Voyage d'étude. 15-18 septembre 1982. Rapport Association des Sédimentologistes Français: 58-61.
ODUM, E.P., 1980. The status of three ecosystem-level hypotheses regarding salt marsh estuaries: tidal subsidy, outwelling and detritus-bassed food chains. In: V.S. KENNEDY. Estuarine perspectives. Academic Press, New York: 485-495.
PIRAZOLLI, P.A., 1986. Secular trends of relative sea level changes indicated by tide gauge records.—J. coastal Res. **1:** 1-26.
RETIÈRE, CH., 1979. Contribution à la connaissance des peuplements benthiques du golfe normanno-breton. Thèse d'Etat, Université de Rennes: 370 pp.
ROBIN, G.Q., 1986. Changing the sea level. In: B. BOLIN, B. DÖÖS, J. JÄGER & R.A. WARRICK. The greenhouse effect, climatic change and ecosystems. S.C.O.P.E. 29.
SCHRICKE, V., 1983. Distribution spatio-temporelle des populations d'Anatidés en transit et en hivernage en baie du Mont-Saint-Michel en relation avec les activités humaines. Thèse de 3ème cycle, Ecologie, Université de Rennes: 299 pp.

CONSEQUENCES OF SEA LEVEL RISE: IMPLICATIONS FROM THE MISSISSIPPI DELTA*

J.W. DAY, JR[1] and P.H. TEMPLET[2]

[1] *Coastal Ecology Institute, Center for Wetland Resources, Louisiana State University, Baton Rouge, Louisiana 70803, U.S.A.*

[2] *Louisiana Department of Environment Quality, Box 44091, Baton Rouge, LA 70804, U.S.A.*
and
Institute for Environmental Studies, Louisiana State University, Baton Rouge, LA 70803, U.S.A.

ABSTRACT

Sea level rise is expected to increase worldwide over the coming decades, and its impacts are beginning to be felt in many areas. Two major direct impacts of sea level rise are submergence and salinity increase. Historically, the Mississippi River Delta has experienced a relative sea level rise (RSLR) and thus serves as an analogy or model for what can be expected elsewhere. Despite long term RSLR primarily due to subsidence, the Mississippi has grown in size over the past several thousand years since eustatic sea level stabilized. Within this century, the net positive growth rate has been reversed and net wetland loss rates as great as 100 km^2 per year have occurred. Much of the wetland loss is associated with human activities that have resulted in a reduction of sediment input to wetlands. Because of this reduction, vertical accretion of the wetland surface is less than RSLR and plants are disappearing due to waterlogging and salinity increase. The resulting loss of wetland plant vigor complicates the problem because the production of plant roots is an important component of soil formation and vertical accretion of the wetland surface.

Two important points to consider in addressing the problem of sea level rise is that there is often a lag time of decades before the response of the natural system to sea level rise becomes evident and that changes in the natural system may be slow at first and then accelerate. The institutional response in Louisiana is complicated, but many of the actions taken may be detrimental in the long run. A common response to rising water levels will be flood control. But, in the Mississippi Delta, dikes along the river have greatly restricted sediment input to wetlands. Additionally, semi-impoundments with water control structures are being considered to protect wetlands from increasing water levels and salinity increases. Many of these depend on gravity drainage, but in a microtidal area such as the Gulf coast, gravity drainage has a finite life span due to rising water levels. Land ownership patterns also complicate a comprehensive approach to the problem due to units selected for management and conflict between short-term and long-term benefits. We conclude that coastal wetlands can be managed to survive rising sea level but that only comprehensive, integrated, long-term planning can effectively deal with the problem of sea level rise. The principle of ecological engineering, where the energies of nature are used as much as possible, should play an integral part of any management plan. Because deltas are probably one of the most threatened of coastal landscapes, an early warning monitoring system is recommended for selected deltas of the world.

1. INTRODUCTION

Numerous reports have recently emphasized the potential impact of global warming trends on future sea level rise in coastal areas (*e.g.* HOFFMANN *et al.*, 1983; Nummedal, 1983). Predictions are that many low lying coastal areas, especially those dominated by wetlands, will be flooded as sea level rises from 1 to 3 meters over the next century (Table 1). Current evidence indicates that sea level rise is leading to wetland loss in a number of coastal areas (CLARK, 1986; HACKNEY & CLEARY, 1987; KANA *et al.*, 1986; STEVENSON *et al.*, 1988). Thus, there is a critical need to consider both the impacts of sea level rise on coastal wetlands and the possible policy responses to these impacts. The Mississippi River Delta area can provide valuable information on both impacts and responses. This area recently has been experiencing a 'relative sea level rise' of about a meter

This article has also been published in Coastal Management* **17: *241-257,*

J.J. Beukema et al. (eds.), Expected Effects of Climatic Change on Marine Coastal Ecosystems, 155–165.

Table 1
Scenarios of Future Sea Level Rise (in cm)[a]

Scenario	Year						
	2000	2020	2025	2050	2075	2080	2100
High[b]	17.1		54.9	16.7	211.5		345.0
Midrange high[b]	13.2		39.3	78.9	136.8		216.6
Midrange low[b]	8.8		26.2	52.6	91.2		144.4
Low[b]	4.8		13.0	23.8	38.0		56.2
Current trends[b] (national)[c]	2.5		5.7	8.8	11.9		15.0
Current trends[d] (Louisiana)[e]	21.3		52.5	83.8	115.0		146.3
Current trends[d] (Louisiana)[f]		72				142	

[a] After Hoffman *et al.*, 1983; Templet and Meyer-Arendt, 1988.
[b] Hoffman *et al.*, 1983.
[c] Midrange of values cited in Hoffman *et al.*, 1983.
[d] Templet and Meyer-Arendt, 1988.
[e] Straight-line projections of relative sea level rise values (1.25 cm/yr) cited in Baumann and DeLaune, 1982.
[f] Midrange of values cited in Nummedal, 1983 (includes a 1-cm/yr eustatic sea level rise for the 1980–2020 period).

per century caused mainly by regional subsidence. In this paper, we will discuss the effects of this rise on coastal ecosystems in Louisiana and some responses to the problems created by this rise.

2. WHAT WILL BE THE IMPACTS OF SEA LEVEL RISE?

In many coastal areas there will be at least two major direct impacts on coastal ecosystems of sea level rise: **submergence** and **salinity** increase. For low lying wetland and terrestrial areas, progressive submergence causes poor drainage and increases waterlogging of soils. These, in turn, lead to lowered plant productivity and ultimately vegetation death. This latter factor is very important because production of biomass by plants plays an important role in vertical accretion (Fig. 1) accounting for approximately half of the biomass in some salt marsh types. In many coastal areas, freshwater input may remain constant while saltwater input increases thereby increasing salinity; also, saltwater intrusion may occur in formerly freshwater areas. These impacts will be most pronounced in large, near-sea level areas, such as deltas and lagoon-like systems. Because these areas are very important for fisheries and agriculture, the impact on local economies may be pronounced.

3. THE MISSISSIPPI RIVER DELTA AS A ANALOGY

The Mississippi River Delta is a large area of lakes, bays, near-sea level wetlands, and low-lying uplands. It serves as an analogy for sea level rise question because it has been subject to a significant historical 'relative' sea level rise. We use the term analogy because, while the sea has not been rising rapidly, the land has been sinking due to regional subsidence. While the average worldwide eustatic rise is about 0.15 cm per year (GORNITZ *et al.*, 1982); the combined relative rate in the Louisiana deltaic region is 1.1 to 1.3 cm per year (BOESCH *et al.*, 1983). Thus, the eustatic component is currently less than 15%, but this percentage may increase rapidly as eustatic sea level rise accelerates. However, we believe that the impacts on natural and social systems are generally the same whether they are caused by the land sinking or sea level rising.

Three major reasons make the Mississippi Delta an excellent place to consider the effects of sea level rise:

- 1. Relative Sea Level Rise (RSLR)

A number of studies have shown that there is a RSLR in the Mississippi deltaic plain, which is currently averaging about 1.2 cm per year (HATTON *et al.*, 1983; DELAUNE *et al.*, 1978; BAUMANN *et al.*, 1984; BAUMANN & DELAUNE, 1982). Deep sedimentary deposits indicate that RSLR has occurred over

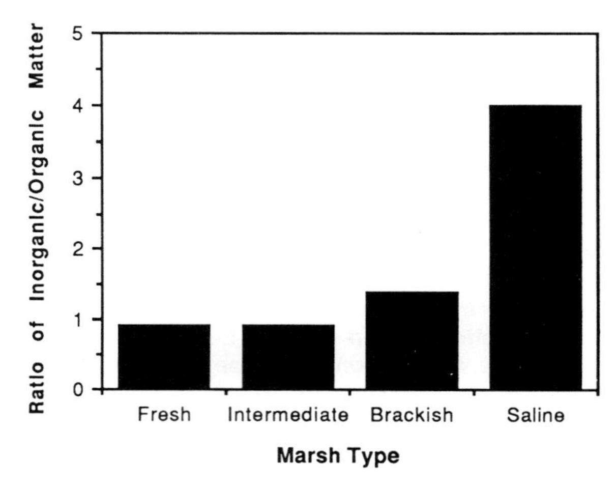

Fig. 1. Ratio of sediment type *vs.* marsh type (HATTON *et al.*, 1983).

Table 2
Summary of Marsh Accretion Rates Measured in Coastal Louisiana and Along the U.S. East Coast[a]

Location	Marsh type	Marsh accretion rate (mm/yr)	Mean sea level rise (mm/yr)	Reference
Louisiana Deltaic Plain	Freshwater		11.0	Hatton *et al.*, 1983
	streamside	10.6		
	backmarsh	6.5		
	Intermediate (*Spartina patens*)			Hatton *et al.*, 1983
	streamside	13.5		
	backmarsh	6.4		
	Brackish (*S. patens*)			Hatton *et al.*, 1983
	streamside	14.0		
	backmarsh	5.9		
	Saline (*S. alterniflora*)		13.0	DeLaune et al., 1978; Baumann, 1980
	streamside	13.5		
	backmarsh	7.5		
Chenier Plain	Salt-brackish (*S. patens*)	7.0	12.0	Baumann and DeLaune, 1982
Georgia	*S. alterniflora*	3-5		Summarized by Hatton *et al.*, 1983
Delaware	*S. alterniflora*	5.0–6.3	3.8	Summarized by Hatton *et al.*, 1983
New York	*S. alterniflora*	2.5–6.3	2.9	Summarized by Hatton *et al.*, 1983
Conn.	*S. alterniflora*	8–10	2.5	Summarized by Hatton *et al.*, 1983
	S. patens	2–5		
Mass.	*S. alterniflora*	2–18	3.4	Redfield, 1972
Maryland	Blackwater	2.6	3.9	Stevenson *et al.*, 1985
	Nanticoke	6.1	3.2	Stevenson *et al.*, 1985

[a] Modified from Boesch *et al.*, 1983 and Stevenson *et al.*, 1986.

geologic time periods. Subsidence accounts for 85 to 90% of the current RSLR, although eustatic sea level rise is predicted to be a larger percentage in the coming decades as true sea level rise accelerates (NUMMEDAL, 1983). The important point, from both ecological and social points of view, is that there has been long term RSLR in the Mississippi River Deltaic Plain.

- 2. Long Term Growth of the Delta

Sea level stabilized at its present level about 5000 to 7000 years ago after being about 100 m lower during the last glaciation (MILLIMAN & EMORY, 1968). Since that time, sedimentation from the Mississippi River has created a large deltaic plain of about $5x10^3$ km^2 of which over $2x10^3$ km^2 are wetlands (KOLB & VAN LOPIK, 1958). The great majority of these wetlands are within 1 m of sea level. The rest of the deltaic plain is made up of shallow aquatic systems and low (up to 5 to 6 m) uplands. Thus, despite continuous RSLR, the delta has experienced an average net areal growth of ~4 km^2 per year over the past 5000 years. The land was formed by series of shifting delta lobes (COLEMAN & GAGLIANO, 1964; FRAZIER, 1967).

- 3. Recent High Land Loss in the Mississippi Delta

Over the past several decades, this long-term pattern of net land gain has been reversed and reports indicate that the rate of land loss accelated to values as high as 100 km^2 per year (GAGLIANO *et al.*, 1981; SASSER *et al.*, 1986). This dramatic reversal is apparently the result of a number of factors, many of which are human activities, all of which contribute to the inability of wetlands to maintain surface elevation in the face of rising water levels (CRAIG *et al.*, 1979; TURNER *et al.*, 1982; DEEGAN *et al.*, 1984; SASSER *et al.*, 1986). In other words, RSLR is greater than annual vertical accretion and is cumulative, leading to the increasing loss rates (HATTON *et al.*, 1983; MITSCH & GOSSELINK, 1986) and a time lag before the affects are evident. There is a time lag because root production by plants can contribute significantly to vertical accretion. This accretion/subsidence imbalance as a reason for land loss is not unique to the Mississippi River Delta. It is happening to a lesser degree on the eastern shore of Chesapeake Bay in Maryland and in other locations along the East Cast of the United States (STEVENSON *et al.*, 1988; Table 2) as well as in the Nile Delta (STANLEY, 1988). STEVENSON *et al.* (1985) concluded that, while most marshes on the East coast were accreting rapidly enough to stay ahead of sea level rise, this was not the case in Louisiana where accretionary balances were negative (subsidence exceeds accretion). The accretion deficit is caused by a number of factors which will be addressed in the next section.

Table 3
Sediment Accretion Deficit[a]

Marsh Type	Vertical Accretion deficit (mm/yr)	Inorganic Areal Accretion Deficit ($g.m^{-2}yr^{-1}$)
Fresh	4.1	409
Intermediate	7.1	1391
Brackish	8.1	2476
Saline	6.0	960

[a] Modified from Hatton, 1983 and Templet and Meyer-Arendt, 1988.

4. CAUSES OF WETLAND DETERIORATION IN THE MISSISSIPPI DELTA

4.1. LACK OF SEDIMENT INPUT

There are two sources of allochthonous sediments that accrete on the surface of wetlands in the Mississippi River Delta. One sources is the direct input of riverine sediments during the annual spring flood. This source has been eliminated for most of the coastal zone by dikes and jetties along the Mississippi River that extend past its mouth. In addition, a number of distributaries that carries sediments to coastal wetlands have been dammed. This virtual elimination of sediment input has led to a vertical accretion deficit of 4.1 to 8.1 mm per year, depending on marsh type. Vertical accretion can be converted to an areal deficit by using the ratio of organic to inorganic (mineral) material and the bulk density. The mineral deficits are in the range of 400 to 2500 $g.m^{-2}.yr^{-1}$ depending on marsh type (Table 3).

Additionally, the Mississippi River carries reduced sediments because of the upstream diversion of sediment by dams, better conservation practices, and the upstream mining of sand and other sediments. In one area of the coast where there is significant sediment input, Atchafalaya Bay, a new sub-delta is growing and loss of existing wetlands has been slowed or reversed (VAN HEERDEN & ROBERTS, 1980).

A second source of sediment is the deposition of resuspended bay-bottom sediments. These are resuspended during high energy events, such as winter frontal passages and tropical storms, and deposited on the surface of wetlands (BAUMANN *et al.*, 1984). These resuspended materials are now the major source of new mineral sediments for much of the coastal zone because riverine sediments have been prevented from reaching Louisiana's coastal wetlands. Resuspended sediments alone do not appear sufficient to maintain surface elevation over a wide area, but they greatly slow wetland loss and may be sufficient to maintain streamside marshes. The Dutch have used brush fencing baffles along the Wadden Sea to encourage settling of resuspended sediments and prevent subsequent sediment loss (WAGRET, 1968; BOUWSEMA *et al.*, 1986).

In the Mississippi River Delta, the input of resuspended sediments has been reduced by canal construction. Over the past 30 to 50 years, over 15,000 km of canals have been constructed for navigation and access and drainage, but mostly for oil and gas exploration and production: 70 to 80% of Louisiana's coastal management permits issued are for oil and gas activities (DNR, 1982). This network of canals has changed the regional hydrology of the coastal zone. Canals are generally straight, with a low spoil bank created by depositing the material excavated during construction along one or both sides of the canal. Spoil banks are generally higher than mean high water; thus, flow across the wetlands surface is altered, and the input of resuspended sediments is reduced. Studies have shown that canals and associated spoil banks alter hydrology and increase waterlogging, reduce productivity of wetlands, cause saltwater intrusion, reduce sediment input to wetlands, and increase the deterioration of natural channels (TURNER *et al.*, 1982; SWENSON & TURNER, 1987; TURNER & CAHOON, 1987). A number of studies have correlated canal density to wetland loss (GRAIG *et al.*, 1980). Because of these impacts of canals, it has been estimated that canals and associated spoil banks directly account for 16 to 20% of direct wetland loss over the past several decades and caused an additional 30 to 45% of wetland loss due to indirect effects.

4.2. SALINITY CHANGES

The dikes along the Mississippi River have stopped most direct fresh water, sediment, and nutrient input into the coastal zone. The canals have allowed more rapid movement of salt water into fresher areas. The intrusion of salt water into abandoned delta lobes is a natural process, but diking, canalization, and subsidence have greatly accelerated it. Salt water can kill or reduce the productivity of fresher vegetation directly causing loss of wetlands, although this is not thought to be the primary cause of plant death or land loss. Salinity increases are a response to subsidence and thus are a symptom of land loss, not a direct cause.

4.3. DEATH OR LOWERED PRODUCTIVITY OF WETLAND VEGETATION

The health of vegetation is a critical factor in maintaining surface elevation because biomass production (roots and above-ground material) can contribute more than 50% of vertical accretion. Fresh, intermediate, and brackish marsh sediments are approximately one-half organic (Fig. 1). Because sediment and water carry nutrients into the wetlands, the dikes also prevent needed nutrients from reaching the vegetation and therefore reduce productivity. In addition, thick stands of vegetation have greater trapping efficiency and more sediment. Studies in the Netherlands have shown that a vertical accretion doubles or triples when vegetation becomes established (DIJKEMA & WOLFF, 1980).

In the Mississippi River Delta a number of factors have combined to reduce the role of wetland vegetation in augmenting vertical accretion. Coastal wetland vegetation grows within a rather narrow elevation range that is related to the local tide range (MITSCH & GOSSELINK, 1986). Thus, if subsidence lowers the surface elevation below this range, vegetation death occurs. There is a large body of literature showing the impact of waterlogging and submergence on wetland plant production and health (*e.g.* MENDELSSOHN *et al.*, 1981, 1982). In the Mississippi River Delta, an important source of new nutrients for salt marsh vegetation accompanies sediment input (DELAUNE *et al.*, 1983). Therefore, a reduction of sediment input lowers productivity and hastens wetland deterioration in at least two ways.

In summary, in the Mississippi River Delta a number of human impacts have interacted with natural processes to cause high rates of wetland loss. These factors contribute to the inability of wetlands to maintain surface elevation in the face of RSLR. Input of both riverine and resuspended sediments to wetlands has been greatly reduced. Wetland productivity has been reduced or wetlands have been killed by submergence and the accompanying saltwater intrusion and this results in a reduction of accretion due to biomass production. It is important to stress that most wetland loss occurs when RSLR is greater than the vertical accretion of the wetland surface and gradual submergence of the wetland takes place. In areas where sediment supply is adequate, wetlands have persisted in the face of RSLR rates greater than 1.0 cm per year (BAUMANN *et al.*, 1984; DELAUNE *et al.*, 1983). Thus an important question is what are the factors which lead to a reduction of sediment input and vertical accretion. As discussed above, activities such as flood control levees and canal construction have significantly reduced sediment input to wetlands and have led to lowered plant productivity.

An extremely important point when considering the effects of sea level rise (and institutional response to it) is that there seems to be a considerable time lag, on the order of decades, before effects become apparent. For example, dikes along the lower Mississippi were substantially completed in the 1930's and widespread canal construction had taken place by the 1960's. Although the effects of subsidence were understood over 30 years ago (KOLB & VAN LOPIK, 1958), it was not until the mid-1970's that the problem of land loss began to be appreciated and not until the mid 1980's that the role of RSLR was beginning to be understood by decision-makers. The rate of wetland loss was low initially but increased very rapidly (GAGLIANO *et al.*, 1981; SASSER *et al.*, 1986).

5. SPATIAL RESPONSES OF THE NATURAL SYSTEM

Recent work indicates that there are distinct spatial patterns of wetland loss in the Mississippi Delta (TURNER & CAHOON, 1987; MAY & BRITSCH, 1987). Some wetland loss occurs as a result of wave erosion of marsh shorelines but this generally accounts for less than 10% of total loss (TURNER & CAHOON, 1987). Most wetland loss occurs in inland wetlands. Areas distant from lakes, bays, and tidal streams as well as in areas of high canal densities deteriorate first because of sediment deficits and waterlogging. Thus, wetland loss is greatest in isolated, stagnant, waterlogged, sediment-starved areas with an ever-increasing accretion deficit. Small ponds open up when vegetation dies. These ponds become larger and coalesce into lakes and bays over time. Such pond formation is especially apparent in areas with high canal density. Thus, the results from the Mississippi Delta indicate that rising sea level will not simply lead to an advance of the edge of the sea. This is especially true in broad near-sea level areas similar to the Mississippi Delta that will likely be most affected by sea level rise. These results suggest that traditional sea defenses such as levees, seawalls, and dikes, will not stop such land loss and, indeed, will probably worsen sediment deficits by eliminating or reducing resuspended sediments.

6. THE INSTITUTIONAL RESPONSE IN LOUISIANA

Coastal and wetland management in Louisiana is very complex. Federal, state, and local public agencies and private groups (landowners, sportsmen, commercial fishermen, trappers, nonrenewable resource exploiters, and conservationists) are involved in the process. In addition, management is performed for a variety of reasons (aquaculture, waterfowl enhancement, land access control, fur mammal management and mineral and timber production). Only recently has management been focused on the problem of land loss and RSLR, and many of the techniques now being used or proposed to address the land loss problem were originally developed for some other wetland management purpose. In addition to activities directed specifically at coastal and wetland management, other activities such as flood protection and minerals extraction also have a profound impact on wetland loss. Because it is likely that some of these activities such as flood control will be used in response to rising water levels, it is instructive to consider their impacts on wetland loss in Louisiana.

Considering the present rate of subsidence and the prospects for future increases in eustatic sea level rise, it appears that many management actions may be contrary to what is necessary to address the wetland loss problem in the long-term. A number of management approaches, however, have become common because they produced short-term benefits or were not immediately or obviously harmful in the short-term (10 to 20 years). This is because the effects of slow submergence on wetlands are gradual and cumulative *i.e.*, the lag time in the response by the natural system to RSLR is on the order of decades. Thus, the effects of deleterious management practices on wetland loss become apparent only slowly. We will now discuss several management practices.

6.1. FLOOD CONTROL

Flood control in the Mississippi River Delta area must contend with the threat of flooding from the River during high flow and from the sea during tropical storms and to the gradual increase in water levels due to RSLR. To protect developed areas, the Mississippi River has been almost completely 'walled in' by dikes. Early flood control along the River was accomplished by ring (encircling) dikes, and the city of New Orleans is still protected by ring dikes because it faces flooding threats from both the River and the sea. The linear dikes bordering the River prevented the introduction of sediments and freshwater into much of the coastal zone and resulted in the problems already discussed. Protection from flooding also stimulated agricultural and urban development to move into low-lying areas. What is needed now is the reintroduction of fresh water and sediments into the coastal areas (VIOSCA, 1927; GAGLIANO *et al.*, 1975, 1976, 1981; COASTAL RESTORATION TECHNICAL COMMITTEE, 1988). However, development makes this difficult, and the widespread network of canals (with attendant spoil banks) within the coastal/wetland system will retard the easy movement of sediment-laden water over wetlands. In retrospect, one could envision a number of ways in which developed areas could have been protected from flooding (such as greater use of encircling dikes and construction of additional floodways) and still provide introduction of freshwater and sediments into the coastal zone. A clear lesson from the Mississippi River Delta is that flood protection in response to sea level rise should not diminish the introduction of freshwater and sediments to coastal wetland areas.

Continuing subsidence and ASLR have caused many low lying wetland areas to experience increase flooding. Gravity drainage systems are still being proposed that involve the construction of many kilometers of canals that make problems of land loss (and water quality) worse. But institutional inertia and reliance on the approaches used in the past make change difficult. For example, Louisiana recently approved a gravity drainage system in the Lake Verret region of the coast, despite the fact that evidence was presented that water levels in the area would likely rise more than a meter (CONNOR *et al.*, 1986) because of the combined effects of subsidence and eustatic sea level rise over the 50-year life of the project. The lesson here is that gravity drainage systems built to drain areas at low water level will become increasingly unworkable as RSLR increases. This has been true for numerous impoundment projects worldwide where localized subsidence has been caused by peat oxidation (DAY *et al.*, 1986).

6.2. IMPOUNDMENT AND SEMI-IMPOUNDMENT

Wetland impoundment (encirclement by dikes for some kind of water management) has been carried out for over 100 years in the Mississippi Delta. New Orleans first developed on the higher natural levees of the River. However, by the late 19th century, the city began to spread into adjacent wetlands. The city is now completely surrounded by dikes and protected by an extensive and expensive drainage system. Despite this, the city experiences severe floodings from extremely heavy short-term rainfall that occurs along the Louisiana coast. During the second half of the 19th century and the early part of this century, extensive areas of wetlands were reclaimed for agriculture. Most of these failed from a combination of subsidence (caused by regional sinking and oxidation of peat) and heavy rains (OKEY, 1918; TURNER & NEILL, 1983). Most are visible today as large rectangular ponds in the coastal marshes.

More recently, semi-impoundment has become common as a technique for wetland management. In this approach, wetland areas are surrounded by low dikes, and water movement is controlled by water control structures, such as weirs, sluices, and gates. This has been done since the 1950's for the management of fur mammals and waterfowl. Recently, semi-impoundment has become much more widespread as a form of marsh management specifically to address the problem of land loss. One of the main objectives is to control saltwater intrusion and prevent death of wetland vegetation. In the past, such plans have been proposed even in areas of salt marsh vegetation where salinity is clearly not a problem. A main question concerning marsh management plans with impoundments is that such features retard the introduction of resuspended sediments and nutrients so important to maintain marsh elevation. Any future plans to introduce riverine sediments will be hampered by the dikes surrounding these management areas. Even though there are questions about the effectiveness of this kind of marsh management, there are plans for projects that include a considerable portion of the coastal wetlands of Louisiana (DAY *et al.*, 1986). The cumulative or long- term impacts of such plans have not been adequately addressed, and it is yet to be demonstrated that such projects will retard land loss. Most of these projects utilize gravity drainage and, thus, will experience loss of effective drainage as RSLR proceeds.

6.3. LAND OWNERSHIP

A complicating problem of wetland management in the Mississippi Delta and of dealing with the affects of sea level rise is land ownership. Land ownership patterns are characterized by a mosaic of private and public properties which rarely coincide with natural drainage basins. Thus, management plans are most often formulated for management units based on ownership, even though most resource managers would agree that management based natural landscape units such as a drainage basin would be more appropriate.

Most of the Mississippi Delta wetlands are privately owned, and the objectives of landowners may not necessarily coincide with those of public agencies. Although wetlands have high ecological value, only part of this may be transformed into economic value for the landowner. Thus a landowner may opt for management for resources which have an immediate economic return (such as waterfowl or furbearers) rather than for resources whose values accrue to society in general (such as water cleansing, export of organic matter, or habitat for fishery species). A particular problem faced by private landowners in Louisiana is revenue from oil and gas production. If wetlands deteriorate into open water, revenues from any mineral production may change from the landowner to the public. Thus a landowner may feel that he is forced to choose between a course of action which ensures recognition of mineral rights and one which is best for wetland conservation. For example, one way a landowner can define property boundaries is with a system of low dikes, but these dikes may interfere with the input of sediments which nourish wetlands. An approach without dikes which maximizes sediment input and maintaines a larger area of wetlands could also lead to wave erosion of wetlands along the wetland-water interface and a net loss of property to the landowner.

Thus, questions of property rights can be an important factor when considering what to do about sea level rise. Even though there is a tendency to formulate management plans based on land ownership, it would be much better to manage based on natural landscape units. As much as possible, landowners should not be put into situations where decisions deleterious to natural resources are in their best interests. Landowners might be given some form of compensation (such as a tax break) for management which is not in their immediate economic self-interest. These points indicate the importance of coordinated planning. Issues related to land ownership have been a complicating factor in wetland management worldwide (DAY *et al.*, 1986), and this should be fully appreciated when seeking approaches to deal with wetland management affected by sea level rise.

7. IMPLICATIONS FROM THE MISSISSIPPI RIVER DELTA FOR OTHER COASTAL REGIONS

The response of wetlands in the Mississippi Delta to RSLR and the effectiveness of different management activities in mitigating the problems caused by rising water levels provide valuable guidance for understanding the effects of submergence due to accelerate eustatic sea level rise on other coastal systems and the most appropriate institutional actions. In this section we discuss and summarize the lessons from the Mississippi Delta, first in terms of the natural systems and then in terms of institutional response.

7.1. NATURAL SYSTEMS

Perhaps the most important finding that comes from a consideration of the Mississippi Delta is that coastal wetlands can withstand considerable sea level rise and not undergo deterioration. This takes place when vertical accretion of the wetland surface is equal to water level rise and for this to happen, there needs to be sediment input. The Mississippi River Delta region has experienced subsidence for the past several thousands years and still built a large deltaic plain. The critical factor is that the system remained open and dynamic, and sediments and water from the river were distributed widely over the deltaic plain. In the dynamic delta system, growth and decay took place with few or no restraints but there was net growth. Over the past several decades, the rate of RSLR has been 1 cm per year or greater. Yet, large areas of the delta have maintained their elevation due to significant inputs of sediments (BAUMANN *et al.*, 1984) and in the Atchafalaya Delta the area of wetlands is actually increasing (VAN HEERDEN & ROBERTS, 1980). Most wetland loss has occurred in areas with low sediment input (DELAUNE *et al.*, 1983). In a number of coastal areas (*e.g.*, the

Netherlands) new sediments have been used to effectively build up the surface of intertidal areas by as much as 4 cm per year (BOUWSEMA *et al.*, 1986).

Two critical problems for wetlands that result from sea level rise are submergence and salinity increase. Submergence occurs when vertical accretion of the wetland surface cannot keep pace with the rate of water level rise. A number of factors contribute to the rate of vertical accretion. Sediment input is an obvious one. In the Mississippi River Delta region, both riverine and resuspended sediments are important. Biomass production by vegetation can contribute as much to vertical accretion as mineral sediment input. Therefore, maintenance or establishment of vegetation is critically important.

Given the present RSLR rates occurring in the Mississippi Delta, there seems to be a time lag of several decades before the effects of sea level rise cause widespread wetland loss. This is because sediment input to wetlands and plant production gives rise to vertical accretion of the wetland surface. The rate of submergence is dependent on how much greater RSLR is than vertical accretion. In general, this means that the rate of submergence is slow so that losses of wetlands are slow first but then accelerate.

The effects of sea level rise will obviously be more critical where there are large near-sea level areas, such as deltas. What is not so widely understood is that coastal areas with low tidal ranges will probably be more quickly affected because the vertical distribution of vegetation is related to local tide range (BAUMANN, 1980). Thus, the impacts of sea level rise (at least on wetland ecosystems) will first become apparent in large flat areas with low tidal range. This suggests that the world's deltas, especially those with low energy, are critical areas for early consideration to deal with the problems caused by rising water levels.

7.2. MANAGEMENT CONSIDERATIONS

When planning for institutional response to counter the impacts of sea level rise on coastal ecosystems, a number of factors need to be considered, especially related to timing and appropriateness of different actions. Those responsible for management should recognize that there is likely to be a considerable time lag (on the order of decades or more) before the effects of sea level rise become clear and that the rate of change will likely accelerate. Thus, action should be taken early, even though neither the need nor the culmination of the sequences of events that lead to land loss may be readily apparent.

Initial responses may be inappropriate unless the problems are very carefully considered over the long-term. Because effects will be minimal and changes occur slowly at first, approaches to addressing the initial problems and symptoms may make the problem worse in the longer-term. As an example, semi-impoundment and reclamation may be suggested as a way to deal with sea level rise, but reduction of sediment input may further lower the rate of vertical accretion of wetlands and gravity drainage will become less and less effective. The solution to the problems caused by short-term piecemeal practices is a comprehensive planning approach which takes a long-term comprehensive view of ASLR. In addition to comprehensive planning, there is a need for strong leadership which can get difficult political and regulatory decisions made. Worldwide, countries, states, or provinces which included ecosystem management within a context of comprehensive planning have had the greatest success in maintaining and managing coastal areas.

Flood protection to combat rising waters should not eliminate widespread sediment and freshwater input to coastal wetlands. Continuous dikes, as have been structured along the Mississippi River, can reduce or eliminate most sediment and freshwater input to coastal wetlands. These inputs are needed to counter submergence and salinity increase. Alternative approaches to protecting developed areas in the coastal zone while ensuring sediment and water input, such as encircling dikes and floodways, need to be considered in comprehensive planning. Finally, it must be realized that flood control which makes use of gravity drainage will become increasingly unworkable unless the surface of the land is built up. If areas where water levels are initially controlled by gravity drainage are to be maintained, then much more expensive pumped drainage will become necessary.

In a number of ways, problems associated with land ownership may complicate efforts to address the effects of sea level rise. Politically and legally, it is much easier to implement management plans based on landscape units defined by ownership. However, ecosystem management, and this especially the case for wetlands, is most effectively done on natural landscape units such as the drainage basin. Different landowners will likely have different objectives in the management of their lands. Private landowners will naturally prefer to manage for resources which provide a direct economic return (such as furbearers or waterfowl) rather than the more nebulous public good. All of this points to the need to include the land ownership issue up front as part of comprehensive planning. And in areas where there is private ownership, these landowners must be involved in comprehensive planning.

Agriculture in low-lying areas presents special problems. Can such areas be maintained with increasing sea level? The construction of dikes and drainage of peat soils will aggravate subsidence and make the problem worse in the long-term. A rotating agricultural system in which riverine sediments are introduced into specific fields might maintain accretion and help fertilize crops. High biomass crops may also be important in maintaining vertical accretion if organic matter is allowed to accumulate. Erosion control techniques, such as cover crops, fallow areas and no-till agriculture, can help reduce erosion and increase accretion. As indicated earlier, gravity drainage is effective for a relatively short period of time where water levels are increasing. Pumped drainage is normally not feasible for agricultural areas. It has been sustained only for highly intensive agriculture in developed countries such as the Netherlands.

Maximum use should be made of natural energies, such as vegetation productivity, winds, river currents, and tides. For example, currents should be used to distribute sediments. Vegetation should be used to enhance vertical accretion. This is the principle of ecological engineering where small amounts of fossil fuel energies are used to channel much larger flows of natural energies (ODUM, 1971). An example of the approach for using natural energies lies in ODUM & DIAMOND'S (1985) energy analysis of the Mississippi River Basin, which showed that, while energy savings accrued to navigation and development caused by dikes on the river, the overall losses of system energy were greater. The implication is that the state and nation would benefit more if the River's water and sediments were used constructively to nourish the wetlands while maintaining navigation and protecting development with ring levees. The Dutch system of building marshes with brush fences mentioned earlier is another example of using natural energies to cause accretion.

Coordinated monitoring and planning should begin in a number of coastal areas of the world to gather data on the effects of sea level rise and to prepare institutional response to it. In this way, the generality of the findings from the Mississippi Delta can be tested. In many cases, there are on-going programs that could be integrated and coordinated. Some possible areas include deltas on the following rivers and areas: Mississippi, Grijalva (Mexico), Ebro, Po, Nile, Camargue, Danube, the Netherlands, Bangladesh, and Maryland (Chesapeake Bay). A number of factors may be measured, including vertical accretion rates, rates of water level increase, vegetation response (productivity, tree ring growth, physiological stress indicators) habitat change over time (mapping), shoreline retreat, and ponding (appearance of new ponds, on the wetlands). A powerful tool, spatial modeling, has evolved over the past several years which can help in predicting the spatial response of low lying areas to sea level rise. We have used spatial modeling to consider the effects of ASLR and the impacts of different management scenarios in coastal Louisiana (COSTANZA *et al.*, 1986).

7.3. LONG-TERM INTEGRATED PLANNING

Because sea level rise is likely to occur over the globe for the forseeable future, it is imperative that it be included in long-term planning at a variety of different scales. This should include international cooperation as well as at the national and local levels. Federal planning efforts should take the impacts of sea level rise into account in all pertinent plans or projects. State level guidelines could be issued to local governments for the regulation of land use through zoning, setbacks, subdivision standards, sewer and other utility placement, and drainage and flood control.

State level coastal zone management agencies could take the lead in these planning efforts because the earliest impacts will be felt in the coastal areas, and state coastal zone agency mandates are generally more comprehensive than traditional single sector agency directives. In addition, existing coastal management agencies could include information and mitigation suggestions in existing permit processes to offset some of the effects of sea level rise (COASTAL SOCIETY, 1986).

8. CONCLUSION

The predicted rise in worldwide sea level will cause a number of impacts on low lying ecosystems and man-made systems. A model for the effects of submergence on these systems is contained in the Mississippi River Delta, which has been experiencing submergence for many years. An important lesson from the Mississippi Delta is that wetlands can continue to exist in the face of rising water levels if there is sufficient sediment input. In many parts of the delta, however, there has been a high loss of land, primarily in interior marshes as sinking exceeds vertical accretion. The cause of the accretion deficit is sediment starvation caused in large part by human activities such as the confinement of the river in the interests of navigation and flood control. There have been a number of institutional actions, some done specifically in response to the problem of wetland loss, which affect the ability of wetlands to maintain themselves. Most of these have taken the form of engineering structures (levees, impoundments, dikes, control structures) and have rarely focused primarily on the natural energies of the River and coastal system to increase sedimentation. Some of these structures yield short-term benefits but are expected to be less effective over the longer term because of the time lag effects. Recommended management strategies drawn from this experience include long-term comprehensive planning to minimize short-term responses, such as agriculture and structures in wetlands, and to maximize sediment and water input and other uses of natural energies. Monitoring the world's deltas should be instituted to provide early warning of the expected impacts and responses.

ACKNOWLEDGEMENTS

This work was supported by the Louisiana Sea Grant College Program and the Louisiana Department of Natural Resources.

9. REFERENCES

ADAMS, R.D., B.B. BARRETT, J.H. BLACKMON, B.W. GANE & W.G. MCINTIRE, 1976. Barataria Basin: geologic processes and framework. Louisiana State University, Center for Wetland Resources, Baton Rouge, LA Sea Grant Publication No. LSU-T-76-006.

BAUMANN, R.H., 1980. Mechanisms of maintaining marsh elevation in a subsiding environment. M.S. Thesis. Louisiana State University, Baton Rouge, Louisiana.

BAUMANN, R.H. & R.D. DELAUNE, 1982. Sedimentation and apparent sea-level rise as factors affecting land loss in coastal Louisiana. In: D. BOESCH. Proceedings of the conference on coastal erosion and wetland modification in Louisiana: causes, consequences and options. FWS/OBS-82/59, Office of Biological Services, U.S. Fish and Wildlife Service, Slidell, LA.: 2-13.

BAUMANN, R.H., J.W. DAY & C.A. MILLER, 1984. Miss. Deltaic Wetland Survival; Sedimentation versus coastal submergence.—Science **224:** 1093-1095.

BOESCH, D.F., D. LEVIN, D. NUMMEDAL & K. BOWLES, 1983. Subsidence in coastal Louisiana: causes, rates and effects on wetlands. FWS/OBS-83/26. US Fish and Wildlife Service, Division of Biological Services, Washington, D.C.

BOUWSEMA, P., J.H. BOSSINADE, K.S. DIJKEMA, J.W.TH.M. MEEGEN, R. VAN REENDERS & W. VRIELING, 1986. De ontwikkeling van de hoogte en de omvang van de kwelders in de landaanwinningswerken in Friesland en Groningen (The progressions in height and areas of the saltwater marshes belonging to the land reclamation projects in Friesland and Groningen).—RIN rapport 86/3.

CLARK, J.S., 1986. Coastal forest tree populations in a changing environment, southeastern Long Island, New York.—Ecol. Monogr. **56:** 259-277.

COASTAL RESTORATION TECHNICAL COMMITTEE, 1988. Report on measures to maintain, enhance, restore, and create vegetated wetlands in coastal Louisiana. Report to the Governor's Coastal Restoration Policy Committee. Office of the Governor, State of Louisiana, Baton Rouge: 39 pp.

COASTAL SOCIETY, 1986. The implications of relative sea level change on coastal decision making. Report of the Coastal Society's Northeast Regional meeting, Oct. 2, 1986, Great Meadows National Wildlife Refuge, Lincoln, MA.

COLEMAN, J.M. & S.M. GAGLIANO, 1964. Cyclic sedimentation in the Mississippi River Deltaic Plain.—Trans. Gulf Coast Ass. Geol. Soc. **14:** 67-80.

CONNER, W., W. SLATER, K. MCKEE, K. FLYNN, I. MENDELSSOHN & J. DAY, 1986. Factors controlling the growth and vigor of commercial wetland forests subject to increased flooding in the Lake Verret, Louisiana watershed. Final Report to the LA. Board of Regents, Baton Rouge, LA: 61 pp.

COSTANZA, R., F.H. SKLAR & J.W. DAY, JR., 1986. Modeling spatial and temporal succession in the Atchafalaya/Terrebonne Marsh/Estuarine complex in S.La. In: DOUGLAS A. WOLFE. Estuarine variability. Academic Press.

COWAN, J.H., JR., R.E. TURNER & D.R. CAHOON, 1986. A preliminary analysis of marsh management plans in coastal Louisiana. Report to Lee Wilson and Associates, Inc., Santa Fe, New Mexico, for the U.S. Environmental Protection Agency: 30 pp + Appendices.

CRAIG, N.J., R.E. TURNER & J.W. DAY, JR., 1980. Wetland losses and their consequences in coastal Louisiana.—Z. Geomorph. Suppl. Bd. **34:** 225-241.

DAY, J.W., JR., R. COSTANZA, K. TEAGUE, N. TAYLOR, G.P. KEMP, R. DAY & R.E. BECKER, 1986. Wetland impoundments: a global survey for comparison with the Louisiana coastal zone. Final Report to Geological Survey Division, Louisiana Department of Natural Resources, Baton Rouge, LA: 140 pp.

DEEGAN, L.A., MCKENNEDY & CHRISTOPHER NEILL, 1984. Natural factors and human modifications contributing to marsh loss in La.'s Mississippi River Deltaic Plain.—Env. Mgt. **8:** 519-528.

DELAUNE, R.D., R.H. BAUMANN & J.G. GOSSELINK, 1983. Relationships among vertical accretion, apparent sea level rise and land loss in a Louisiana Gulf Coast Marsh.—J. sediment. Petrol. **53:** 147-157.

DELAUNE, R.D., W.H. PATRICK, JR. & R.J. BURESH, 1978. Sedimentation rates determined by ^{137}Cs dating in a rapidly accreting salt marsh.—Nature **275:** 532-533.

DELAUNE, R.D., W.H. PATRICK, JR. & S.R. PEZESHKI. Survival of coastal wetlands forests: sedimentation *vs.* submergence. New Forest.

DIJKEMA, K.S. & W.J. WOLFF, 1983. Flora and vegetation of the Wadden Sea Islands and coastal areas. Report 9 of the Wadden Sea working group.

DNR, 1986. Report on Status of CZM projects, Nov. 1986 and proposed coastal protection. Master Plan, DNR/LA. Geol. Survey.

——, 1982. Cote de la Louisiane. Coastal Resources Program, Jan. 1982. DNR, Baton Rouge, Louisiana.

FRAZIER, D.E., 1967. Recent Deltaic deposits of the Mississippi River: Their development and chronology.—Trans. Gulf coast Ass. Geol. Soc. **17:** 287-315.

GAGLIANO, S.M. & J.L. VAN BEEK, 1975. An approach to multiuse management in the Mississippi Delta system. In: M.L. BROUSSARD. Deltas, models for exploration. Houston Geological Society, reprint.

——, 1976. Mississippi River sediment as a resource. In:

R.S. SAXENA. Modern Mississippi Delta - depositional environments and processes. A guide book for the AAPG/SEPM field trip, May 23-26, reprint.

GAGLIANO, S.M., K.J. MEYER-ARENDT & K.M. WICKER, 1981. Land loss in the Mississippi River Deltaic Plain.—Trans. Gulf Coast Ass. Geol. Soc. **31:** 295-300.

GORNITZ, V., S. LEBEDEFF & J. HANSEN, 1982. Global sea level trend in the past century.—Science **215:** 1611-1614.

HACKNEY, C.T. & W.J. CLEARY, 1987. Salt marsh loss in southeastern North Carolina lagoons: importance of sea level rise and inlet dredging.—J. Coast. Res. **3** (10): 93-97.

HATTON, R.S., R.D. DELAUNE & W.H. PATRICK, JR., 1983. Sedimentation, accretion and subsidence in marshes of Barataria Basin, Louisiana.—Limnol. Oceanogr. **28:** 494-502.

HEERDEN, I. VAN & H.H. ROBERTS, 1980. The Atchafalaya Delta-Louisiana's new prograding coast.—Trans. Gulf. Coast Ass. Geol. soc. **30:** 497-506.

HICKS, S.D., 1978. An average geopotential sea level series for the U.S.—J. Geophys. Res. **83:** 1377-1379.

HOFFMAN, J.S., 1983. Projecting sea level rise to the year 2100. In: Coastal Zone '83. Third Symposium on Coastal and Ocean Management, American Society of Civil Engineers, June 1-4, San Diego, CA.: 2784-2795.

HOFFMAN, J.S., D. KEYES & H.G. TITUS, 1983. Projecting future sea level rise: methodology, estimates to the year 2100, and research needs. Washington, D.C.: Office of policy and Resource Management, U.S. EPA, 2nd Ed.

KANA, T.W., B.J. BACA & M.L. WILLIAMS, 1986. Potential impacts of sea level rise on wetlands around Charleston, South Carolina. U.S. Environmental Protection Agency, Washington, D.C., EPA 230-10-85-014.

KOLB, C.R. & VAN LOPIK, 1958. Geology of the Mississippi Deltaic Plain, Southeastern La. Report to the U.S. Army Engineer Waterways Experiment Station CE, Vicksburg, Miss., Tech. Rept.: 3-483, 2 vols.

MAY, J.R. & L.D. BRITSCH, 1987. Geological investigation of the Mississippi River deltaic plain, land loss and land accretion. Waterways Experiment Station, Corps of Engineers, Vicksburg, Ms. Tech. Rept. GL-87-13: 53 pp.

MENDELSOHN, I.A., K.L. MCKEE & W.H. PATRICK, 1981. Oxygen deficiency in *Spartina alterniflora* roots; metabolic adaptation to anoxia.—Science **214:** 439-441.

MENDELSSOHN, I.A., K.L. MCKEE & M.T. POSTEK, 1982. Sublethal stresses controlling *Spartina alterniflora* productivity. Wetlands Ecology and Management. Int. Sc. Publ., Jaipur, India: 223-242.

MILLIMAN, J. & K.O. EMERY, 1968. Sea levels during the past 35,000 years.—Science **162:** 1121-1123.

MITSCH, W.J. & J.G. GOSSELINK, 1986. Wetlands. Van Nostrand Reinhold Co., Inc. N.Y., N.Y.: 178-181.

NUMMEDAL, D., 1983. Future sea level changes along the Louisiana coast.—Shore and Beach **51:** 10-15.

ODUM, H.T., 1971. Env. power and society. John Wiley and Sons, N.Y., N.Y.

ODUM, H.T. & C. DIAMOND, 1985. Energy systems overview of the Miss. R. Basin. Report to the Cousteau Society.

OKEY, C.W., 1918. The subsidence of muck and peat soils in southern Louisiana and Florida.—Trans. Am. Soc. Civ. Eng. **82:** 396-422.

RAMSEY, KAREN E., THOMAS F. MOSLOW & SHEA PENLAND, 1985. Sea level rise and subsidence in coastal Louisiana. Proc. Ass. State Floodplain Manager's Conference.

REDFIELD, A.C., 1972. Development of a New England salt marsh.—Ecol. Monogr. **42:** 201-237.

SASSER, C.E., M.D. DOZIER, J.G. GOSSELINK & J.M. HILL, 1986. Spatial and temporal changes in La.'s Barataria Basin Marshes.—Envir. Mgt. **10:** 671-680.

STANLEY, D.J., 1988. Subsidence in northeastern Nile Delta: rapid rates, possible causes, and consequences.—Science **240:** 497-500.

STEVENSON, J.C., L.G. WARD & M.S. KEARNEY, 1985. Vertical accretion in marshes with varying rates of sea level rise. In: D.A. WOLFE. Estuarine variability. Academic Press, Orlando, Fla.

———,1988. Sediment transport and trapping in marsh systems: implications from tidal flux studies.—Mar. Geol. **80:** 37-59.

SWENSON, E.M. & R.E. TURNER, 1987. Spoil banks: effects on a coastal marsh water-level regime.—Est. Coast. Shelf Sci. **24:** 599-609.

TEMPLET, P., 1987. The policy roots of La.'s land loss crisis. Coastal Zone '87, Proceedings of the Fifth Symposium on Coastal and Ocean Management, V1: 714 pp.

TEMPLET, P.H. & K.J. MEYER-ARENDT, 1988. Louisiana wetland loss: a regional water management approach to the problem.—Env. Mgt. **12:** 181-192.

TITUS, J.G., 1986. Greenhouse effect, sea level rise, and coastal zone management.—CZMJ. **14:** 147-171.

TURNER, R.E. & C. NEILL, 1983. Revisiting the marsh after 70 years of impoundment. In: R.J. VARNELL. Water Quality and Wetland Management Conference Proceedings. New Orleans. LA.: 309-332.

TURNER, R.E., R. COSTANZA & W. SCAIFE, 1982. Canals and wetland erosion rates in coastal Louisiana. In: D. BOESCH. Proceedings of the Conference on Coastal Erosion and Wetland Modification in Louisiana: Causes, Consequences, and Options. FWS/OBS-82/59, Office of Biological Services, U.S. Fish and Wildlife Service, Slidell, LA.: 73-84.

TURNER, R. & D. CAHOON, 1987. Causes of wetland loss in the coastal central Gulf of Mexico. Final Rept. submitted to Minerals Management Service, New Orleans, LA. Contract No. 14-12-0001-30252. OCS Study/MMS 87-0120. 3 Vol.

VIOSCA, P., JR., 1927. Flood control in the Mississippi Valley in it's relation to La. Fisheries.—Trans. Am. Fish. Soc. 57.

WAGRET, P., 1968. Polderlands. Methuen and Co., London: 288 pp.

POSSIBLE EFFECTS OF SEA LEVEL CHANGES ON SALT-MARSH VEGETATION*

A.H.L. HUISKES

Delta Institute for Hydrobiological Research, Vierstraat 28, 4401 EA Yerseke, The Netherlands

ABSTRACT

The composition of salt-marsh vegetations is strongly influenced by tidal movements. The flooding frequency influences the zonation of the vegetation. In turn, this zonation pattern is modified by sedimentation and erosion processes brought about by the tidal movements. Within the vegetation intra- and interspecific processes due to tidal action both on the population and on the individual level may also modify the vegetation pattern. Changes in the pattern of the tidal movements will affect the three aforementioned processes and will either directly or indirectly alter the pattern of the salt-marsh vegetation.

1. INTRODUCTION

The composition of the vegetation of salt marshes is generally the result of three patterns, closely related to the tidal movements of the water they border. Firstly a zonal pattern caused by the frequency of flooding and related to the tolerance of the different plant species to flooding. Secondly a pattern related to the composition of the salt marsh soil, also influenced by the tides through *e.g.* sedimentation and erosion, resulting in a modification of the zonation pattern of the vegetation. Thirdly a pattern caused by the inter- and intraspecific relationships of salt-marsh species which in turn is modifying the other two patterns.

These inter- and intraspecific relationships are influenced by the tides. Any change in the tidal movements (as a result from a rise in the general sea level) will influence directly or indirectly the vegetation pattern of a salt marsh.

2. INFLUENCE OF THE TIDAL AMPLITUDE ON ZONATION OF VEGETATION

The salt marsh vegetation shows in general a zonal pattern which is commonly ascribed to the direct influence of the inundation frequency (BEEFTINK, 1965; CHAPMAN, 1974, 1976). This zonation is not the result of a gradient in salinity (GILLHAM, 1957), which is also present in the marsh. Beeftink states in his standard work on salt marsh vegetation (BEEFTINK, 1965) that the periodic character of the ebb and flood movements is responsible for the characteristic zonation pattern in the salt marsh vegetation. This pattern is clearly seen only in those situations where the increase in height of the salt marsh soil is gradual and the soil pattern is also not disturbed otherwise by the tides.

Such a nearly ideal situation occured in the Markiezaat marsh South of the town of Bergen op Zoom (Fig. 1). Due to the endikement of the area in connection with the Delta Project, this area was turned into a freshwater marsh in 1985.

Fig. 2 shows a map of the central part of this salt marsh. The map was drawn by Schat in 1978 as a part of a study on the population dynamics of some salt marsh species. The vegetation zones, running parallel to each other and perpendicular to the main direction of the ebb and flood currents are apparent. Schat related the vegetation zones to the level of the marsh with respect to N.A.P. (Dutch Ordnance Datum, about to Mean Sea Level). Schats study followed an earlier study by BOS & SIMONS (1964) in the same marsh describing the same pattern as shown by Schat but modified in certain parts of the marsh, ascribed to differences in soil pattern. Also KOOISTRA (1978) related the vegetation pattern in general to the tidal dynamics.

The flooding frequency in the aforementioned studies can be calculated from the recorded high water levels and the height of the marsh according to the Dutch Ordnance Datum (N.A.P.). SIEREVELD *et al.* (1979) give a relationship between the number of floodings per year and the height alone (N.A.P.) (Fig. 3). A point at 1 m above N.A.P. is flooded every tide, being every tide in the semidiurnal cycle. A point at 2.50 m above N.A.P. however is flooded about 30 times a year in the Rattekaai marsh (Fig. 1) but only 5 times a year in the Stavenisse marsh. This is caused by the difference in tidal amplitude at the various marshes due to the funneling effect of the tidal inlet (as the Oosterschelde factually is). If the flooding frequency curve for the Rattekaai marsh is applied to the vegetation zones in the Markiezaat marsh (SCHAT, 1978), a marsh with almost the same tidal

* *Communication no 433, Delta Institute for Hydrobiological Research.*

J.J. Beukema et al. (eds.), Expected Effects of Climatic Change on Marine Coastal Ecosystems, 167–172.

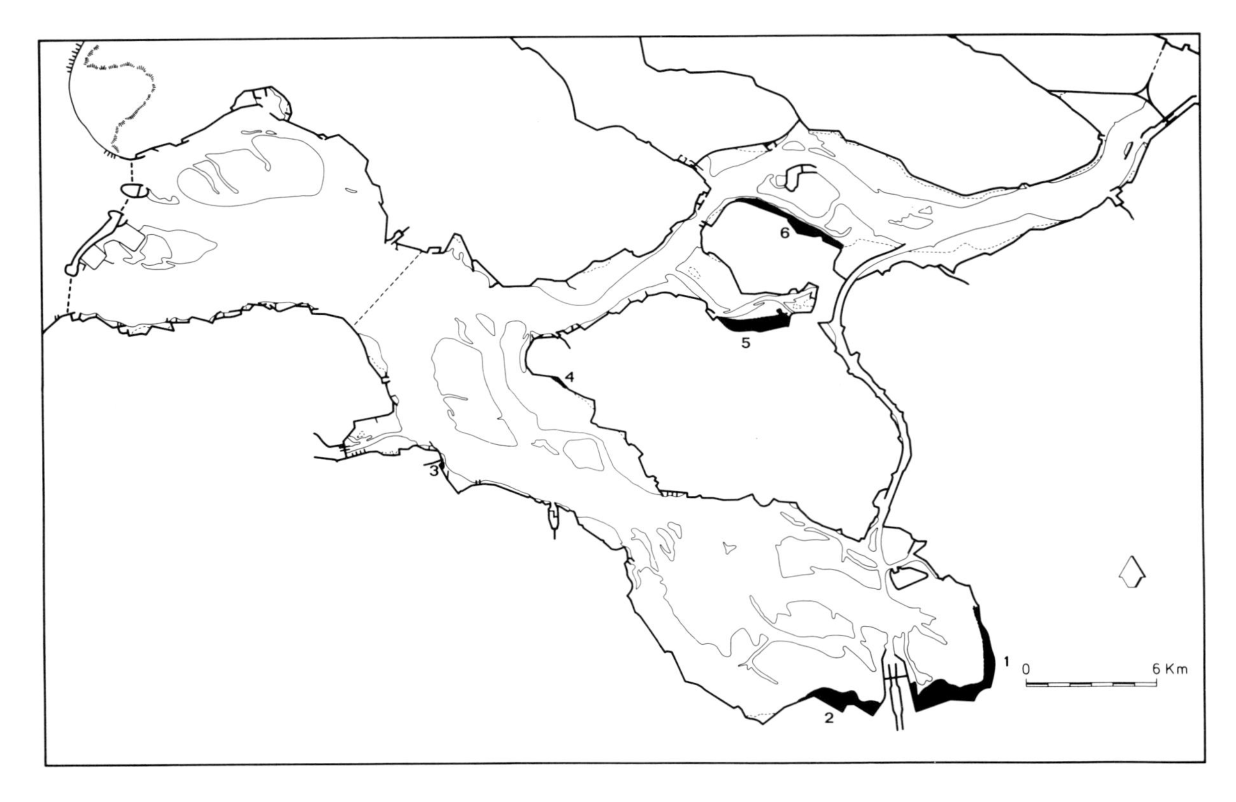

Fig. 1. Map of the Oosterschelde area in the S.W. Netherlands indicating the salt marshes mentioned in the present paper: 1. = Markiezaat marsh, South of Bergen op Zoom; 2. = Rattekaai marsh; 3. = Goese Sas; 4. = Stavenisse marsh; 5. = Krabbenkreek marsh; 6. = marshes near Philipsdam.

amplitude, the vegetation zones described by Schat are flooded with a frequency as shown in Table 1.

3. MODIFICATIONS OF THE ZONAL VEGETATION PATTERN

Not every salt marsh has a vegetation like the Markiezaat marsh. As was stated before this marsh has a very gradual slope, it is situated perpendicular to the prevailing wind which results in an unidirectional vector of tides and wave force, and it has a relatively firm sandy sediment. Other marshes have a steeper slope, are more subjected to wind and/or wave attack and may have places where the sediment may easily become mobile again. These factors may be responsible for local differences in erosion and sedimentation patterns, resulting in a pattern of creeks, marsh flats, salt pans and other geomor-

TABLE 1

Inundation frequencies (number of floodings per year) of different vegetation zones in the salt marsh South of Bergen op Zoom. (N.A.P. = Dutch Ordnance Level, which is close to mean tide level).

	level (*cm + N.A.P.*)	*floodings* ($n.y^{-1}$)
Salicornia-Zostera zone	110 - 130	600 - 700
Transition zone with *Spartina*	130 - 150	630 - 670
Spartina zone	150 - 195	350 - 630
Puccinellia zone	195 - 225	90 - 350
Limonium zone	225 - 250	25 - 90
Elytrigia zone	250 - 325	1 - 25

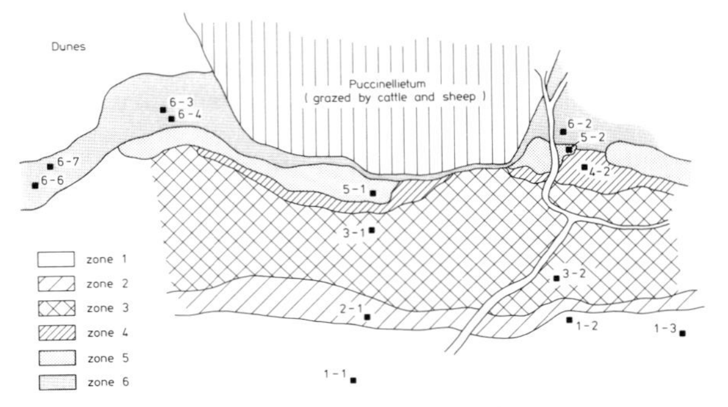

Fig. 2. Different vegetation zones in the Markiezaat marsh, South of Bergen op Zoom. The black squares are the locations of research plots.
Zone 1: sparse vegetation of *Zostera marina* var. *stenophylla*, *Zostera noltii*, small tufts of *Spartina anglica* and *Salicornia dolichostachya*;
Zone 2: vegetation dominated by larger *Spartina anglica* tussocks;
Zone 3: vegetation dominated by *Spartina anglica* and *Aster tripolium*;
Zone 4: vegetation dominated by *Puccinellia maritima*, with *Spartina anglica* and *Aster tripolium*;
Zone 5: vegetation dominated by *Limonium vulgare*, *Plantago maritima*, *Triglochin maritima*;
Zone 6: strandline zone dominated by *Elytrigia pungens*. (after Huiskes *et al.*, 1985).

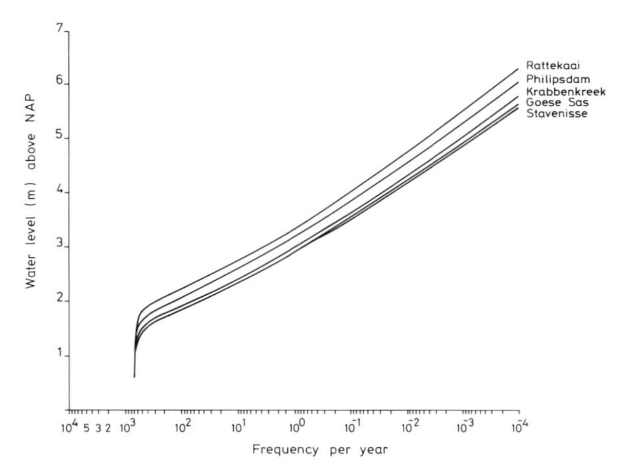

Fig. 3. Number of floodings per year at different levels above N.A.P. (Dutch Ordnance Level) in a number of salt marshes in the Oosterschelde area.
(after SIEREVELD *et al.*, 1979).

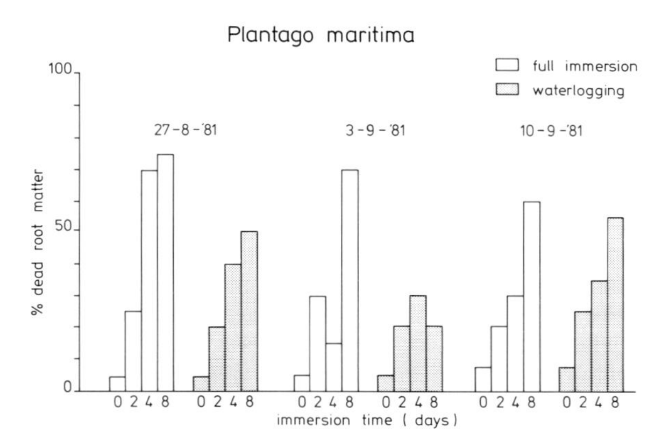

Fig. 4. The change in percentage dead root matter after 0, 2, 4, and 8 days of full immersion of the plants or of waterlogging the root system only.
(after GROENENDIJK, 1984).

phological differences. Local differences in geomorphology are caused by and are causing in turn differences in soil structure. These differences in soil structure may locally modify the vegetation pattern, thus changing the zonal pattern of the vegetation. Several authors describe for instance the differences between the vegetation on the creek banks and the adjacent salt marsh flat (BEEFTINK, 1965; CHAPMAN, 1974; RANWELL, 1972; and others).

Another example is the occurence of places with very bad drainage in the higher salt marsh. These areas, called salt pans (LONG & MASON, 1983), have a vegetation which is normally found in lower areas of the marsh (*e.g.* a *Spartina anglica* and/or *Salicornia* vegetation). In cases of pans that contain water for prolonged periods, the vegetation usually consists of algae (NIENHUIS, 1978).

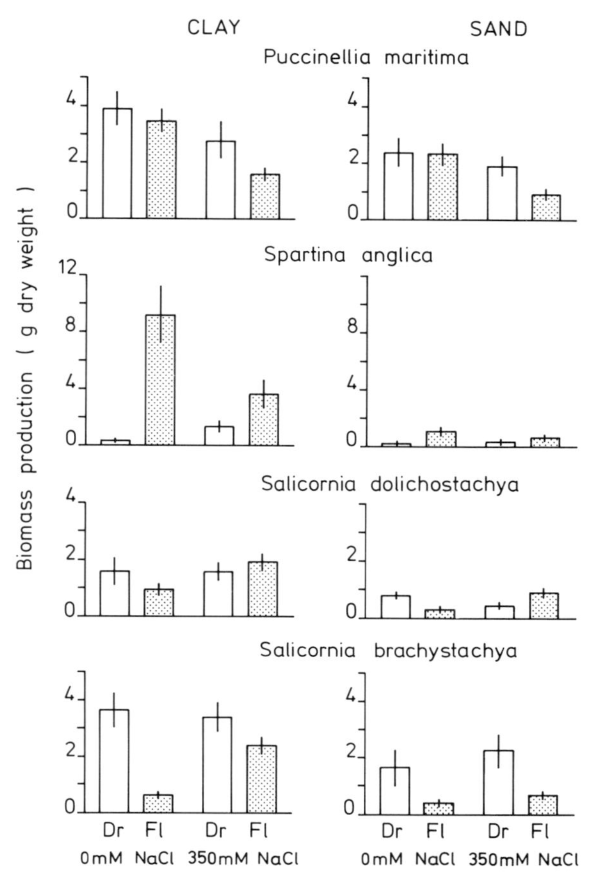

Fig. 5. Biomass production of four salt-marsh species, grown on sand or clay, drained or flooded, with either 0 or 350 mmol.dm^{-3} NaCl soil moisture. Mean and SE of 3-12 plants grown on drained soil and 9-21 plants grown on flooded soils. (redrawn from: VAN DIGGELEN, 1988).

The existence of areas in high marsh with a vegetation characteristic for the lower areas of the marsh is generally caused by the structure of the soil, which in turn is the result of the duration of the inundation.

4. RESPONSES OF THE VEGETATION TO INUNDATION

Having established the fact that the pattern in salt marsh vegetation is directly or indirectly caused by the pattern of the tidal movement, it is important to understand what flooding does to individual plant species. GROENENDIJK (1984) divides inundation in immersion (flooding of the complete plant) and waterlogging (flooding of the root system only). Both types of inundation may have a different effect on certain plant species, as may be seen in Fig. 4: Flooding causes an increase in the amount of dead roots, proportional to the duration of the flooding. It is

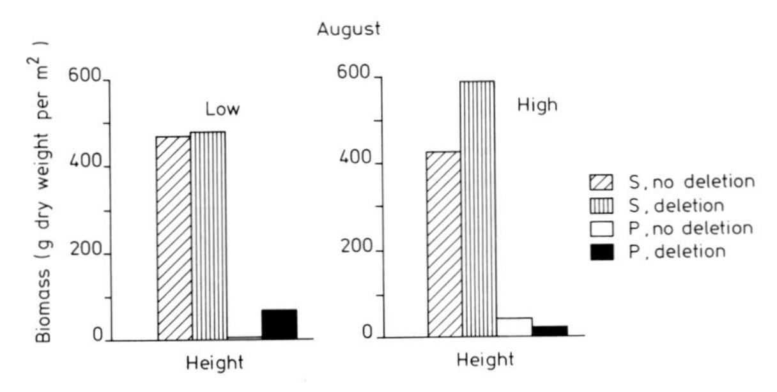

Fig. 6. Biomass of *Spartina anglica* and *Puccinellia maritima* (in gram dry weight per m^2) at two sites of a mixed vegetation at the end of the growing season. (redrawn with permission from SCHOLTEN *et al.*, in press).

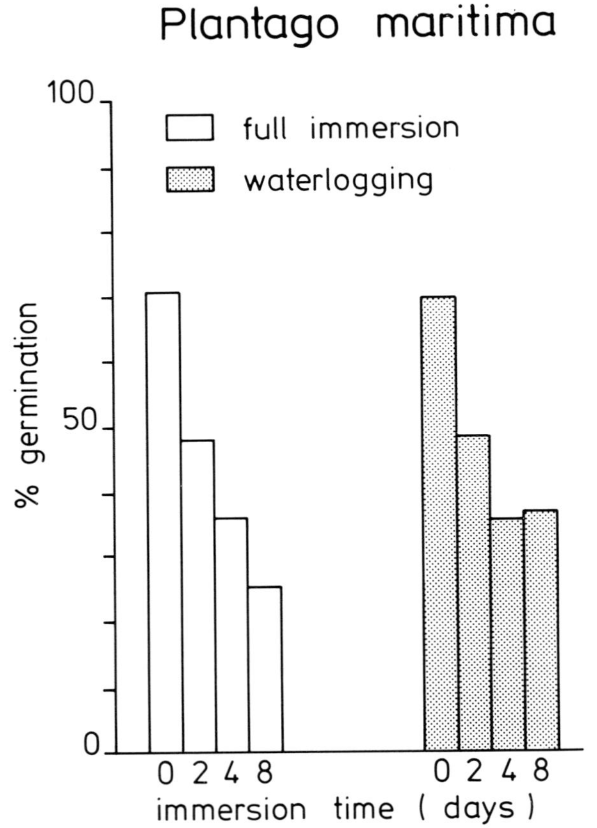

Fig. 7. The change in percentage germination of seeds ripened on the plants while subject to the treatment, after 0, 2, 4, or 8 days immersion or waterlogging. (after GROENENDIJK, 1984).

also shown that a full immersion of the plants results in a higher amount of dead root matter than waterlogging of the soil. This may be ascribed to the aeration of the roots. Many salt marsh plants have the possibility of translocating oxygen from the air via the shoot towards the roots (ARMSTRONG, 1975 and others). Groenendijks experiments show that flooding may have an influence on the functioning of the plants in the marsh. In his thesis VAN DIGGELEN (1988) reports on the biomass production of salt marsh plants when grown in different substrata and flooded or drained. He performed this experiment with different species. The results of the experiment are summarized in Fig. 5. The plants produce more biomass in a clayey soil than in a sandy soil. This occurs in the natural marshes as well (De Leeuw, pers. comm.). A comparison of the biomass of plants growing on the same soil type but under different flooding regimes shows differences within and between the species studied. *Puccinellia maritima* grows better in non-flooded, non-saline situations. *Spartina anglica* however grows better in flooded situations but seems to prefer non-saline soils. The two *Salicornia* species however tend to prefer a saline situation over a freshwater situation. *S. dolichostachya* produces in that situation more biomass when flooded, while *S. brachystachya* seems to have a higher biomass in a drained situation.

Van Diggelens experiment shows in which situation individual plant species grow better. Following this study on single-species reactions, SCHOLTEN *et al.* (in press) studied the interrelationships between plant species, growing in different levels of the salt marsh. The results of a field experiment in which a mixture of *Spartina anglica* and *Puccinellia maritima* was studied under high- and low-marsh situations is shown in Fig. 6. In both marsh areas, situations with either one of the species removed were compared with unperturbed plots. In the low marsh, where *Spartina* is the more common species, *Puccinellia* will disappear when *Spartina* is present in the plots, while *Spartina* performs equally well whether *Puc-*

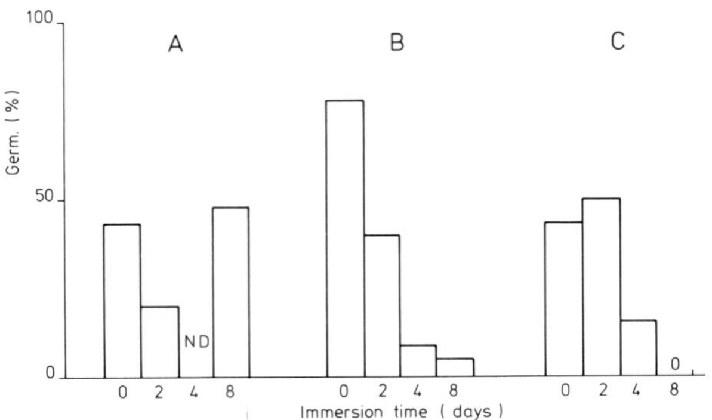

Fig. 8. Effect of prolonged immersion on the germination rate of seed from *Aster tripolium* motherplants which were subjected to (A) immersion before flowering; (B) immersion during flowering; (C) immersion after flowering. The results are averages of two batches of 100 seeds.

TABLE 2

Increase of high water level and calculated increase in number of floodings above which the vegetation will not be able to exist any longer (after Beeftink, 1987).

Community	*Salt-marsh level*	*Increase HWL (in cm)*	*Increase floodings ($n.y^{-1}$)*
Lolio-Potentillion	upper edge	1 - 3	+ 1
Armerion maritimae	upper	3 - 5 (10)	+ 10
Puccinellion maritimae	middle	10 - 15 (40)	+ 100
Spartinion	low	30 - 50	+ 300

cinellia is present or not. In the higher marsh *Puccinellia* grows even better when *Spartina* is present in the plot (which may be attributed to the shelter the *Spartina*-plants give to *Puccinellia*). *Spartina* however grows better when *Puccinellia* is removed. The result of the interactions between two species in the salt marsh seems to depend on the place in the marsh which is likely to be connected with the influence of the flooding regime at the applicable level of the marsh.

The vegetation pattern in salt marshes is not governed by the influence on the competitive power of individual plants only. Flooding influences plant species also on the species population level as GROENENDIJK (1984) has shown. In Fig. 7 the germination percentages of seed batches from plants grown under different inundation regimes are shown. Germination decreased proportional to the duration of the flooding. Waterlogging or full immersion did not make a difference in germination percentages. The seeds that did not germinate were all dead; decrease in germination was not counteracted by an increase in dormancy of the seed batches. In Fig. 8 the same trend in germination is shown for *Aster tripolium* (GROENENDIJK, 1985). The results shown in Fig. 7 were obtained from seed batches that ripened on the plants during the flooding treatments. In the experiment of which the resuls are shown in Fig. 8 the treatment was given at three different periods, before flowering, during flowering and during seed ripening. It seems that a long period of flooding during seed ripening has a stronger negative effect on the subsequent germination than the same period of flooding during flowering. There is however a substantial variation which makes the drawing of firm conclusions difficult.

5. CHANGE IN INUNDATION REGIME

In the former section it was shown that biomass production, competitive ability and reproduction may change when the inundation regime is altered as will happen with an increase in mean sea level. BEEFTINK (1979) showed that an increase in sea level will have a different effect on the various vegetation types in the marsh. A *Spartina*-vegetation is able to withstand an increase in sea level of 30 to 40 cm while an Armerion of the higher marsh will change when the increase in sea level is exceeding 3 to 5 cm. The highest parts of the marsh, where a Lolio - Potentillion is found will even change when the increase in flood level is as little as 1 to 3 cm. Table 2 converts the increase in flood level to an increase in flooding frequency. This change or disappearance of certain vegetation types may directly be ascribed to the performance of individual species as explained before.

A rise in sea level of more than 10 to 15 cm will theoretically change the differentiated salt-marsh vegetation with low marsh, middle marsh and high marsh habitats into a low marsh habitat which has a far less differentiated vegetation. However the change of the sedimentation processes due to the change in mean sea level may counteract the aforementioned alterations. Enhanced sedimentation will increase the height of the marsh and will also alter certain soil characteristics like the salinity and the aeration of the soil. To what extent these processes will balance each other is unclear.

6. POSSIBLE RESPONSES OF PLANT SPECIES TO INCREASE OF SEA LEVEL

Plant species in the salt marsh may become adapted to a higher inundation frequency. It is known that a number of salt marsh species have polyploid series on the genus level which indicates the ability of differentiation within a species. This phenomenon of population differentiation within a salt-marsh species was described for *Salicornia* (SCOTT, 1977; JEFFERIES *et al.*, 1981; HUISKES *et al.*, 1985); *Suaeda maritima* (JOENJE, 1978) and *Aster tripolium* (HUISKES & VAN SOELEN, 1987; GRAY, 1987). The findings by these authors indicate that, given time, certain plant species might be able to become adjusted to higher inundation frequencies, thus resulting in a new equilibrium in the vegetation.

Acknowledgements—I am indebted to Prof. C.H.R. Heip, who made a large number of very useful remarks for improving the text. I wish to thank Jaap van Diggelen and Martin Scholten for allowing me to use a figure from as yet unpublished work. I am grateful to Mr. A.A. Bolsius for the quick and accurate drawing of the figures.

7. REFERENCES

ARMSTRONG, W., 1975. Waterlogged soils. In: J.R. ETHERINGTON. Environment and plant ecology. Wiley, New York: 181- 218.

BEEFTINK, W.G., 1965. De zoutvegetatie van ZW-Nederland beschouwd in Europees verband.—Meded. Landbouwhogeschool Wageningen **65:** 1-167.

——, 1979. The structure of salt-marsh communities in relation to environmental disturbances. In: R.L. JEFFERIES & A.J. DAVY. Ecological Processes in Coastal Environments. Blackwell, Oxford: 77-93.

——, 1987. Vegetation responses to changes in tidal inundation of salt marshes. In: J. VAN ANDEL *et al.* Disturbance in grasslands. Junk Publishers, Dordrecht: 97-117.

BOS, E.S. & H.H.J. SIMONS, 1964. Vegetatieonderzoek van het schorrencomplex ten zuiden van Bergen op Zoom. Student report, Delta Institute, Yerseke.

CHAPMAN, V.J., 1974. Salt Marshes and Salt Deserts of the World. Cramer, Lehre: 1-392.

——, 1976. Coastal Vegetation, 2nd ed. Pergamon Press, Oxford.

DIGGELEN, J. VAN, 1988. A comparative study on the ecophysiology of salt marsh halophytes. Thesis, Amsterdam.

GILLHAM, M.E., 1957. Coastal vegetation of Mull and Iona in relation to salinity and soil reaction.—J. Ecol. **45:** 757-778.

GRAY, A.J., 1987. Genetic change during succession and stability. Blackwell, Oxford: 273-293.

GROENENDIJK, A.M., 1984. Tidal management: Consequences for the salt-marsh vegetation.—Wat. Sci. Tech. **16:** 79-86.

——, 1985. Ecological consequences of tidal management for the salt-marsh vegetation.—Vegetatio **62:** 415-424.

HUISKES, A.H.L. & J. VAN SOELEN, 1987. Seed productivity and seed polymorphism in *Aster tripolium*. In: A.H.L. HUISKES, C.W.P.M. BLOM & J. ROZEMA. Vegetation between land and sea. Junk Publishers, Dordrecht: 202-211.

HUISKES, A.H.L., H. SCHAT & P.F.M. ELENBAAS, 1985. Cytotaxonomic status and morphological characterisation of *Salicornia dolichostachya* and *Salicornia brachystachya*.—Acta bot. neerl. **34:** 271-282.

JEFFERIES, R.L., A.J. DAVY & T. RUDMIK, 1981. Population biology of the salt marsh annual *Salicornia europaea* agg.—J. Ecol. **69:** 17-31.

JOENJE, W., 1978. Plant colonisation and succession on embanked sandflats. Thesis, Groningen.

KOOISTRA, M.J., 1978. Soil development in recent marine sediments of the intertidal zone in the Oosterschelde, The Netherlands. A soil micromorphological approach. Thesis, Amsterdam.

LONG, S.P. & C.F. MASON, 1983. Salt Marsh Ecology. Blackie, Glasgow.

NIENHUIS, P.H., 1978. Dynamics of benthic algal vegetation and environment in Dutch estuarine salt marshes, studied by means of permanent quadrats.—Vegetatio **38:** 103-112.

RANWELL, D.S., 1972. Ecology of salt marshes and sand dunes. Chapman and Hall, London: 112-131.

SCHAT, H.M. 1978. Populatiebiologie van *Salicornia stricta* en *Salicornia brachystachya* en van enkele andere soorten op de schorren then zuiden van Bergen op Zoom. Student report, Delta Institute, Yerseke.

SCHOLTEN, M., K. KAAG, M. WEBER, P. BLAAUW & J. ROZEMA, 1989. The impact of competitive interactions on seedling distribution on salt marshes.—Aquatic Bot., in press.

SCOTT, A.J., 1977. Reinstatement and revision of *Salicorniaceae* J. Agardh (Caryophyllales).—Bot. J. Linn. Soc. **75:** 357-374.

SIEREVELD, J.J. & J.P. AL, 1979. Uitbreiding Schorren. Notitie DDMI-79.419. Rijkswaterstaat, Deltadienst, Middelburg.

SALT MARSHES IN THE NETHERLANDS WADDEN SEA: RISING HIGH-TIDE LEVELS AND ACCRETION ENHANCEMENT

K.S. DIJKEMA[1], J.H. BOSSINADE[2], P. BOUWSEMA[3] and R.J. DE GLOPPER[4]

[1] *Research Institute for Nature Management, Dept. of Estuarine Ecologie, P.O. Box 59, 1790 AB Den Burg, Texel, The Netherlands*

[2] *Department of Public Works, Groningen Department, P.O. Box 30041, 9700 RM Groningen, The Netherlands*

[3] *Department of Public Works, Groningen Department, P.O. Box 20003, 9930 PA Delfzijl, The Netherlands*

[4] *Department of Public Works, Flevoland Department, P.O. Box 600, 8200 AP Lelystad, The Netherlands*

ABSTRACT

Sea-level rise will become a worldwide threat to coastal marshes by affecting the marsh vegetation through an increased number of tidal floodings and an increase in wave energy. The survival of salt marshes depends on the accretionary balances in both the marsh zone itself and the pioneer zone in front of the marsh.

The pioneer zone, which is transitional to the tidal flats, is situated on a level which is most affected by wave action. Moreover, it lacks the protection of a closed vegetation cover. In the Netherlands Wadden Sea more than half of the foreland salt marshes (which are man-made) showed an accretional deficit in this pioneer zone during a period with rising high-tide levels. This will lead to cliff formation and marsh erosion from the seaward edge. Sea-edge erosion has been observed for most of the barrier-island salt marshes in the Wadden Sea.

In the salt-marsh zone itself mostly a positive accretionary balance occurs due to the effect of the vegetation cover on sedimentation and erosion protection. The vertical accretion rate of the man-made foreland marshes is high enough to compensate for a future sea-level rise of 1 to 2 cm per y. The outlook for the barrier-island salt marshes is bad if the sea-level rise will increase to values of 0.5 to 1.0 cm per y, which might be expected for the next century. The response of marsh vegetation and sedimentation to year-to-year changes in mean high-tide levels complicates the effects of a long-term sea-level rise and needs more study.

A general conclusion is that management techniques to prevent negative effects of sea-level rise have to direct most attention to the pioneer zone where marsh growth starts. Further field studies on the processes of sedimentation and plant establishment in both man-made and natural pioneer zones may provide a base for future management.

1. INTRODUCTION

Salt marshes are a transitional sedimentary belt between the sea and terrestrial habitats which were more extensive in former times and provided a continuous landscape along many low-lying coasts (BEEFTINK, 1977; DIJKEMA, 1987a). Up till now, the areal extent of salt marshes in Europe has suffered most from major losses due to embankments (BEEFTINK, 1975; VERHOEVEN *et al.*, 1980; DIJKEMA *et al.*, 1984; DIJKEMA, 1987b; KÖNIG, 1987). In the future our attention must be focused on the effects of sea-level rise, which will become a worldwide threat to coastal marshes (BROUNS, 1988; BOORMAN *et al.*, 1989).

Salt marshes are natural or semi-natural ecosystems. The plant and animal communities develop in a close interaction with the hydrological and geomorphological processes. Most plants and invertebrate animals are highly specialized organisms. Disappearance of the salt-marsh habitats will mean the loss of a highly evaluated genetic resource. Salt marshes also provide resting, breeding and feeding grounds for substantial numbers of birds, many of them migratory. And, with respect to sea-level rise, salt marshes perform an important function in coastal defence, absorbing wave energy, especially during storm surges (NIEMEYER, 1977; DIECKMANN, 1987; ERCHINGER, 1987).

The objective of this paper is to discuss the chances for the salt marshes in the northwest European Wadden Sea to survive future sea-level rise (sections 2 and 5) and the possibilities for accretion-enhancement techniques to prevent marsh loss (section 6). To reach that objective the formation of salt marshes is discussed in relation to mean high tide

J.J. Beukema et al. (eds.), Expected Effects of Climatic Change on Marine Coastal Ecosystems, 173–188.

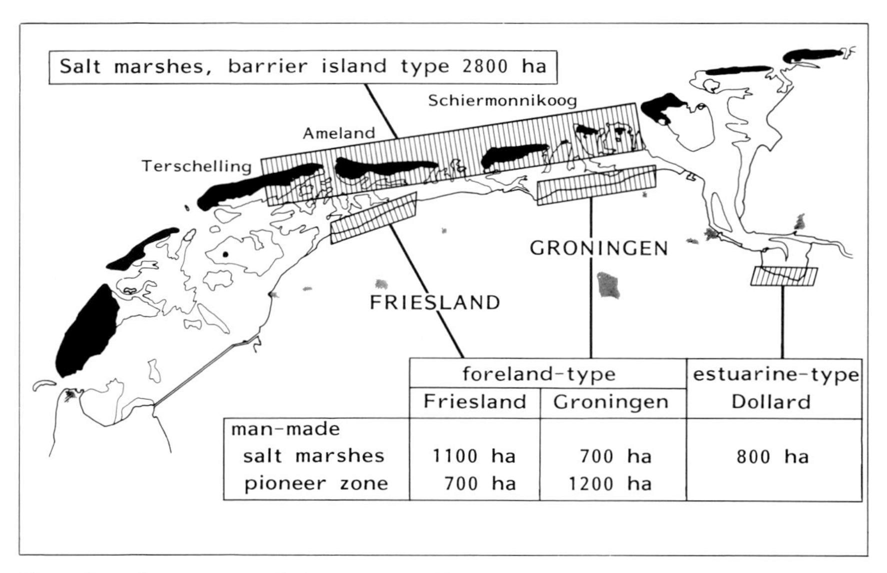

Fig. 1. Locations, geomorphological types (DIJKEMA, 1987a) and areal extent of salt marshes in the Netherlands Wadden Sea.

(MHT) level (section 3). A case study on the development of man-made salt marshes (section 4) stresses the impact of MHT-rise on both surface erosion and sea-edge erosion of salt marshes.

2. ENVIRONMENTAL SETTING

2.1. THE SALT MARSHES

The Netherlands, German and Danish Wadden Sea is a marine sedimentary shore of 450 km length protected by a chain of barrier islands (WOLFF, 1983). The Wadden Sea is mesotidal and two thirds of the 8,000 km^2 area is emergent at low tide. Except for local *Zostera* beds the emergent flats are bare; most of them consist of sand, but along the mainland, in estuaries and locally on the tidal divides mud flats occur. Salt marshes are mostly located along the edges of the Wadden Sea; the areal extent is $\sim$ 350 km^2. About 200 km^2 is more or less man-made with help of marsh accretion-enhancement techniques (DIJKEMA *et al.*, 1984).

The man-made salt marshes in the Netherlands Wadden Sea belong to the foreland-type of salt marsh (DIJKEMA, 1987a); they develop along the mainland of the provinces of Friesland and Groningen in front of an alluvial coastal plain (Fig. 1). Often the sedimentation rate is much higher and the sediment layer of a greater thickness than in the barrier-connected salt marshes. The barrier-type develops naturally at the lee side of the barrier islands (Fig. 1) on high sand flats, often enhanced by the construction of sand dikes. The tidal range of the discussed sites increases from 2.0 m in the west to 2.5 m in the east.

2.2. RELATIVE SEA-LEVEL RISE

Relative sea level in the Wadden Sea is rising due to a combination of factors:

- (a). Eustatic sea-level rise, mostly due to melting of mountain glaciers and thermal expansion of the ocean upper water masses. It is an effect of global warming. GORNITZ *et al.* (1987) estimated by two methods global values of 0.10 $\pm$ 0.01 and 0.12 $\pm$ 0.03 $cm.y^{-1}$ for the period 1880–1980.
- (b). Tectonic subsidence as counterpart for the tectonic uplift of Scandinavia. This seems of little importance as the Netherlands Wadden Sea is at the margin of the zone of influence of isostatic rebound (AUBREY, 1984).
- (c). Regional variation in the tectonic movements (NOOMEN, in prep.). Relative to selected stable stations for the entire Netherlands a significant and regular subsidence of 0.04 $cm.y^{-1}$ for Texel and for the Groningen north coast occurs. For the Friesland coast there are not enough data, but the north coast seems to be stable. For the Frisian barrier islands and the Southwest-Netherlands there are no data at all.
- (d). Compaction of older Holocene peat and clay layers with a theoretical average rate of 0.025 $cm.y^{-1}$ during the last 10 000 years (VEENSTRA, 1980). The process of compaction, however, occurs rather quickly at first and then gradually slows down. Present-day compaction is negligible, apart from that of recent sediments.
- (e). Subsidence due to gas extraction occurring around production sites. From 1964 till now a mean subsidence rate of 0.1 to 0.2 $cm.y^{-1}$ has been

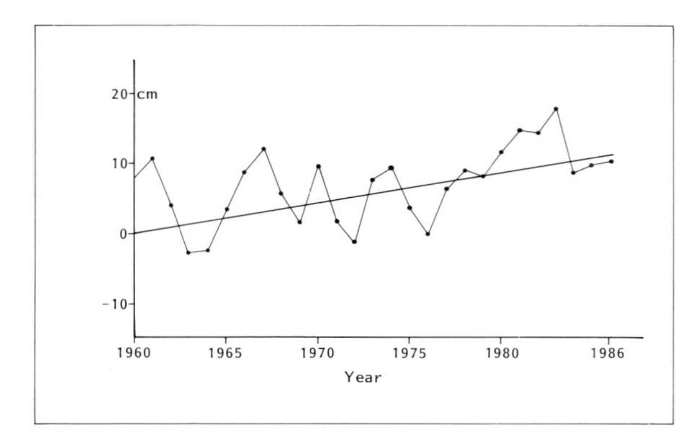

Fig. 2. Trend in MHT-levels for the period 1960-1986. Based on records of the tidal gauges at Harlingen, Holwerd and Schiermonnikoog in the Wadden Sea.

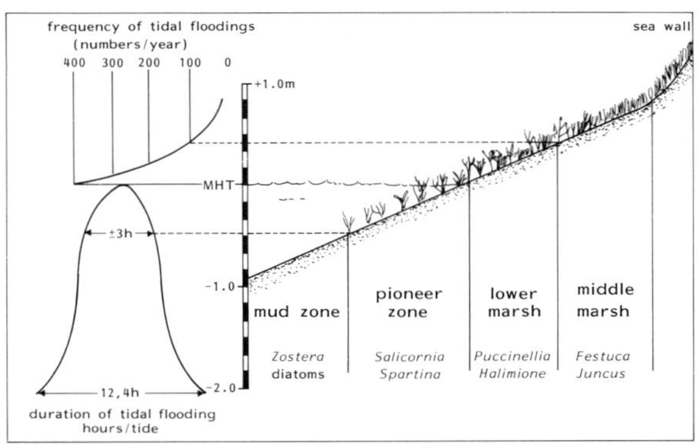

Fig. 3. Zonation of salt marshes in relation to duration and frequency of tidal floodings. After ERCHINGER (1985).

measured for the eastern man-made marshes along the Groningen coast. For the period 1987–2025 this rate is expected to increase to an average of 0.1 to 0.6 $cm.y^{-1}$. For a barrier salt marsh on Ameland a subsidence rate of 0.8 to 1.5 $cm.y^{-1}$ is predicted for the period 1987-2003.

To compare the estimations with the measured increase in mean relative sea level factors (d) and (e) are irrelevant because levelings are carried out from fixed stations. The eustatic rise (a) and the tectonic subsidence (b + c) together can be estimated at a + b + c = 0.10 to 0.16 $cm.y^{-1}$, being somewhat lower than the registration of tide gauges for the Netherlands Wadden Sea (0.15 to 0.17 $cm.y^{-1}$ for the period 1933–1980; VAN MALDE, 1984).

For the past 50 years an increase in tidal amplitudes has lead to a much faster rise in relative MHT-levels as compared to mean relative sea levels (VAN MALDE, 1984). DIJKEMA *et al.* (1988) report a mean value of 0.44 $cm.y^{-1}$ for the period 1960–1986 in the man-made salt marshes, calculated as the trendline from the year to year fluctuations in MHT (method of least squares; Fig. 2). For the nearby barrier islands Terschelling, Ameland and Schiermonnikoog the MHT-rise for the period 1960–1988 is 0.4, 0.5 and 0.3 $cm.y^{-1}$ respectively. In the German Wadden Sea comparable values for a rise in MHT-levels have been found: 0.64 $cm.y^{-1}$ for the entire area in the period 1958–1983 (FÜHRBÖTER, 1986) and 0.43 to 0.46 $cm.y^{-1}$ for stations near the Netherlands Wadden Sea in the period 1954–1986 (JENSEN *et al.*, 1988). This increase may be due to changes in the wind direction to northwest and north (LAMB & WEISS, 1979) and locally to dredging of the estuaries (VAN MALDE, 1984). Estimations of future MHT-levels are not available.

Sea levels (factor a) are expected to rise faster in the near future due to the 'Greenhouse Effect' which will lead to a warmer climate. Estimations of future sea levels are only rough estimates. ROBIN (1986) presented a sea level-temperature regression based on an analysis of all data and results of models. A prediction of global warming of 3.5°C + 2.0°C due to CO_2-doubling (Villach-I Conference) will lead to a global sea-level rise of 80 cm (possible range 20 to 165 cm) in the next hundred years. OERLEMANS (in press) has estimated the various contribution to sea-level rise separately, based on a comparable temperature scenario (Villach-II Conference) and a lower contribution from thermal expansion (WIGLEY & RAPER, 1987). His most likely values of global sea level rise relative to 1985 are 6.2 cm in 2000, 20.5 cm in 2025, 33.0 cm in 2050 and 65.6 cm in 2100. The contributions from glaciers and thermal expansion are quite comparable. The values for Antarctica and Greenland are smaller and neutralize each other. The standard deviation in Oerlemans' estimation is in the same order as the predicted sea-level rise itself. The trends in tectonic subsidence (factors b and c), compaction (d, negligible) and subsidence due to gas extraction (e) should be added to these figures.

3. IMPACTS OF SEA-LEVEL CHANGES ON SALT MARSHES

3.1. FORMATION AND ZONATION OF NATURAL SALT MARSHES

To understand the impact of sea-level fluctuations and marsh accretion-enhancement techniques on salt marshes it is best to start with the development of natural salt marshes. Conditions for salt-marsh formation are best on a gently sloping shoreline with little wave energy and sufficient sediment supply (DIJKEMA, 1987a). Secondly, the surface elevation, tidal amplitude and drainage must be sufficient to allow periods of soil aeration necessary for plant growth (ARMSTRONG *et al.*, 1985). Pioneer plants (*Spartina anglica* and *Salicornia dolichostachya*) and further sedimentation will do the rest, creating an environment which promotes a closed cover of perennial halophytic plants (KAMPS, 1962; DIJKEMA, 1983a).

Puccinellia maritima is an important species in the Wadden Sea during this stage of salt-marsh formation. Around the MHT-level this perennial grass reaches sufficient coverage to enhance sedimentation to the highest rate of all stages of salt-marsh formation (WOHLENBERG, 1933; JAKOBSEN, 1954; BOUWSEMA *et al.*, 1986; DIJKEMA *et al.*, 1988), to favour the development of a natural creek system and to prevent erosion of their newly created habitat (WOHLENBERG, 1953; KAMPS, 1962; VON WEIHE, 1979). The drainage system is an important stimulus for the growth of most plants and promotes succession to the next vegetation types in salt-marsh development.

In this way halophytic plants play a dominant role in salt-marsh formation and also create the environments for subsequent steps in marsh development, thus leading to a striking zonation pattern of vegetation types (Fig. 3). This zonation is related to MHT (the duration and frequency of tidal floodings; RANWELL, 1972; BEEFTINK, 1965, 1977; DIJKEMA, 1983a; WESTHOFF, 1987), the water and soil salinity as well as the combination of these factors with soil aeration (BEEFTINK, 1977; COOPER, 1982; VINCE & SNOW, 1984; ARMSTRONG *et al.*, 1985; VAN DIGGELEN, 1987).

ARMSTRONG *et al.* (1985) found three main zones in soil aeration corresponding with our salt-marsh zones (Fig. 3): (1) reduced conditions dominate, phases of oxidation occur only near the surface ($\leqslant 5$ cm) and at neap-tide periods, conditions characteristic of the pioneer zone with *Spartina anglica*; (2) aerated conditions dominate, but are lowered monthly for a few days by high tides at spring tide, characteristic of the lower marsh zone with *Puccinellia maritima, Halimione portulacoides* and *Elymus pycnanthus*; and (3) the longest period of oxidation, free from fluctuations except for the very high spring tides, characteristic of the middle marsh zone with *Festuca rubra*.

Soil aeration can be considered a key-factor for the initial stages in salt-marsh formation, whereas soil salinity is more important with increasing elevation and for brackish marshes (estuaries, deltas). As the tolerances of halophyte species for edaphic factors overlap each other, competitive interactions between the plant species also play an important role in determining the zonation pattern (SNOW & VINCE, 1984; BERTNESS & ELLISON, 1987). A fast growth rate in spring is especially important and gives a species an advantage in the competition for light during the rest of the growing season (SCHOLTEN *et al.*, 1987).

TABLE 1

Maximum rise in MHT-level at which the vegetation community recovers (for sites with a tidal amplitude of roughly 3 to 4 m). Based on observations of the effects of an increase in tdal amplitudes through barrage building in the Volkerak, Southwest-Netherlands (modified after Beeftink, 1979).

Salt-marsh zone	*Vegetation community*	*MHT-rise (in cm)*
Pioneer zone	Spartinetum	30-40
Lower marsh zone	Puccinellietum maritimae	15
	Plantagini-Limonietum	10
	Halimionetum portulacoides	10
	Atriplici-Elytrigietum pungentis	10
	Puccinellietum maritimae, with middle marsh elements	3-5
Middle marsh zone	Artimisietum maritimae	3
	Juncetum gerardii	3
	Juncetum gerardii, with upper marsh elements	1-3
Upper marsh zone	Lolio-Potentillion anserinae	0-1

3.2. EFFECTS OF CHANGES IN MHT-LEVEL ON VEGETATION

Starting from the above described relation between vegetation zones and the elevation relative to MHT, it might be expected that changes in MHT-levels lead to changes in the vegetation of the zones. From examples of the effects of artificial MHT-rises and natural MHT- fluctuations in the Wadden Sea and the Southwest-Netherlands this statement will be discussed.

3.2.1. EFFECTS OF ARTIFICIAL MHT-RISES

As a consequence of the construction of the Afsluitdijk in 1932 a sudden rise in MHT of 0.15 m occurred near the salt-marsh island Griend in the Netherlands Wadden Sea (FEEKES, 1950). Before 1932 there was a stable zonation of mainly *Puccinellia maritima, Limonium vulgare, Festuca rubra* and *Elymus pycnanthus* for decades. In the period 1932-1934 (or 1935) the grass species disappeared completely, with *Elymus* as the last one. From the remaining species *Aster tripolium, Plantago maritima, Juncus gerardii* and *Armeria maritima* disappeared, *Artemisia maritima* being the last one in 1936. In the period 1934-1940 the barren sites were occupied by the annuals *Salicornia europaea, Suaeda maritima* and *Atriplex littoralis*. In this period the island was damaged by severe gales and sand which blew in the remaining salt marsh. In the period 1940-1946 the vegetation regenerated, starting with *Puccinellia maritima*.

Comparable sudden rises in MHT occurred recently in the Southwest-Netherlands as an effect of the Delta Works, where parts of estuaries were cut off from the tides by enclosure dams (BEEFTINK, 1979, 1987a). A rise of about 0.30 m in MHT had different effects on the increase in flooding frequency for each salt-marsh zone: for the pioneer and lower marsh zone flooding was about twice often, but for the middle and upper marsh zones it was sevenfold. The vegetation of the upper and middle marsh zones was much more vulnerable to the abrupt rise of MHT compared to the vegetation of the lower marsh and pioneer zones (Table 1).

3.2.2. EFFECTS OF NATURAL MHT-FLUCTUATIONS

Smaller changes in the year-to-year MHT-levels, up to about 0.10 $m.y^{-1}$, occur as natural tidal variations (Fig. 2). BEEFTINK (1987b) has related vegetation responses in the lower marsh zone on a foreland type of salt marsh to clusters of years with either an increased or decreased MHT-level. An increase in the flooding frequency caused an increase in the cover of the annual species *Salicornia europaea* and *Suaeda maritima*, probably by a better dispersion of the seeds and the availability of enough bare soil. Also some perennial species of the pioneer and lower marsh zones, *Spartina anglica, Triglochin maritima* and *Limonium vulgare* increased in cover. On the other hand, the lower and middle marsh species *Puccinellia maritima, Glaux maritima, Festuca rubra, Artemisia maritima* and *Elymus pycnanthus* decreased in cover, for which decreased soil aeration could have been responsible. The behaviour of some species such as *Aster tripolium* and *Halimione portulacoides* suggests a shift of the populations parallel to the change in flooding frequency (BEEFTINK, 1987b).

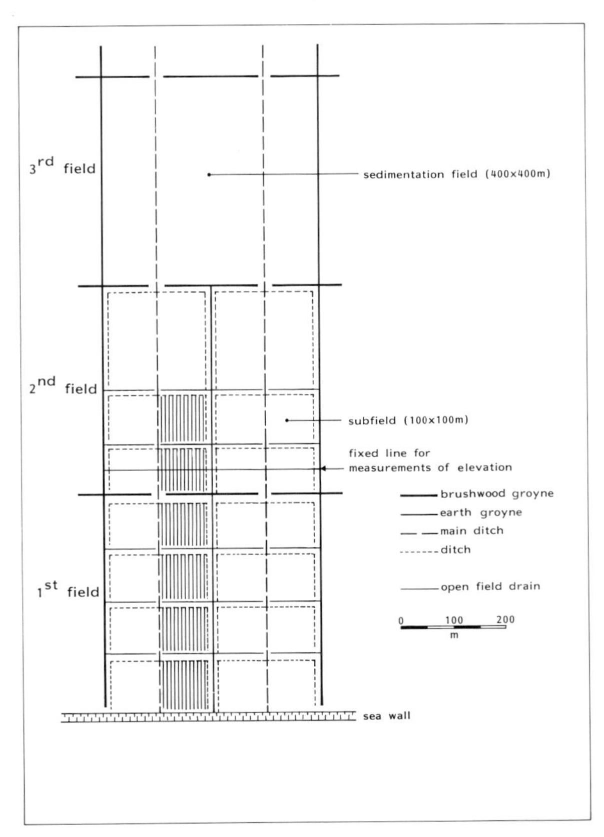

Fig. 4. Lay-out of a series of three sedimentation fields from the seawall to the tidal flats.

On the barrier island of Schiermonnikoog, OLFF *et al.* (1988) studied the relation between annual changes in the cover of species and fluctuations in the monthly frequency of tidal floodings over a period of 15 years. Most of their sample plots were situated in the middle marsh zone. It appeared that for the lower part of the plots higher annual MHT-levels had a negative effect on the cover of most marsh species. Most negative correlations were found with the inundation frequency in summer, indicating the harmful effect of summer inundations, probably on the soil aeration. For the higher part of their marsh the numbers of species with a positive and a negative relation with increased tidal floodings were about equal, but the positive correlations had much higher values. Among the species with high positive correlations with raised MHT-levels, *Festuca rubra, Juncus maritimus, Juncus gerardii* and *Artemisia maritima* frequently occurred in the seed bank of the salt marsh. These species may have the advantage of quick re-establishment during vegetation regeneration in the gaps. The other species with high positive correlations to raised MHT-levels, *Triglochin maritima* and *Puccinellia maritima*, are among the most salt tolerant ones (OLFF *et al.*, 1988).

3.3. FLUCTUATIONS AND LONG-TERM CHANGES IN VEGETATION

The examples show that deviation in the year-to-year MHT-levels leads to changes in the species composition of the salt-marsh zones. Factors controlling these changes are soil aeration, the dissemination mechanisms of the species and competitive interactions between the plants. DE LEEUW *et al.* (in prep.) found the soil salinity to be correlated to rainfall deficit and not to the frequency of tidal floodings, however. From areas with a high subsidence rate in the Mississippi delta there is considerable evidence that an accretionary deficit can cause a deterioration in the salt-marsh vegetation through waterlogging (MENDELSSOHN *et al.*, 1981; ORSON *et al.*, 1985; MENDELSSOHN *et al.*, 1988). Readers are referred to HUISKES (this issue), ROZEMA *et al.* (1985), GROENENDIJK *et al.* (1987) and VAN DIGGELEN (1988) for ecophysiological background.

It may be expected that, if MHT-deviations follow a trend in one direction, succession or regression of vegetation to communities of higher or lower marsh zones will occur (DANKERS *et al.*, 1987). This has been shown for brackish marshes in the Baltic Sea, where the long-term trend of relative sea level is decreasing due to land-uplift. This caused the vegetation zones to move seaward in a long-term process. On a short term (< 10 year), however, year-to-year changes in water level appeared to be more important (ERICSON, 1980; CRAMER & HYTTEBORN, 1987). BEEFTINK (1987a, 1987b) found that even one year with a tide deviation of more than 0.05 to 0.10 m can lead to changes in the cover of some plant species. Changes occur within the same year when the tides are lower than normal. When the tides are higher than normal there is a delay of one or more years. Therefore, salt-marsh accretion is a step-by-step process, progressing during periods with low MHT-levels (*cf.* ERICSON, 1980; BEEFTINK, 1987a).

Historical data on the start of marsh growth on different sites in relation to MHT-level are needed to gain more insight into the hypothesis on marsh accretion. In studying the impacts of relative sea-level rise on salt marshes it is important to separate the fluctuations and long-term changes in the marsh zones.

4. DEVELOPMENT OF MAN-MADE SALT MARSHES AND MUD FLATS DURING A PERIOD WITH RISING HIGH-TIDE LEVELS

4.1. HISTORY AND PRESENT RESEARCH

Techniques to enhance marsh accretion have been based on the processes of natural salt-marsh formation. The oldest and most simple technique is drainage of the surface of the marsh along with a narrow strip of the adjoining mud flat (farmers method). This was performed by local farmers by digging small, open field drains spaced about 5 to 15 m apart. The drains need to be cleaned from sediment regularly. The technique was used as early as 1740 in the Dollard estuary (STRATINGH & VENEMA, 1855). By this artificial drainage, soil aeration is improved and the aerated zone is shifted to a lower level relative to MHT, resulting in horizontal marsh extension.

A newer technique uses sedimentation fields surrounded by brushwood groynes of ~ 0.30 m in height above MHT (Fig. 4). By this technique the most critical factor in salt-marsh formation, wave energy, is reduced to enhance the sedimentation directly. The sedimentation field technique was developed in Schleswig-Holstein, West-Germany, and applied in the Netherlands in 1935. After 1945 it was applied on a large scale by the Department of Public Works (KAMPS, 1962). In Germany the standard size of the sedimentation fields is 200 x 400 m, but smaller fields often occur there (100 x 200 m, 200 x 200 m, 200 x 300 m). In the Netherlands the standard size is larger, 400 x 400 m, but sedimentation fields of other size may occur (DIJKEMA *et al.*, 1988).

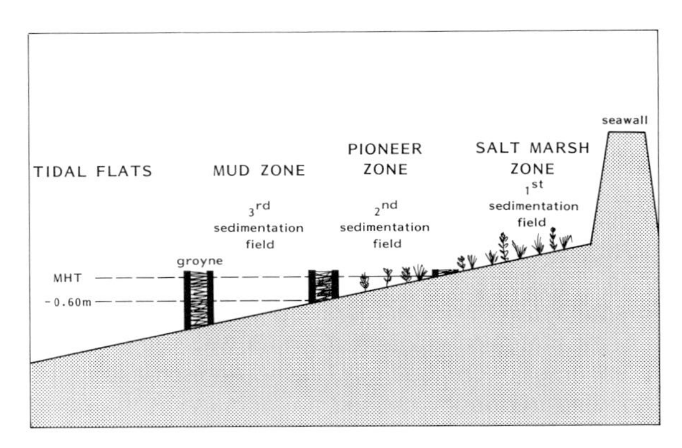

Fig. 5. Cross-section of the zones in man-made salt marshes and mud flats.

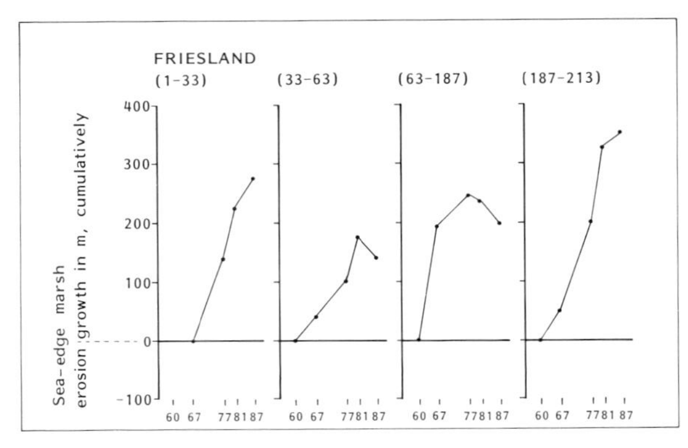

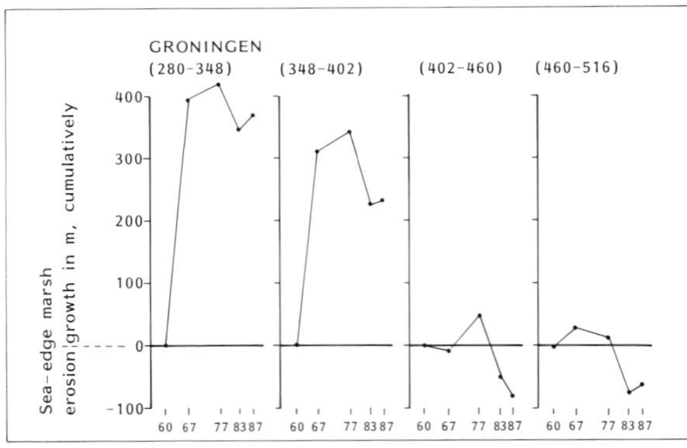

Fig. 6. Seaward marsh growth in the Netherlands man-made marshes, cumulatively relative to 1960. After DIJKEMA *et al.*, 1988.

In the Netherlands a research programme is now in operation to develop modified accretion techniques, being more based on enhancement of the natural processes of salt-marsh formation. The programme is carried out cooperatively by the Department of Public Works (RWS, which manages the area, is responsible for the soil research and takes care of recording and processing field data) and the Research Institute for Nature Management (RIN, responsible for the ecological research and the coordination of the study).

The present research programme started in 1982 and consists of two parts:
- 1. Trend analysis of existing data. Data on surface elevation, soil composition, vegetation, and management practices have been recorded in 33 observation fields ($\sim 50.10^4$ m^2 each) from about 1950 (sometimes 1937) onwards, each of them representing 2 km coastline.
- 2. Field experiments with drainage techniques. In 1982 five experimental fields ($\sim 30.10^4$ m^2 each) were selected in locations with different sedimentation rates and grazing intensities, including an experimental drainage system. In 1987, 8 more experimental fields were added to put smaller types of field drains and newly developed drain-digging machines into practice.

The man-made salt marshes along the north coast of Friesland and Groningen show a main zonation similar to that in natural salt marshes (Figs 3 and 5). This zonation occurs in the observation fields and the experimental fields as well.

4.2. HORIZONTAL ACCRETION

To compare the areal changes of salt marsh for different sites and time periods the seaward growth of the marsh edge can be used. Fig. 6 shows the accretion of the man-made salt marshes with the year 1960 as starting point. The data are from the observation fields and averaged for coastal stretches with a comparable development. In our study the marsh edge is defined by a 5% cover of *Puccinellia maritima* and includes all vegetation communities of the alliances Puccinellion maritimae and Armerion maritimae.

The average horizontal accretion for the period 1960-1987 is 8.2 m. y^{-1} for Friesland and 4.7 m.y^{-1} for Groningen (DIJKEMA *et al.*, 1988). Those rates are comparable to the horizontal accretion in German man-made salt marshes: 7.1 m.y^{-1} for the period 1830-1978 and 4.7 m.y^{-1} for the period 1940-1979 (DIECKMANN, 1988). Only along the western and eastern flanks of the Frisian coast the accretion progresses consistently for the entire period (more than 10 m.y^{-1} on average). For the major coastal stretch in Friesland (central part) and the western half of the Groningen coast, however, almost all accretion occurred in the sixties with high accretion rates of 28-49 m.y^{-1}; lateron the marsh area stabilized (+ 3 m.y^{-1}) and for the past 10 years erosion has occurred (−9 m.y^{-1}). The eastern part of the Groningen coast even shows a net reduction of the salt marsh area after 1960.

To find an explanation for the stagnation of salt-marsh growth it is important to look at the relation between the vegetation zones and the surface elevation

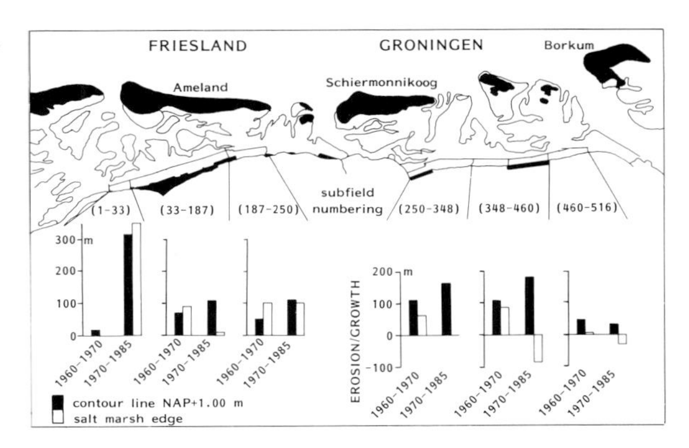

Fig. 7. Seaward growth of the contour line of NAP + 1.00 m, compared to the salt-marsh edge. After BOUWSEMA *et al.*, 1986.

of the marsh (Fig. 3). It might be expected that the seaward marsh edge would follow the horizontal growth or retreat of a certain contour line. As a 5% cover of *Puccinellia maritima* is normally reached at a level of ~1.00 m above Dutch Ordnance Level (NAP), the seaward extension of this contour line has been compared to that of the salt marsh edge (Fig. 7). For the first period (1960-1970) the relation is rather close as the salt marsh extension follows the seaward shift of the NAP + 1.00 m contour line. After 1970 the NAP + 1.00 m contour line keeps moving seaward due to continued vertical accretion, but the salt-marsh growth falls far behind in most areas. For the period 1965-1985 this stagnation corresponds with a vertical retreat of the lower marsh limit of 0.16 m for the central part of the Frisian coast and with even 0.30 m for the entire Groningen coast (BOUWSEMA *et al.*, 1986). Only on the western and eastern flanks of the Frisian coast this problem does not occur.

There should have been changes in one or more environmental factors causing the disturbed relation between the lower salt-marsh limit and the surface elevation. DIJKEMA *et al.* (1988) mention three changes:
- 1. The rise in MHT-level of ~11 cm for the period 1960-1986 (section 2.2), which covers about most of the problem in Friesland and about half of the problem in Groningen. Therefore, all data on elevation have been transformed relative to the trendline in annual MHT-levels (increase of 0.44 cm per year; Fig. 2). Summer and winter values for MHT are not significantly different from each other (95% confidence limits). In this way net results of the accretion are shown.
- 2. An increase in wave energy. The rise in MHT-level caused an increase in the frequency of overflowings of the brushwood groynes from about 100 to 200 tides a year along all man-made salt marshes. The wave energy may be further increased by

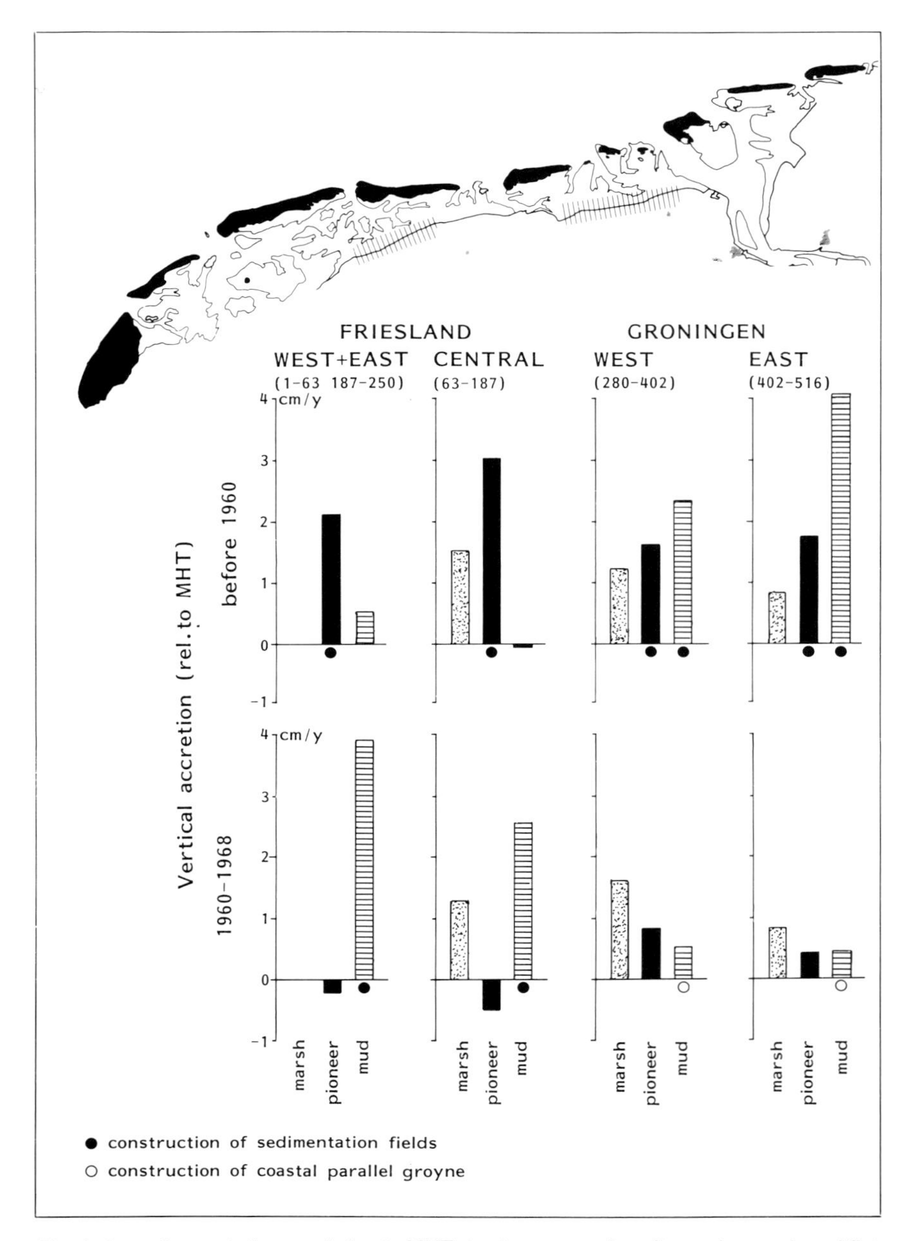

Fig. 8. Accretionary balance relative to MHT-rise in man-made salt marshes and mud flats in the Netherlands Wadden Sea for the period in which the sedimentation fields were constructed.

a tendency to formation of terraces in the seaward part of the salt marsh, causing a steeper slope of the marsh - mud flat transition. Both changes must have negative effects on the settlement of marsh plants. For example, the establishment of *Spartina anglica* is limited by wave energy (KÖNIG, 1948; DIJKEMA, 1983a; VAN EERDT, 1985b; GROENENDIJK, 1986). This point needs attention in further studies.

- 3. Changes in the method of drainage. In field experiments this point is now being tested.

4.3. VERTICAL ACCRETION

Data on elevation are available for fixed lines in the observation fields (Fig. 4). These transects are situated parallel to the coastline, ~ 100 m apart from each other from the seawall towards the tidal flats of the Wadden Sea (DIJKEMA *et al.*, 1988). The annual vertical accretion rate relative to MHT-rise (increase of 0.44 cm.y^{-1}; Fig. 2) has been averaged for the zones of Fig. 5, for 4 time periods and for coastal

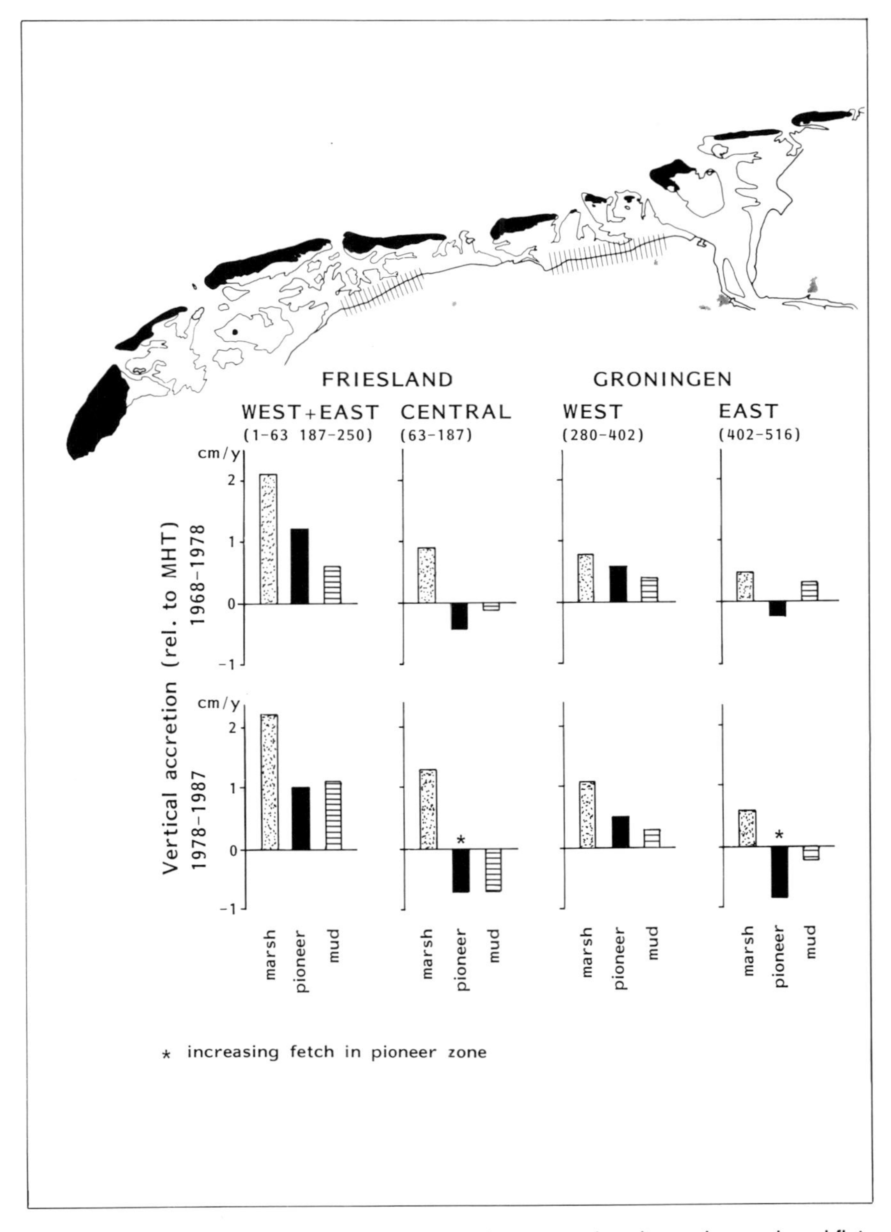

Fig. 9. Accretionary balance relative to MHT-rise in man-made salt marshes and mud flats in the Netherlands Wadden Sea for the period after construction of the sedimentation fields.

stretches with high and low vertical accretion. The accuracy of these values is at least ± 0.1 cm. Figs 8 and 9 show the spatial and temporal variability of the vertical accretion.

In the first periods (before 1960 and 1960-1968) there is a much higher vertical accretion rate as compared to later years (Fig. 8). The high accretion occurs directly after the construction of the sedimentation fields. The average duration of the period with raised accretion levels is 4 years and the total net accretion in that period is 0.10 to 0.20 m (DIJKEMA *et al.*, 1988).

In the following periods (1968-1978 and 1978-1987), the vertical accretion slows down and differentiates related to the surface elevation, accretionary conditions and vegetation (Figs 9 and 10):

- 1. The outer sedimentation fields includes the bare mud zone lower than 0.60 m below MHT. There is a positive accretionary balance relative to the MHT-rise for 70% of the coastal length presently.

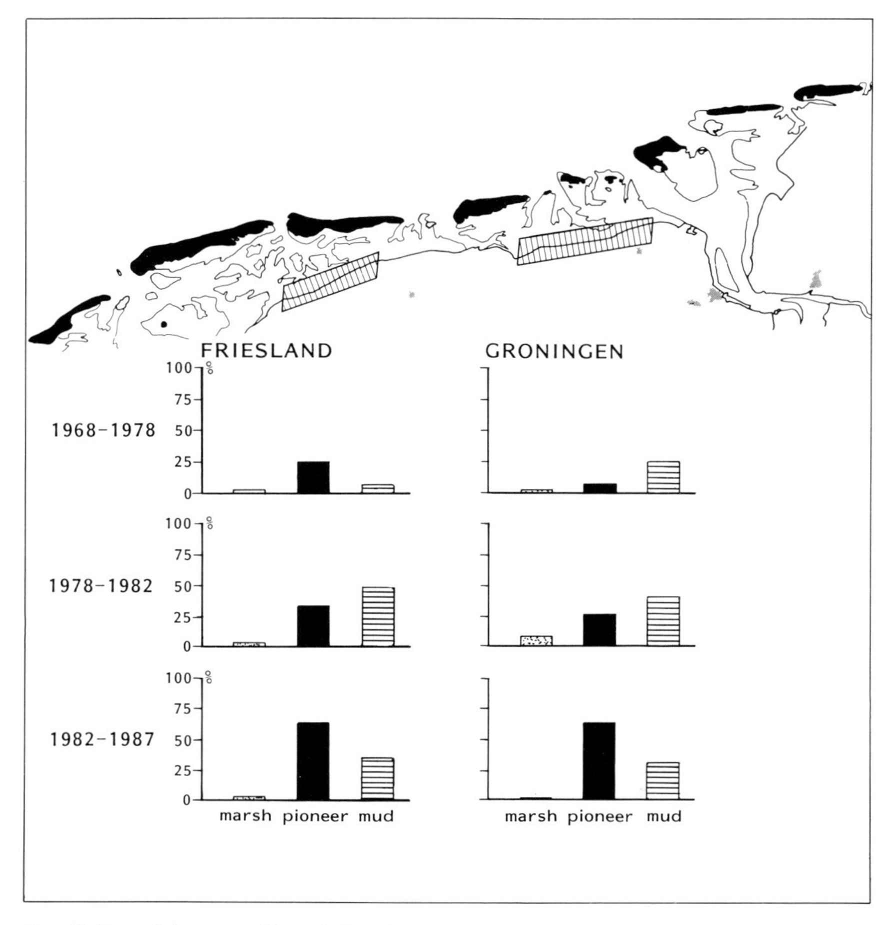

Fig. 10. Part of the coastal length (in %) with an accretionary deficit relative to MHT for the man-made salt marshes and mud flats in the Netherlands Wadden Sea.

The mean vertical accretion in Friesland is very close to the rates on the adjacent tidal flats (a 1-km wide strip of tidal flats in front of the sedimentation fields; DE BOER & KOOMEN, 1983). However, local variation in the two zones are not correlated. The variation in size of the sedimentation fields has no effect on the vertical accretion (DIJKEMA *et al.*, 1988). Even the construction of a coastal parallel groyne (Groningen 1960-1968; Fig. 8) did not improve the sedimentation conditions in the outer sedimentation fields.

It may be concluded that the accretion on the adjacent tidal flats is more important for the accretion in the outer sedimentation fields than the lay-out of the groynes. Nevertheless, the groynes are responsible for a large initial accretion.

- 2. The pioneer zone includes the sedimentation fields between 0.60 m and about 0.20 m below MHT, partly with an open pioneer vegetation of *Spartina anglica, Salicornia dolichostachya*, or both. This zone is transitional between the mud and the overgrown marsh and has a width of about 400 m. The accretionary balances for the coastal stretches with high and low accretion are completely different (Fig. 9). In the higher accretionary parts (Friesland-west and -east flanks and Groningen-west) the vertical accretion regularly increases from the mud zone through the pioneer zone towards the salt-marsh zone. The accretionary balance relative to MHT-rise is positive for each zone (even twice as much or more than the present MHT-rise).

In the lower accretionary parts (Friesland-central and Groningen-east) a decline is found in the accretionary sequence in the pioneer zone, reaching a negative accretionary balance of 0.7 to 0.8 $cm.y^{-1}$ for the period 1978-1987 and covering more than 60% of the coastal length presently (Fig. 10). This accretionary deficit is more than the annual MHT-rise of 0.44 $cm.y^{-1}$, thus constitutes real erosion. The erosion is mainly found in observation fields in the pioneer zone where the fetch has been increased due to bad up-keep of the groynes (Fig. 9).

It is concluded that, contrary to the outer sedimentation fields, vertical accretion in the pioneer zone is dependent on the size of the sedimentation fields. This is due to the large effect of wave energy in this zone especially (OLSEN, 1959; DRONKERS, 1984; SCHOOT & VAN EERDT, 1985). Waves are an important obstacle for natural salt-marsh formation. The groynes are most effective in affecting the wave factor in this zone, providing that the fetch for the main wind direction is not longer than ~200 m.

- 3. The salt-marsh zone includes the landward row of sedimentation fields above a level of about 0.10 m below MHT with a sward of mostly perennial salt-marsh species (DIJKEMA, 1983a; BOUWSEMA, 1986, 1988). Groynes no longer occur at this elevation. The accretionary balance is positive relative to MHT-rise in all observation fields and all time periods (except for one case where the accretion equals MHT-rise). The vertical accretion doubles in the higher accretionary parts or turns from erosion into accretion in the lower parts due to the protection and the drying of the sediment by a closed vegetation cover. The maximum vertical accretion rate of all zones is found from MHT to MHT + 0.20 m. With a further increase in elevation the vertical accretion decreases again due to the decreased number of tidal floodings (DE GLOPPER, 1981; BOUWSEMA *et al.*, 1986).

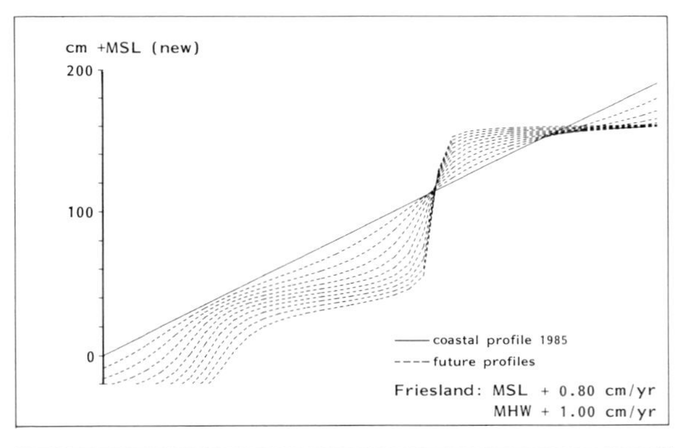

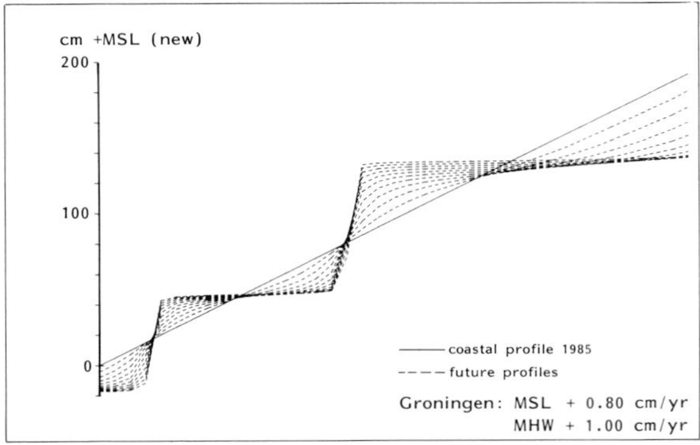

Fig. 11. The future coastal profiles in the man-made salt marshes and mud flats in the Netherlands Wadden Sea, calculated as 10-year intervals from 1985 on. Adapted after BROUNS, 1988.

BOUWSEMA *et al.* (1986), DE GLOPPER (1986) and DIJKEMA *et al.* (1988) could not find a direct relation between changes in drainage techniques and vertical accretion. Because in the long run a closed vegetation cover enhances the vertical accretion more than sedimentation fields, draining surely has an indirect effect on the vertical accretion: when the elevation is sufficient, draining promotes a closed vegetation cover.

The accretionary balance has, on an average, mean positive values of 1.6 cm.y^{-1} for Friesland and 1.1 cm.y^{-1} for Groningen. The highest mean values occur in Friesland (western part) with ~4 cm.y^{-1}. Gross vertical accretion values are 0.44 cm.y^{-1} more and are comparable to other foreland-type salt marshes. With marker horizon techniques DE GLOPPER (1981) measured accretion rates of 1.0 to 2.3 cm.y^{-1} for low salt marshes near our man-made salt marshes. DIECKMANN (1988) found averages of 0.6 to 2.5 cm.y^{-1} for Schleswig-Holstein (German Wadden Sea). On natural foreland salt marshes in the Braakman (Southwest-Netherlands), VERHOEVEN & AKKERMAN (1967) measured rates of 1 to 5 cm.y^{-1} for the creek levees and 0.7 to 1.7 cm.y^{-1} for the basins.

5. CAN COASTAL MARSHES SURVIVE FUTURE SEA-LEVEL RISE ?

5.1. INTRODUCTION

As explained in section 3 the answer to this question firstly depends on the long-term rate of MHT-rise. Apart from the large uncertainties in mean sea-level rise, predictions of the future MHT-levels are lacking. This point needs attention in further studies.

The MHT-rise has to be compared to both the marsh zone and to the tidal flats in front of it. The zones in front of the salt marsh hardly get any attention in papers on accretion and erosion of salt marshes; the only criterion for survival is a positive accretionary balance for the marsh zone itself (*e.g.* HATTON *et al.*, 1983; ORSON *et al.*, 1985; STEVENSON *et al.*, 1986). However, the possible loss of salt marsh through erosion from the seaward edge depends on the accretionary processes in the zones in front of the marsh (DIJKEMA *et al.*, 1988; BOORMAN *et al.*, 1989).

5.2. PIONEER ZONE

Sea-edge erosion is what has to be expected for more than half of the Netherlands man-made salt marshes. If the present accretionary deficit in the pioneer zone continues (Figs 9 and 10), cliff formation on a large scale will be the result (Figs 11 and 12). Cliffs show a rate of horizontal retreat from 0.02 to 0.40 m.y^{-1} along the Groningen foreland (DE

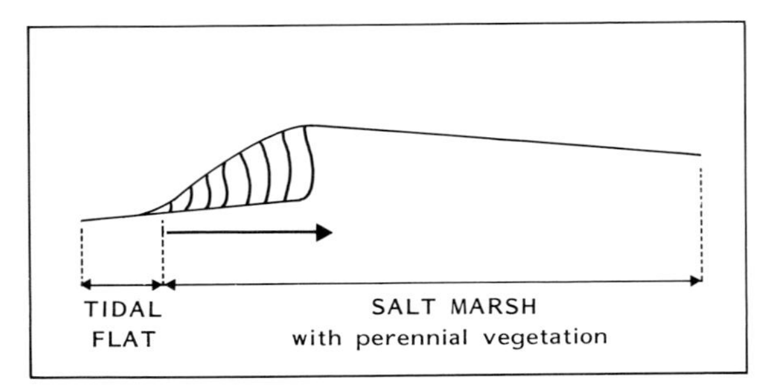

Fig. 12. Cross-section of a salt marsh with the successive stages of sea-edge erosion through cliff formation. After DIJKEMA *et al.*, 1988.

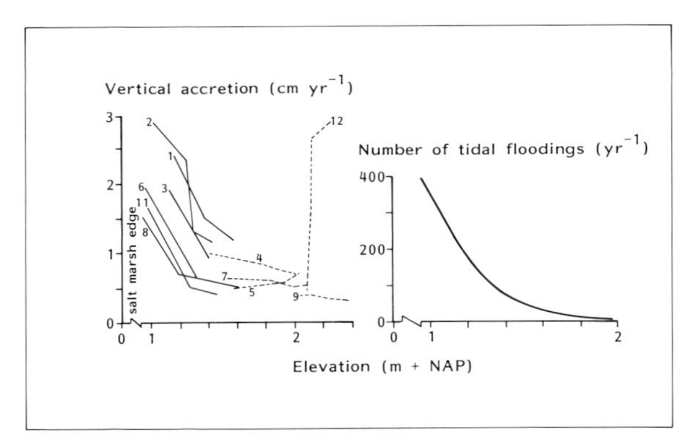

Fig. 13. Relation between the annual vertical accretion rate and the elevation of the marsh, compared to the annual number of submergions. Data collected for the period 1962-1973 with help of the marker horizon technique on artificially drained salt marshes of the foreland type in Friesland (1-3) and Groningen (4-12). After DE GLOPPER (1981).

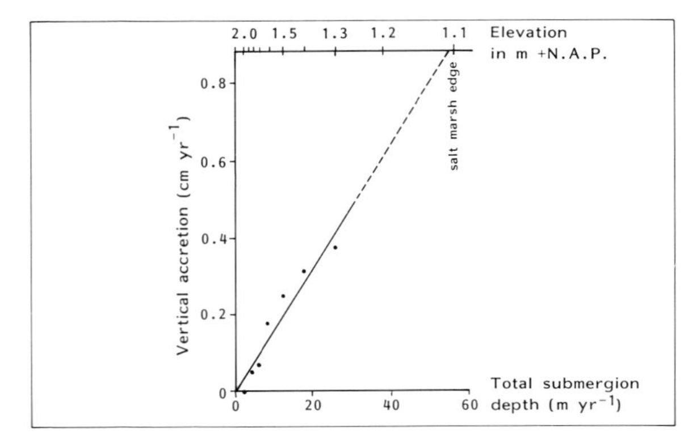

Fig. 14. Relation between the annual vertical accretion rate and the elevation of the marsh, compared to the total submergion depth. Data based on contour maps of 1962 and 1986 of the barrier type of salt marsh Nieuwlandsrijd on Ameland. After EYSINK (1987).

GLOPPER, 1981), 0.40-0.80 m.y^{-1} along the Oosterschelde foreland (VAN EERDT, 1985a), 1.93 m.y^{-1} along the Ems estuary (DE GLOPPER, 1981), 6 m.y^{-1} on former man-made salt marshes in Germany (ERCHINGER, 1987) to 3 to 11 m.y^{-1} along the Dengie foreland (Essex, Great Britain, HARMSWORTH & LONG, 1986).

Sea-edge erosion of salt marshes has been observed from the middle of the 19th century for most of the barrier islands in the Wadden Sea (EHLERS, 1988). The horizontal retreat was about 0.5 m.y^{-1}. Perhaps the conditions are not adequate for vertical accretion on both the salt marshes and the adjacent tidal flats on the barrier islands; the concentrations of suspended mud in front of the Groningen man-made marshes appeared to be four times the concentrations south of the barrier islands (KAMPS, 1962).

5.3. MARSH ZONE

It is unlikely that an increased sea-level rise will directly affect the areal extent of the foreland type of salt marshes in the Wadden Sea through regressive succession of vegetation in the lower marsh zone. The vertical accretion (Fig. 9) is high enough to compensate for a future sea-level rise of 1 to 2 cm.y^{-1} which is more than the expected global sea-level rise for the next century. For the higher marsh zones this is not true with respect to their present vertical accretion (DE GLOPPER, 1981). However, increase of the number of tidal floodings due to an accretionary deficit will increase the accretion to the rate of the lower marsh zone again (Fig. 13; CHAPMAN, 1976; DE GLOPPER, 1981; PETHICK, 1981; BOUWSEMA *et al.*, 1986; EYSINK, 1987). In general, this mechanism will provide for marshes to accrete in vertical direction in response to relative sea-level rise (STEVENSON *et al.*, 1986).

The outlook for the barrier type of salt marshes in the Wadden Sea is bad if the sea-level rise increases to values of 0.5 to 1.0 cm.y^{-1} or more. There is evidence that there is a balance between the present MHT-rise and accretion of the marsh. For a low marsh site on the barrier island of Terschelling a vertical accretion of 12 cm between 1958 and 1979 has been recorded by marker horizon techniques (0.6 cm.y^{-1}; ROOZEN, 1985). EYSINK (1987) found accretion rates of 0.1 to 0.4 cm.y^{-1} on the barrier island of Ameland, which may increase to maximal 0.3 to 0.8 cm.y^{-1} if the number of tidal floodings should increase to that of the lower marsh (Fig. 14).

If sea-level rise will lead to a long-term accretionary deficit in the marsh zone, the survival of the salt marsh depends on the response of the vegetation on the increased number of tidal floodings. For the barrier island of Ameland DANKERS *et al.* (1987) predicted that at a subsidence rate of 0.8 to 1.5

cm.y^{-1} due to gas extraction, regression to communities of lower marsh zones or to bare tidal flats will occur, even if the increased accretion is taken into account. However, the response of vegetation to year-to-year changes in MHT-levels may disturb this process. It is hypothesized that a delay in the regression of vegetation (section 3.3) may allow an increased vertical accretion during years with raised MHT-levels and thus may compensate for (part of) the sea-level rise.

6. COULD MARSH ACCRETION-ENHANCEMENT TECHNIQUES BE A TOOL IN PREVENTING MARSH LOSS?

6.1. NEGATIVE ACCRETIONARY BALANCE IN THE PIONEER ZONE

A general conclusion is that management techniques in order to prevent negative effects of increased sea-level rise have to direct most attention to the pioneer zone where marsh growth starts. If this zone keeps a positive accretionary balance there is no problem at all. If the balance becomes negative, techniques that enhance natural accretion might be a better management tool, from the viewpoint of nature preservation, than traditional solutions like revetments, sediment nourishment, etc.

The man-made salt marshes in the Netherlands Wadden Sea have shown sea-edge erosion during a period with rising high-tide levels. The marsh zone itself had a positive accretionary balance. In a management proposal for the Department of Public Works (DIJKEMA *et al.*, 1988) the authors propose modifications in the lay-out of the sedimentation fields in the pioneer zone (the zone in front of the salt marsh which is most affected by wave energy) to counteract the sea-edge erosion. In accordance with these proposels, the fetch between the main groynes will be reduced to 200 m in this zone. Also the elevation of the groynes will be raised and adapted to recent MHT-levels and to subsidence due to gas extraction. (On the other hand, seaward coast parallel groynes in the outer sedimentation fields could be abandoned.)

Further research in the pioneer zone is needed to develop accretion-enhancement techniques to prevent future marsh loss through increased sea-level rise. The study should include the processes of sedimentation and plant establishment in front of both man-made and natural salt marshes:

- a. Vegetation dynamics in the pioneer zone as a function of:
- 1. year to year fluctuations in MHT (effect on the aeration pattern of Armstrong);
- 2. drainage systems (effects of natural creek systems and man-made drains through soil aeration or soil sulphide);
- 3. wave energy levels (effect of the fetch in sedimentation fields);
- 4. soil fertility (effects of sediment composition and accretion rate).

- b. The accretional balance in the pioneer zone as a function of:
- 1. vegetation structure (effect on sedimentation and protection against erosion);
- 2. year to year fluctuations in MHT (effects on sediment input);
- 3. wave energy levels (effects of modified sedimentation fields);
- 4. drainage techniques (effects on sediment input, sedimentation and protection against erosion).

6.2. NEGATIVE ACCRETIONARY BALANCE IN THE MARSH ZONE

The possibilities to prevent marsh loss are small for salt marshes with a negative accretionary balance. The conditions for vertical accretion have to be optimized by appropriate management (*e.g.* low intensity of grazing; in some cases drainage may be helpful).

Field studies in the marsh zone should provide more insight into the effects of sea-level fluctuations during a long-term trend of relative sea-level rise. In the Wadden Sea the authors conducted a monitoring study at field sites with a relatively high vertical accretion rate (rising high-tide levels in man-made salt marshes; section 4), while the first author has started such a study at a field site with a low vertical accretion rate (subsidence due to gas extraction on the barrier island Ameland; section 5). As a base for future management additional experimental studies are needed on the processes of sedimentation and vegetation dynamics at these field sites:
- a. Vegetation dynamics in the marsh zone during a long-term trend of relative MHT-rise (effects of year-to-year fluctuations in MHT through soil aeration, soil sulphide or soil salinity; does the aeration pattern of Armstrong follow the MHT differences or is it dependent on the drainage system?).
- b. The accretionary balance in the marsh zone during years with raised MHT-levels (the hypothesis in section 5.3 states that a delay in the regression of vegetation may allow an increased vertical accretion).

Above all, however, to prevent marsh loss it is important to eliminate subsidence and any negative impacts on sediment supply. The Mississippi Delta is a good example of a coastal area where sediment starvation caused by human impact leads to enormous marsh losses (DAY & TEMPLET, this issue; BAUMANN *et al.*, 1984). When there is no real chance to prevent the marsh loss by accretion-enhancement (when the sediment supply is insufficient), it might be better to influence the large-scale dynamic processes by sediment nourishment or allow transgression processes to rebuild barriers, tidal flats and salt marshes. For the barrier islands in the Wadden Sea with their low vertical accretion rate these might be the only options in the long run.

ACKNOWLEDGEMENTS

We would like to thank the management of our departments for the confidence in our work and the willingness to employ the results. The manuscript was critically reviewed by Dr. Wim Beeftink of the Delta Institute for Hydrobiological Research, Yerseke, Dr. Joop J.W.M. Brouns of the Research Institute for Nature Management, Texel (RIN) and Gail L. Chmura of the Center for Wetland Resources, Louisiana State University. Text processing was performed by Michaela Scholl of the RIN and drawings by Arjan J. Griffioen of the RIN.

7. REFERENCES

ARMSTRONG, W., E.J. WRIGHT, S. LYTHE & J.T. GAYNARD, 1985. Plant zonation and the effects of the spring-neap tidal cycle on soil aeration in a Humber salt marsh.—J. Ecol. **73:** 323-339.

AUBREY, D.G., 1984. Recent sea levels from tide gauges: problems and prognosis. In: Glaciers, ice sheets and sea level: effect of a CO_2-induced climatic change. United States Department of Energy, Washington: 73-91.

BAUMANN, R.H., J.W. DAY & C.A. MILLER, 1984. Mississippi deltaic wetland survival: Sedimentation versus coastal submergence.—Science **224:** 1093-1095.

BEEFTINK, W.G., 1965. De zoutvegetatie van ZW-Nederland beschouwd in Europees verband. Med. Landbouwhogeschool 65-1. Veenman, Wageningen: 1-167.

——, 1975. The ecological significance of embankment and drainage with respect to the vegetation of the south-west Netherlands.—J. Ecol. **63:** 423-458.

——, 1977. The coastal salt marshes of western and northern Europe: an ecological and phytosociological approach. In: V.J. CHAPMAN. Wet coastal ecosystems. Elsevier, Amsterdam, Oxford, New York: 109-155.

——, 1979. The structure of salt-marsh communities in relation to environmental disturbances. In: R.L. JEFFERIES & A.J. DAVY. Ecological processes in coastal environments. Blackwell, Oxford, London, Edinburgh, Melbourne: 77-93.

——, 1987a. Vegetation responses to changes in tidal inundation of salt marshes. In: J. VAN ANDEL *et al.* Disturbance in grasslands. Junk, Dordrecht: 97-117.

——, 1987b. De betekenis van de factor getij voor de schorrevegetatie. In: J. ROZEMA. Oecologie van Estuariene Vegetatie. Vrije Universiteit Amsterdam; Delta Instituut voor Hydrobiologisch Onderzoek, Yerseke: 1-45.

BERTNESS, M.D. & A.M. ELLISON, 1987. Determinants of pattern in a New England salt marsh plant community.—Ecological Monographs **57:** 129-147.

BOER, M. DE & M.A.J. KOMEN, 1983. Erosie en sedimentatie in de landaanwinningswerken op het Friese wad 1966-1978.—Rijkswaterstaat, Adviesdienst Hoorn, Nota WWKZ-82.H003: 1-22.

BOORMAN, L.A., J.D. GOSS-CUSTARD & S. MC GRORTY, 1989. Climatic change, rising sea level and the British coast.—HMSO, London. Natural Environment Research Council. ITE research publication **1:** 1-24.

BOUWSEMA, P., 1986. Vegetatiekartering in de landaanwinningswerken. In: R.J. DE GLOPPER. Wadden en landaanwinning: voordrachten, gehouden voor de wetenschappelijke bijeenkomst in mei 1985. Flevobericht 252, Rijksdienst voor de IJsselmeerpolders, Lelystad: 23-32.

——, 1988. Vegetatiekarteringen van de Friese en Groninger noordkust (1960, 1965, 1970, 1975, 1980 en 1981/83). Rijkswaterstaat Directie Groningen, Dienstkring Baflo: 1-38 + 6 maps.

BOUWSEMA, P., J.H. BOSSINADE, K.S. DIJKEMA, J.W.TH.M. VAN MEEGEN, R. REENDERS & W. VRIELING, 1986. De ontwikkeling van de hoogte en van de omvang van de kwelders in de landaanwinningswerken in Friesland en Groningen.—Rijkswaterstaat Directie Groningen, Nota ANA-86.05 Rijksinstituut voor Natuurbeheer, RIN-rapport 86/3: 1-58.

BROUNS, J.J.W.M., 1988. The impact of sea level rise on the Dutch coastal ecosystems.—Research Institute for Nature Management, Texel, The Netherlands. RIN-report 88/60. Netherlands Institute for Sea Research. NIOZ-report 1988/8: 1-102.

CHAPMAN, V.J., 1976. Coastal vegetation. Pergamon Press, Oxford etc.: 1-292.

COOPER, A., 1982. The effects of salinity and waterlogging on the growth and cation uptake of salt marsh plants.—New Phytol. **90:** 263-275.

CRAMER, W. & H. HYTTEBORN, 1987. The separation of fluctuation and long-term change in vegetation dynamics of a rising seashore.—Vegetatio **69:** 157-167.

N. DANKERS, K.S. DIJKEMA, G. LONDO, & P.A. SLIM, 1987. De ecologische effecten van bodemdaling op Ameland.—Rijksinstituut voor Natuurbeheer, Texel, RIN-rapport 87/14: 1-90.

DIECKMANN, B., 1987. Bedeutung und Wirkung des Deichvorlandes für den Küstenschutz. In: N. KEMPF, J. LAMP & P. PROKOSCH. Salzwiesen: Geformt von Kustenschutz, Landwirtschaft oder Natur?—WWF-Deutschland, Tagungsbericht 1, Husum: 163-187.

DIECKMANN, R., 1988. Entwicklung der Vorländer an der nordfriesischen Festlandküste.—Wasser und Boden **3:** 146-150.

DIGGELEN, J. VAN, 1988. A comparative study on the ecophysiology of salt marsh halophytes. Thesis Free University, Amsterdam: 1-208.

DRONKERS, J., 1984. Import of fine marine sediment in tidal bassins.—Netherlands Institute for Sea Research, Publication Series **10:** 83-105.

DIJKEMA, K.S., 1983a. The salt-marsh vegetation of the mainland coast, estuaries and Halligen. In: K.S. DIJKEMA & W.J. WOLFF. Flora and vegetation of the Wadden Sea islands and coastal areas. Balkema, Rotterdam: 185-220.

——, 1983b. Use and management of mainland salt marshes and Halligen. In: K.S. DIJKEMA & W.J. WOLFF. Flora and vegetation of the Wadden Sea islands and coastal areas. Balkema, Rotterdam: 302-312.

——, 1987a. Geography of salt marshes in Europe.—Z. Geomorph. N.F. **31** (4): 489-499.

——, 1987b. Changes in salt-marsh area in the Netherlands Wadden Sea after 1600. In: A.H.L. HUISKES, C.W.P.M. BLOM & J. ROZEMA. Vegetation between land and sea. Junk, Dordrecht, Boston, Lancaster: 42-49.

DIJKEMA, K.S., W.G. BEEFTINK, J.P. DOODY, J.M. GEHU, B. HEYDEMANN & S. RIVAS MARTINEZ, 1984. Salt marshes in Europe. Council of Europe, Strasbourg: 1-178.

DIJKEMA, K.S., J. VAN DEN BERGS, J.H. BOSSINADE, P. BOUWSEMA, R.J. DE GLOPPER & J.W.TH.M. VAN MEEGEN, 1988. Effecten van rijzendammen op de opslibbing en de omvang van de vegetatiezones in de Friese en Groninger landaanwinningswerken.—Rijkswaterstaat Directie Groningen, Nota GRAN 1988-2010; Rijksinstituut voor Natuurbeheer, Texel, RIN-rapport 88/66; Rijksdienst voor de IJsselmeerpolders, Lelystad, RIJP-rapport 1988-33 Cbw: 1-119.

EERDT, M.M. VAN, 1985a. Schorkliferosie in de Oosterschelde.—Landschap **2:** 99-108.

——, 1985b. The influence of vegetation on erosion and accretion in salt marshes of the Oosterschelde, The Netherlands.—Vegetatio **62:** 367-373.

EHLERS, J., 1988. Morphologische Veränderungen auf der Wattseite der Barriere-Inseln des Wattenmeeres.—Die Küste **47:** 3-30.

ERCHINGER, H.F., 1985. Dünen, Watt und Salzwiesen. Der Niedersachsische Minister für Ernährung, Landwirtschaft und Forsten, Hannover: 1-59.

——, 1987. Salzwiesenbildung und -erhaltung — Lahnungsbau und Begruppung für den Küstenschutz. In: N. KEMPF, J. LAMP & P. PROKOSCH. Salzwiesen: Geformt von Kustenschutz, Landwirtschaft oder Natur? WWF-Deutschland, Tagungsbericht 1, Husum: 279-296.

ERICSON, L., 1980. The downward migration of plants on a rising Bothnian seashore.—Acta Phytogeographica Suecica **68:** 61-72.

EYSINK, W.D., 1987. Gaswinning op Ameland-oost, effecten van de bodemdaling.—Delfts Hydraulic Laboratory, Report H114: 1-53.

FEEKES, W., 1950. Bouw- en plantengroei. In: G.A. BROUWER *et al.* Griend, het vogeleiland in de Waddenzee. Martinus Nijhoff, 's Gravenhage: 82-187.

FÜHRBÖTER, A., 1986. Veranderungen des Sakularanstieges an der deutschen Nordseeküste.—Wasser und Boden **9:** 456-460.

GLOPPER, R.J. DE, 1981. De snelheid van de opslibbing en van de terugschrijdende erosie op de kwelders langs de noordkust van Friesland en Groningen. In: 50 jaar onderzoek. Flevobericht 163. Rijksdienst voor de IJsselmeerpolders, Lelystad: 43-51.

——, 1986. De Rijkslandaanwinningswerken; werkwijzen en enkele onderzoeksresultaten. In: R.J. DE GLOPPER. Wadden en landaanwinning: voordrachten, gehouden voor de Wetenschappelijke Bijeenkomst in mei 1985. Flevobericht 252. Rijksdienst voor de IJsselmeerpolders, Lelystad: 23-32.

GORNITZ, V. & S. LEBEDEFF, 1987. Global sea-level changes during the past century. In: D. NUMMEDAL, O.H. PILKEY & J.D. HOWARD. Sea-level fluctuations and coastal evolution. Society of Economic Paleontologists and Mineralogists Spec. Pub. 41, Tulsa, Oklahoma: 3-16.

GROENENDIJK, A.M., 1986. Establishment of a *Spartina anglica* population on a tidal mudflat: a field experiment.—J. Environmental Management **22:** 1-12.

GROENENDIJK, A.M., J.G.J. SPIEKSMA & M.A. VINK-LIEVAART, 1987. Growth and interactions of salt-marsh species under different flooding regimes. In: A.H.L. HUISKES, C.W.P.M. BLOM & J. ROZEMA. Vegetation between land and sea. Junk, Dordrecht, Boston, Lancaster: 236-258.

HARMSWORTH, G.C. & S.P. LONG, 1986. An assessment of saltmarsh erosion in Essex, England, with reference to the Dengie Peninsula.—Biological Conservation **35:** 377-387.

HATTON, R.S., R.D. DELAUNE & W.H. PATRIC, 1983. Sedimentation, accretion, and subsidence in marshes of Baratavia Basin, Louisiana.—Linmol. Oceanogr. **28:** 494-502.

HAYES, M.O., 1979. Barrier island morphology as a function of tidal and wave regime. In: S.P. LEATHERMAN. Barrier Islands, from the Gulf of St. Lawrence to the Gulf of Mexico. Academic Press, New York: 1-27.

JAKOBSEN, B., 1954. The tidal area in south-western Jutland and the process of the salt marsh formation.—Geogr. Tidsskr. **53:** 49-61.

JENSEN, J., H.-E. MÜGGE & G. VISSCHER, 1988. Untersuchungen zur Wasserstandsentwicklung in der Deutschen Bucht.—Die Küste **47:** 135-161.

KAMPS, L.F., 1962. Mud distribution and land reclamation in the eastern wadden shallows.—Rijkswaterstaat comm. 4, Den Haag: 1-73.

KÖNIG, D., 1948. *Spartina townsendii* an der Westkuste von von Schleswig-Holstein.—Planta **36:** 34-70.

——, 1987. Historisches über Wattenmeer-Salzwiesen. - In: N. KEMPF, J. LAMP & P. PROKOSCH. Salzwiesen: Geformt von Küstenschutz, Landwirtschaft oder Natur? WWF-Deutschland. Tagungsbericht 1, Husum: 31-70.

LAFRENZ, P., 1957. Über die Pflege und Nutzung des Anwachses und der Deiche an der Dithmarscher Küste.—Die Küste **6:** 94-129.

LAMB, H.H. & I. WEISS, 1979. On recent changes of the wind and wave regime of the North Sea and outlook.—Fachliche Mitteilungen, Amt für Wehrgeophysik **194:** 1-108.

LEEUW, J.E., H. OLFF & J.P. BAKKER, in prep. Year-to-year variation in salt-marsh production as related to inundation and rainfall deficit.

MALDE, J. VAN, 1984. Voorlopige uitkomsten van voortgezet onderzoek naar de gemiddelde zeeniveaus in de Nederlandse kustwateren.—Rijkswaterstaat, Directie Waterhuishouding en Waterbeweging, Nota WW-WH 84.08: 1-20.

MENDELSSOHN, I.A & K.L. MC KEE, 1988. *Spartina alterniflora* die-back in Louisiana: time-course investigation of soil waterlogging effects.—J. Ecol. **76:** 509-521.

MENDELSSOHN, I.A., K.L. MC KEE & W.H. PATRIC, 1981. Oxygen defiency in *Spartina alterniflora* roots: metabolic adaptation to anoxia.—Science **214:** 439-441.

NIEMEYER, H.D., 1977. Seegangsmessungen auf Deichvorländer.—Forschungsstelle Norderney, Jahresbericht 1976, **28:** 113-139.

NOOMEN, P., in prep. Bodembeweging in Nederland. Proceedings, Symposium KIvI, RWS e.a.

OERLEMANS, J., 1989. A projection of future sea level.—Climatic Change: in press.

OLFF, H., J.P. BAKKER & L.F.M. FRESCO, 1988. The effect of fluctuations in tidal inundation frequency on a saltmarsh vegetation.—Vegetatio **78:** 13-19.

OLSEN, H.A., 1959. The influence of the Rømø Dam on the sedimentation in the adjacent part of the Danish Wadden Sea.—Geogr. Tidsskr. **58:** 119-140.

ORSON, R., W. PANAGEOTOU & S.P. LEATHERMAN, 1985. Response of tidal salt marshes of the US Atlantic and Gulf coast to rising sea level.—J. Coastal Res. **1:** 29-37.

PETHIC, J.S., 1981. Long-term accretion rates on tidal salt marshes.—J. Sedim. Petrology **51:** 571-577.

RANWELL, D.E., 1972. Ecology of salt marshes and sand dunes. Chapman and Hall, London: 1-258.

ROBIN, G. DE Q., 1986. Changing the sea level. In: B. BOLIN, B.R. DÖÖS, J. JÄGER & R.A. WARRICK. The greenhouse effect, climatic change and ecosystems. SCOPE 29, Wiley, London.

ROOZEN, A.J.M., 1985. Een kwart eeuw onderzoek aan vegetatiesuccessie op de Boschplaat van Terschelling.—De Levende Natuur **86:** 74-80.

ROZEMA, J., P. BIJWAARD, G. PRAST & R. BROEKMAN, 1985. Ecophysiological adaptations coastal halophytes from foredunes and salt marshes. —Vegetatio **62:** 499-521.

SCHOLTEN, M., P.A. BLAAUW, M. STROETENGA & J. ROZEMA, 1987. The impact of competitive interactions on the growth and distribution of plant species in salt marshes. In: A.H.L. HUISKES, C.W.P.M. BLOM & J. ROZEMA. Vegetation between land and sea. Junk, Dordrecht, Boston, Lancaster: 270-281.

SCHOOT, P.M. & M.M. VAN EERDT, 1985. Toekomstige ontwikkeling van de schorgebieden in de Oosterschelde. Rijksuniversiteit Utrecht, Geografisch Instituut; Rijkswaterstaat Deltadienst, Hoofd afd. Milieu en Inrichting. Nota DDMI 85.23.

SNOW, A.A. & S.W. VINCE, 1984. Plant zonation in an Alaskan salt marsh II. An experimental study of the role of edaphic conditions.—J. Ecol. **72:** 669-684.

STEVENSON, J.C., L.G. WARD & M.S. KEARNEY, 1986. Vertical accretion in marshes with varying rates of sea level rise. In: D.A. WOLFE. Estuarine variability. Academic Press, San Diego, New York etc.: 241-259.

STRATINGH, G.A. & C.A. VENEMA, 1855. De Dollard of Geschied-, Aardrijk- en Natuurkundige beschrijving van dezen boezem der Eems. Oomkens, Zoon & Schierbeek, Groningen: 1-329.

VEENSTRA, H.J., 1980. Introduction to the geomorphology of the Wadden Sea area. In: K.S. DIJKEMA, H.-E. REINECK & W.J. WOLFF. Geomorphology of the Wadden sea area. Balkema, Rotterdam: 8-19.

VERHOEVEN, B. & J. AKKERMAN, 1967. Buitendijkse mariene gronden, hun opbouw, bedijking en ontginning.—Van Zee tot Land 45, Tjeenk Willink, Zwolle: 1-93.

VERHOEVEN, B., M. JESPERSEN, D. KÖNIG & E. RASMUSSEN, 1980. Human influences on the landscape of the Wadden Sea area. In: K.S. DIJKEMA, H.-E. REINECK & W.J. WOLFF. Geomorphology of the Wadden sea area. Balkema, Rotterdam: 110-135.

VINCE, S.W. & A.A. SNOW, 1984. Plant zonation in an Alaskan salt marsh. I. Distribution, abundance and environmental factors.—J. Ecol. **72:** 651-667.

WEIHE, K. VON, 1979. Morphologische und ökologische Grundlagen der Vorlandsicherung durch *Puccinellia maritima* (Gramineae).—Helgol. Wiss. Meeresunters. **32:** 239-254.

WESTHOFF, V., 1987. Salt marsh communities of three Westfrisian Islands, with some notes on their long-term succession during half a century. In: A.H.L. HUISKES, C.W.P.M. BLOM & J. ROZEMA. Vegetation between land and sea. Junk, Dordrecht, Boston, Lancaster: 16-40.

WIGLEY, T.M.L. & S.C.B. RAPER, 1987. Thermal expansion of sea water associated with global warming. —Nature **330:** 127-131.

WOHLENBERG, E., 1933. Das Andelpolster und die Entstehung einer charakteristischen Abrasionsform im Wattenmeer.—Wiss. Meeresunters. NF. Abt. Helgoland **19** (4): 1-3.

——, 1953. Sinkstoff, Sediment und Anwachs am Hindenburgdam.—Die Küste **2:** 31-94.

WOLFF, W.J., 1983. Ecology of the Wadden Sea. Balkema, Rotterdam: 1-2050.

INSHORE BIRDS OF THE SOFT COASTS AND SEA-LEVEL RISE

J.D. GOSS-CUSTARD[1], S. MCGRORTY[1] AND R. KIRBY[2]

[1] *Institute of Terrestrial Ecology, Furzebrook Research Station, Wareham, Dorset, BH20 5AS, U.K.*

[2] *Ravensrodd Consultants, 6 Queens Drive, Taunton, Somerset, TA1 4XW, U.K.*

ABSTRACT

This paper discusses the possible effect of a rise in sea level of about 1 m in 100 years on the birds that utilise inshore waters and the sedimentary intertidal zone in North-West Europe. In open coastal areas, such an increase in sea level could especially affect the sediments, marshes and invertebrates. These changes in turn could affect the feeding and breeding of many species of birds through the greater erosion of the intertidal flats and marshes, and the consequent reduction in the quality and area of habitat available for both feeding and breeding, and possibly through an increase in the levels of turbidity of inshore waters. In more sheltered areas, the effects are more difficult to predict because of uncertainty about how the sediment regime may change, relative to the rate of rise of the sea level.

1. INTRODUCTION

The mudflats, sandflats and marshes of the intertidal zone in North-West Europe, along with the adjacent inshore waters, provide feeding and/or breeding grounds for millions of birds. Most of these birds are migratory. Many breed as far north as the Arctic and pass along the coasts of Europe on their way to, or back from, wintering areas further to the south and west. For these birds, the inshore waters and flats provide a vital source of food for replenishing their energy reserves for onward migration and, frequently, safe places where the annual moult of the flight feathers can take place. Others remain to winter in these areas, though this is less common in the eastern regions where the arrival of ice in winter drives any remaining birds further southwards and westwards. Some birds also breed in the salt marshes that flank the intertidal flats, or use the flats to obtain food to feed their young being reared inland. For many birds, these areas of coast provide important resources that are used in one way or another throughout the year.

This paper reviews the possible consequences for these birds of a rise in sea level of 1 m over the next 100 years. It discusses how such a change might affect the sediment and shore profiles, how these in turn might affect the birds' resources and thus the numbers of birds themselves.

2. GENERAL ENVIRONMENTAL CONSEQUENCES OF A RISE IN SEA LEVEL

Predicting the effects of sea-level rise, with all the variables which will interact, is a difficult task. Moreover, the effect is likely to vary according to whether the shore in question is an open coastal flat or large estuary, or, alternatively, whether it is an estuary well protected by a narrow entrance.

2.1. OPEN COASTAL FLATS AND LARGE ESTUARIES

Without sea walls or any other kind of barrier to the advance of the sea, a rise in sea level should lead to the classic 'sea-level transgression' over land. In this, erosion of the coast occurs and the sediments thereby derived are pushed inshore and redeposited. Provided the slope of the shore remained unchanged, no coastline should be lost and the shore would retain its existing width. It would take many years for a marsh to be eroded and for new mudflats to develop, and thus it would be expected that the rate of change to the new sediment regime would be slow. In other words, there would be an inevitable lag.

However, along low-lying shorelines, this transgression is prevented at present by sea walls. With a rise in sea level, the overall width of the intertidal zone would decrease as low water mark rises. It is expected that erosion would increase at all levels of the shore, mainly as a consequence of the increased rates of 'wave attack'. The derived sediments would be transported by the currents but would not be able to reach the hinterland because of the sea walls. Much of the sediment would be deposited in 'sinks' elsewhere, probably below the gradually rising low tide mark.

Gently-sloping coasts should develop cliffs at the

J.J. Beukema et al. (eds.), Expected Effects of Climatic Change on Marine Coastal Ecosystems, 189–193.

seaward edge of the marsh as erosion occurs. The marshes would fragment and deeper creeks would develop, thus further decreasing the total area of vegetation in the marsh.

Seawards of the marsh edge, the increased rates of erosion would increase sediment instability. More 'winnowing' of the sediments would probably occur. That is, because the sediments would be more frequently in suspension in an environment with generally higher levels of energy, more of the finer particles should be removed and carried off elsewhere. Depending on the nature of the sediment already present, this could produce two trends in sediment change. One is that in the higher tidal flats especially, the sediments would become increasingly composed of 'lag deposits' of coarser materials, such as gravel or the shells of large dead , molluscs. The other is that, with the finer sediments gradually removed, only the underlying layers of rock or hard clay would remain. Soft coasts would tend to be replaced by harder and/or coarser ones. So although the erosion of the marsh at the upper levels of the shore would compensate to some extent for the loss of mudflats that would occur through inundation at the bottom of the shore, the flats that remain would be changed in kind.

In addition to becoming narrower and steeper, it is also possible that, as the downshore areas erode further, the profile would become more concave so that the length of the exposure time at low tide would decrease. However, this is speculation, as there is no well-established link between wave climate and the shape of the shore profile. But even so, exposure time would decrease because, as a result of winnowing, the level of the surface would in any case be lowered. Because of this 'degradation' of the mudflats, the rise in sea level would have a greater impact on exposure time than would be apparent from the rise in sea level alone.

2.2. AREAS PROTECTED BY NARROW ENTRANCES

In estuaries and inlets well protected from the vigorous wave climate of the open coasts, the postulated increase in sediment supply in the sea should, in the first instance, lead to accretion, the reverse of what would happen along the open coasts. It is not yet possible to predict the probable net rate of change in the levels of the flats in these areas, nor would it be easy to do. The extent to which it happened would depend on the unpredictabele relative rates of rise in the level of the sea and of the flats. But any increase in the level of the tidal flats relative to sea level would lead to an encroachment by marsh vegetation onto the flats. At some stage, as the sea level rose still further, the coastal dunes and sand bars that often protect such sheltered inlets would themselves be eroded and over-topped; the mouth of the estuary would be widened. When this occurred, the estuary or bay would no longer be protected, so that erosion should set in and the erosional trends predicted for the open coasts would occur in these sheltered estuaries and bays too.

2.3. GENERAL ECOSYSTEM EFFECTS

The increased sediment instability in the coastal waters would probably have a number of consequences. If the turbidity of the water arising from the increased rates of erosion were sufficient to decrease light penetration, primary productivity due to phytoplankton might decrease. Were this to happen, coastal areas in general could be less biologically productive. Second, the general mobility of the sediments should increase so that changes in sediment type and level would be expected to occur more often, as may happen already in some highly dynamic systems, such as the Severn estuary. Third, unless they were prevented by improved coastal defences, storm surges would become more damaging and breach the sea defences more often. Finally, the ground water inland would become more saline and salt water would penetrate further upstream in rivers; both are occurring now in Louisiana (SALINAS *et al.*, 1986).

3. INVERTEBRATES OF THE INTERTIDAL ZONE

Most of the birds that inhabit the intertidal areas feed on the macro-invertebrates that live in the sediments. Therefore most attention is given to this group of organisms.

3.1. OPEN COASTAL FLATS

Were there no sea walls or other barriers to the sea, the shore would in effect simply move inland. However, the rate of change may have implications for the invertebrates. Since in general the rate of change would be slow in relation to the life cycles of the invertebrates, the fauna overall is unlikely to show any changes other than 'normal' population fluctuations. However, eroding sediments usually have poor faunas compared with areas of deposition, presumably because larvae have difficulties in setttling.

The most realistic scenario is that the sea walls will remain and be improved. The changes in the environment detailed above could have many implications for the invertebrates.

Any narrowing of the beach would lead to a reduction in the total area available at the levels of the shore that particular species live. This would reduce the total abundance of invertebrates.

Any concentration of the erosion in the upper mudflats would favour those species that live in the lower flats. However, in several of these, the larvae settle in the siltier sediments at the top of the beach. The predicted changes in the upper mudflats could thus adversely affect species such as *Arenicola marina* and *Macoma balthica* which behave this way.

As the proportion of fine particles in the sediment is reduced by erosion, there will be a corresponding reduction in the organic matter content of the sediment (WOLFF, 1973). This in turn would lead to a reduction in the productivity of deposit feeding invertebrates, such as *Arenicola marina* (LONGBOTTOM, 1970).

If the reduction in fine particles and the associated organic matter were to continue, deposit feeders could be replaced by suspension feeders. However, if the concentration of particles in the water column rose above a critical threshold, the suspension feeders would also be reduced because their feeding efficiency declines as their filtration mechanisms become blocked (WIDDOWS *et al.*, 1979). Any reduction in the phytoplankton productivity in the inshore waters would disadvantage both deposit and suspension feeders alike.

If erosion of the surfase sediments causes the substrate to become increasingly covered with lag deposits of shells and gravel, some protection might be afforded to any finer deposits lying beneath. Such areas may have good populations of shellfish and worms, but may equally have only an impoverished fauna. Without any underlying fine sediments, coarser substrates consisting of sand, shells or stones have very poor faunas because of the grinding action of the particles when moved by waves and currents.

The hard, clay base that would be the ultimate result of continued erosion supports just a few burrowing animals.

Overall, the trend in the open coasts would be towards less productive coarse sediments or hard clay substrates. A greater proportion of the intertidal area of estuaries would also become marine and sandy and the brackish section, with its very productive mudflats, would be reduced in area and located further inshore.

3.2. AREAS PROTECTED BY NARROW ENTRANCES

The dominant sedimentary process in such places would initially be deposition: low-level flats would be eventually replaced by higher-level flats. This will lead to a change in the species composition of the invertebrates, but the flats would remain productive until they reached a level at which they were not covered by water on Neap tides. Colonisation by salt marsh plants would then occur, and the area of suitable intertidal flats for the major invertebrate food species of birds would decrease. Once the sea-level rise overtops any protective bar at the mouth of the estuary, erosion will begin and the changes noted in the preceding section would be expected to take place.

4. BIRDS

The effects on the birds would depend on the resource and the habitat they exploit.

4.1. BIRDS FEEDING IN THE INSHORE WATERS

A number of birds use the shallow waters as fishing grounds during both the breeding and the non-breeding season. They include, for example, several species of terns, saw-billed ducks such as the red-breasted merganser, *Mergus serrator,* cormorants, *Phalacrocorax carbo,* and shags, *P. aristotelis.*

Any increase in the turbidity of the inshore waters would make it more difficult for these birds to catch fish. Furthermore, the fish stocks themselves might be reduced by the increased turbidity and the stronger wave attack which could act directly on the fish themselves or indirectly on their food organisms.

4.2. BIRDS FEEDING ON THE INTERTIDAL FLATS

Many species of birds exploit the invertebrates that live in and on the intertidal sediments. They include a number of species of wildfowl, notably the shelduck, *Tadorna tadorna*, and eider duck, *Somateria mellisima,* along with several species of gulls, notably the black-headed gull, *Larus ridibundus,* and about 20 species of wading birds. Conservation concern focusses at present on those species for which estuaries are the main source of food, and this includes the shelduck and most waders. Although some of these species breed near the shore and exploit its resources, most of them occur in their largest numbers in North-West Europe outside the breeding season while either on migration or overwintering.

Some wildfowl, notably the brent goose, *Branta bernicla* and the wigeon, *Anas penelope,* also feed in winter on the vegetation living on the surface of the intertidal flats. In parts of their winter range, herbivorous wildfowl, regularly eat out the food supply on the flats and have to move onto other sources. The increased rates of erosion and levels of turbidity expected to occur from a rise in sea level seem likely to decrease still further the food supply of these birds.

Amongst carnivorous shorebirds, there is increasing evidence that the food supply is critical to determining the numbers of many of them. As bird density increases, various forms of interference that occur between birds as they forage become more intense and the rate of depletion of the prey also increases. Such competition could set a limit to the numbers that can use an intertidal area (GOSS-CUSTARD, 1985). Recent studies on dunlin, *Calidris alpina,* and grey plover, *Pluvialis squatarola,* suggest that, for some species, many estuaries may already be at capacity (GOSS-CUSTARD & MOSER, 1988; MOSER, 1988).

Food abundance is thought to interact with the time available for feeding to determine the numbers of birds that could be supported by a particular intertidal flat. The birds that forage on the exposed flats are able to feed only at low water so the exposure time of the flat affects how long they can forage. When the feeding time is reduced by the removal of the upper levels of the beach by reclamation for industrial purposes, the numbers of some species decrease sharply (EVANS, 1981).

Another factor which may affect the numbers of birds using an intertidal flat for feeding is the width of the shore. Perhaps because many avian predators attack from the direction of the shore (WHITFIELD, 1985), some waders seem to avoid narrow shores (BRYANT, 1979). Many wildfowl also seem to avoid narrow shores, presumably because of the greater risk of disturbance from people.

Most of the changes in the nature of the sediments and invertebrates of the intertidal zone anticipated earlier in this paper would be disadvantageous for most of the birds. First, the abundance of many of the food organisms would be reduced because of the unfavourable change in the nature and the stability of the sediments, the greater turbidity of the water and the reduced productivity of the inshore waters. Second, the reduced width, and therefore area, of the shore would force birds to feed at higher densities with a consequent increase in both interference and depletion competition. Third, the reduced exposure time following the degradation of the flats should remove some of the time available for feeding. Finally, the narrower shores might increase the rates of predation and disturbance on vulnerable species. Since waders are thought to die already from food shortage, either directly through starvation or indirectly because hungry birds are easier for predators to catch, a further reduction in the quality of their feeding conditions would probably lead to a decrease in population size (GOSS-CUSTARD, 1980). Any reduction in the abundance of suspension feeding mussels, *Mytilus edulis,* due to increased turbidity would be expected also to disadvantage eider ducks. Only the turnstone, *Arenaria interpres,* might benefit because harder and coarser substrates, with the associated algae, provide places for their prey to live.

4.3. BIRDS FEEDING ON SALT MARSHES

Some waders feed in the marsh creeks and muddy patches, notably redshank, *Tringa tonanus,* and grey plover. Many wildfowl feed in the marsh, notably the brent goose and wigeon, which concentrate in the grazed areas. These resources may not only be important for overwintering survival and for preparing for migration, but also affect the subsequent breeding succes (TEUNISSEN *et al.*, 1985). Several passerines also forage in the marsh and for the twite *Acanthis flavirostris,* the salt marsh is its main wintering habitat. According to a study by DAVIES (1987) on the Wash, reed buntings, *Emberiza schoeniclus,* preferred the vegetation communities at the higher levels of the marsh whereas rock pipits, *Anthus spinoletta,* and skylarks, *Alauda arvensis,* were widely scattered. The twites were found primarily in the Salicornia and Aster zones at the lower levels of the marsh.

Though it is expected that the lower levels of the marsh would be eroded, it is not clear to us what the subsequent changes in the floristic communities in the mid and upper levels might be. Clearly, with the greater amount of sediment in suspension, there would be the opportunity for marsh levels to rise, but at what rate relative to sea-level rise, and what level of the marsh, is unclear. It is, therefore, difficult to predict how birds feeding on marsh vegetation might be affected because of uncertainty about which zones in the marsh would be most affected.

4.4. BIRDS BREEDING ON SALT MARSHES

A number of species nest in salt marshes, including waders, notably redshank, oystercatcher, *Haematopus ostralegus,* lapwing, *Vanellus vanellus*, ringed plover, *Charadrius hiaticula,* and occasionally other species as well gulls, notably the black-headed gull and some terns, such as the common tern, *Sterna hirundo;* wildfowl, for example the mallard, *Anas platyrhynchos* and shelduck and a number of passerines, notably the skylark, reed bunting and meadow pipit, *Anthus pratensis.* The nests of these birds are vulnerable to flooding on high spring tides, and this can be a major cause of nest loss (HALE, 1980).

For nests to avoid being flooded, they must be placed in the mid or upper levels of the marsh. If the additional sediment in the water settles in these two zones at a rate equal or in excess of the rise in sea level, the nests of these birds should be secure, at least in the mid term. But if this is not so, then increased losses of nests from flooding would be predicted. Whether displaced birds could find alternative areas is also uncertain but, in Britain, the drainage of wetlands and coastal disturbance seems already to have reduced the numbers of many breeding birds, implying that spare breeding habitat does not exist.

5. DISCUSSION

With such a speculative paper, further discussion is hardly required. However, there are a number of additional points to be made.

First, tidal ranges vary considerably around the coasts of North-West Europe. The smallest occur along the Wadden Sea and it is here that the earliest, and largest, effects might be expected to be seen. For many birds, this is an extremely important area and, were its carrying capacity to be reduced, the pressure on other intertidal areas further to the west and south would in turn be increased. Indeed, a reduction in the food supplies in the Wadden Sea could make it very difficult for many birds to complete their migrations, so important a staging post is the area.

Second, the rise in sea level would be accompanied by a rise in ambient temperature. This would be helpful to many of the wintering birds because their energy demands would be lower, their food organisms would be more active and so accessible to the birds and the feeding grounds would be less frequently covered by ice. The wide geographic range of most of the invertebrates eaten by birds suggest that the species composition of the food supply would be unlikely to change much, but the increased temperatures might increase their growth rates and this too would favour some of the larger-bodied bird species which eat large prey. However, no means yet exist for calculating the extent to which any such improvements in the food supply, resulting from the increased temperature, would offset the reduction in the food supplies due to the rise in sea level itself.

Finally, no allowance has been made for possible responses by man to the sea-level rise, such as the construction of storm surge barriers or the encouragement of accretion at high shore levels. If such steps are taken to protect the coast, further consideration needs to be given to the possible effects of sea-level rise on inshore bird communities.

6. REFERENCES

BRYANT, D.M., 1979. Effects of prey density and site characteristics on estuary usage by overwintering waders (Charadrii).—Estuar. coast. mar. Sci. **9:** 369-385.

DAVIES, M., 1987. Twite and other wintering passerines on the Wash salt marshes. In: P. DOODY & B. BARNETT. Research and Survey in Nature Conservation: The Wash and its environment. NCC, Peterborough: 123-132.

EVANS, P.R., 1981. Reclamation of intertidal land: some effects on Shelduck an wader populations in the Tees estuary.—Verh. orn. Ges. Bayern **23:** 147-168.

GOSS-CUSTARD, J.D., 1980. Competition for food and interference among waders.—Ardea **68:** 31-52.

——, 1985. Foraging behaviour of wading birds and the carrying capacity of estuaries. In: R.M. SIBLY & R.H. SMITH. Behavioural Ecology: ecological consequences of adaptive behaviour. Blackwells, Oxford: 169-188.

GOSS-CUSTARD, J.D. & M.E. MOSER, 1988. Rates of change in the numbers of dunlin, *Calidris alpina,* wintering in British estuaries in relation to the spread of *Spartina anglica.*—J. Appl. Ecol. **25:** 95-109.

HALE, W.G., 1980. Waders. New Naturalist Series. Collins, London: 1-320.

LONGBOTTOM, M.R., 1970. The distribution of *Arenicola marina* (L.) with particular reference to the effects of particle size and organic matter in the sediments.—J. exp. mar. Biol. Ecol. **5:** 138-157.

MOSER, M.E., 1988. Limits to the numbers of grey plovers, *Pluvialis squatarola,* wintering on British estuaries: an analysis of long-term population trends.—J. Appl. Ecol. **25:** 473-486.

SALINAS, L.M., R.D. DELAUNE & W.H. PATRICK, 1986. Changes occurring along a rapidly submerging coastal area: Louisiana, USA.—J. Coast. Res. **2:** 269-284.

TEUNISSEN, W., B. SPAANS & R. DRENT, 1985. Breeding success in Brent in relation to individual feeding opportunities during spring staging in the Wadden Sea.—Ardea **73:** 109-119.

WHITFIELD, D.P., 1985. Raptor predation on wintering waders in South-East Scotland.—Ibis **127:** 544-558.

WIDDOWS, J., P. FIETH & C.M. WORRALL, 1979. Relationships between seston, available food and feeding activity in the common mussel, *Mytilus edulis* L.—Mar. Biol. **50:** 195-207.

WOLFF, W.J., 1973. The estuary as a habitat. An analysis of data on soft-bottom macrofauna of the estuarine area of the rivers Rhine, Meuse and Scheldt.—Zool. Verh., Leiden **126:** 1-242.

EFFECTS OF INCREASED SOLAR UV-B RADIATION ON COASTAL MARINE ECOSYSTEMS: AN OVERVIEW

KEES J.M. KRAMER

Laboratory for Applied Marine Research, MT-TNO, P.O.Box 57, 1780 AB Den Helder, The Netherlands

ABSTRACT

The objective of this overview is to estimate the possible biological effects on the Dutch coastal marine ecosystem of increased UV-B irradiation caused by a thinning of the ozone layer. As no relevant UV-B data or biological effect data are available for the Dutch coastal waters, extrapolations from the literature for other areas are necessary. This paper is based on an extensive literature survey (KRAMER, 1987).

Many processes affect the transmission of UV-B in the atmosphere and in seawater. From the various physical, (photo)chemical and biological processes that interfere with the transmission of UV-B and the biological (adverse) effects that have been observed in laboratory experiments and in the natural environment, predictions are made of the possible effects on organisms in the water column and on the tidal flats.

It is concluded that the relatively high concentrations of suspended particulate matter and yellow substance in the Dutch coastal waters limit the transmission of UV-B to the surface layers (<0.5–1 m) of the water. Accordingly, no biological effects are expected in the water column. On the tidal flats, however, effects on benthic organisms are expected due to the shallow water depth and the low turbidity of water at low tide. Experiments are necessary to confirm these predictions.

1. INTRODUCTION

The electromagnetic radiation emitted by the sun covers a large range of wavelengths. Radiation of wavelengths between 100 and 400 nm is called ultraviolet light. This range is somewhat arbitrarily subdivided into three radiation bands: UV-C (100–280 nm), UV-B (280-315 nm) and UV-A (315–400 nm). The biological effects (see below) of UV-A are much less marked than those of UV-B and UV-C.

Interference by atmospheric constituents results in a depletion of the spectrum emitted by the sun (Fig. 1). Ozone (O_3), for example, is particularly important at wavelengths below 330 nm. The shortest UV wavelength detectable at sea level is about 290 nm. Several authors have described the amount and nature of the UV reaching ground (*e.g.* NAGARAJA RAO *et al.*, 1984).

The (present) total amount of ozone, integrated over the total atmospheric column may be taken to be 0.340 cm for a typical mid-year value for mid-latitude locations (DÜTSCH, 1971; NAGARAJA RAO *et al.*, 1984). In calculations a value of 0.32 cm ozone depth is often used (ZANEVELD, 1975).

The potential depletion of ozone, mainly due to interference by man-made substances such as chlorofluorcarbons (CFC's; CLYNE, 1976), has been reviewed by HIDALGO (1975) and ROWLAND (1982).

The increase in UV-B flux reaching the earth's surface as a result of ozone depletion is a cause of concern because of its ecological impact. Although the additional UV radiation reaching the earth's surface as a result of a 10% ozone reduction would amount to less than a 0.5% increase of the total sunlight energy (and a 1% increase of the total UV component), a serious biological problem will arise if the particular spectral changes coincide with spectral characteristics of biologically damaging reactions. This is the case for UV-B, a relatively small increase of which has particularly damaging consequences (see below).

The impact of such an increase on marine ecosystems can be considerable, because seas cover about 70% of the earth's surface, and because marine ecosystems depend strongly on what happens in the upper sea layers, the euphotic zone.

Several reviews deal with the various aspectes of irradiance in general (BAINBRIDGE *et al.*, 1966), and, in particular, those of increased UV-B upon the marine system (*e.g.* CALKINS, 1982A; TITUS, 1986), the physical aspects of light penetration (ZANEVELD, 1975; JERLOV, 1976), the biological effects (CALKINS, 1975A, 1982A; NACHTWEY *et al.*, 1975; WORREST, 1982, 1986) and the photochemical aspects (ZAFIRIOU *et al.*, 1984). An introduction to and survey

J.J. Beukema et al. (eds.), Expected Effects of Climatic Change on Marine Coastal Ecosystems, 195–210.

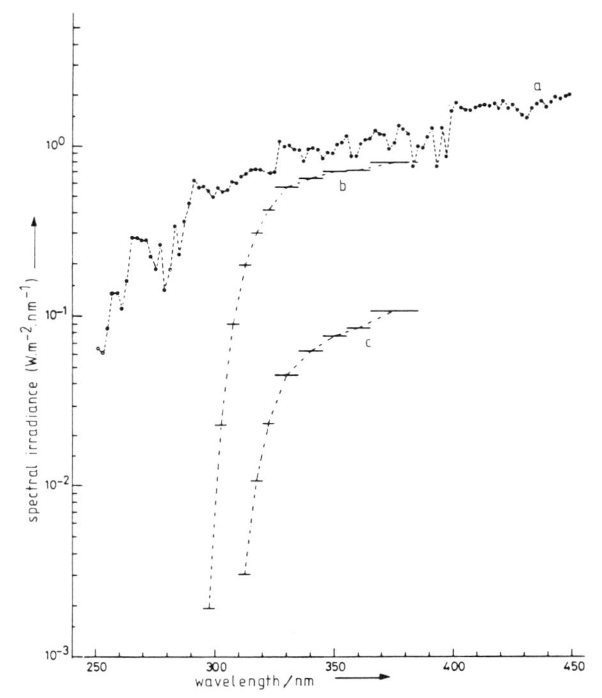

Fig. 1. Solar irradiance, without interfering atmosphere (a) and at the sea surface at a solar altitude of 60° (b) and of 10° (c), equivalent to the situation in the Netherlands at noon on a clear day in June and December, respectively (from: GEZONDHEIDSRAAD, 1986).

of the UV-B problem can be found in CALDWELL & NACHTWEY (1975).

No qualitative or quantitative data were found for the Dutch coastal ecosystem. Furthermore, most data have been obtained under laboratory or semi-natural conditions. Therefore, the various physical, biological and chemical processes that affect the behaviour of UV-B irradiation in the marine environment will be given and the findings extrapolated to the natural environment and in particular to the specific situation in the Dutch coastal waters of the North Sea and the Wadden Sea.

Some limitations should be allowed. The paper only deals with the aquatic environment, as (semi)terrestrial systems (salt marshes) can easily be treated with terrestrial systems. Atmospheric processes (meteorological, photochemical, etc.) are not treated extensively, only those processes that have a direct influence on the marine system being mentioned. It is assumed that a reduction in the ozone thickness occurs, and hence an increase in UV-B. Where and how this happens is beyond the scope of this paper.

2. PHYSICAL AND OPTICAL CONSIDERATIONS

2.1. ATMOSPHERIC PROPERTIES

The primary energy incident on the surface of the sea is the global radiation from sun and sky. Interference by atmospheric constituents, for example scattering by air molecules and absorption by atmospheric gasses (ozone, oxygen, water vapour, and others) and absorption and scattering by particles (including water droplets) removes a considerable part of the shortwave component of sunlight (Fig. 1). The effects of the atmosphere upon the sun's irradiance (clouds, haze) have been reviewed by SHETTLE *et al.* (1975). The spectral distribution varies only slightly with solar elevation above 15°; at low sun heights the skylight becomes dominant over direct sunlight. Although the visible component of the sunlight is not greatly dependent on the thickness of the atmosphere penetrated, the UV-B portion is strongly absorbed by ozone, and is thus highly dependent on the solar angle (CALKINS & THORDARDOTTIR, 1980).

The spectral distribution of solar radiation at sea-level is calculated in Fig. 2, note the steep gradient in the UV region of the spectrum. The amount of UV reaching the sea surface and the spectral distribution is given by GREEN *et al.* (1974, 1975), BAKER *et al.* (1980) and NAGARAJA RAO *et al.* (1984). The ozone depth (the integrated amount of ozone at a given place) in the stratosphere varies with location and time. An average value for this parameter at present is 0.32 cm (GREEN *et al.*, 1975). A reduction of 25% and 50% in the ozone layer thickness thus gives an ozone depth of 0.24 and 0.16 cm respectively: these figures are usually applied here. The spectral irradiance at the ocean surface for several ozone thicknesses is presented in Fig. 3.

2.2. AIR-SEA INTERFACE

Reflection is another process at the air-sea interface that is important for the actual downward UV-B irradiance; it depends on solar elevation and on the roughness of the sea surface (wind). Large effects of these parameters can be observed, especially at lower solar elevations. This means that on a summer's day a maximum transmittance will be observed for less than six hours in the Dutch coastal ecosystem (about 53°N). Another, more chemical, phenomenon at the interface is the sea surface microlayer. This is the very top water layer (<1 mm), where many organic molecules such as surfactants, hydrophobic (lipids and sterols) and hydrophilic (long chain proteinaceous substances) concentrate (GARRETT, 1967; MACINTYRE, 1974). The presence of these compounds might result in relatively large absorption of radiation, especially in coastal, polluted (hydrocarbons) areas; no quantitative data are, however, available.

2.3. OPTICAL PROPERTIES OF NATURAL WATERS

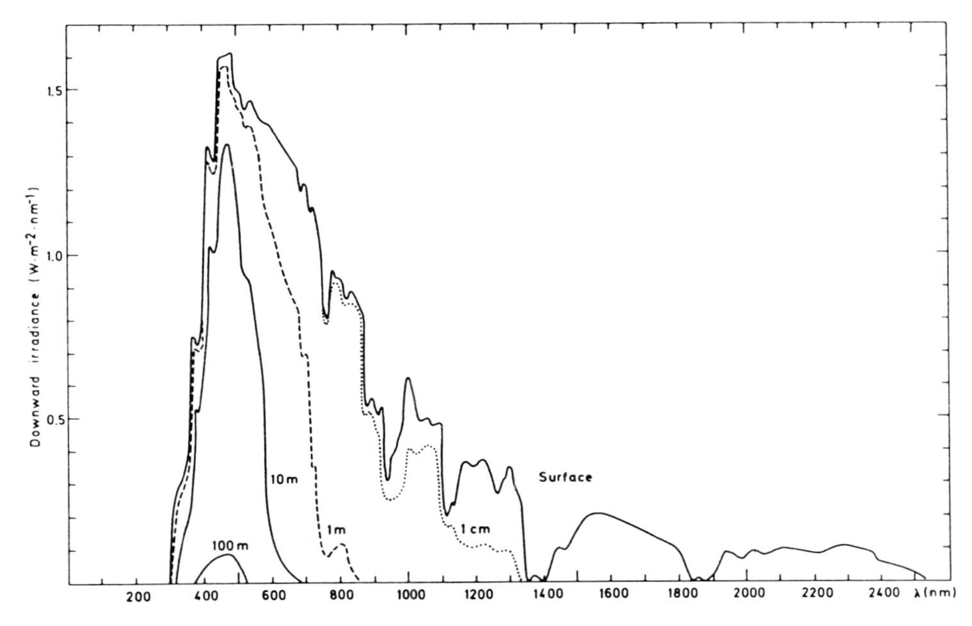

Fig. 2. The complete spectrum of downward irradiance in the sea (from: JERLOV, 1976).

2.3.1. ABSORPTION AND SCATTERING IN THE WATERCOLUMN

Several compounds influence the absorption and scattering of UV-B in seawater: dissolved and particulate matter.

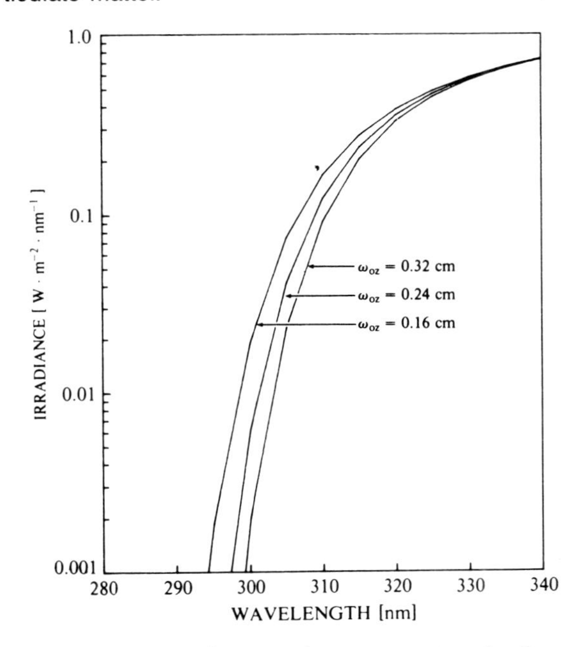

Fig. 3. Spectral irradiance at the ocean surface for three ozone depths: 0.32 cm, 0.24 cm and 0.16 cm (from: BAKER *et al.*, 1982).

For pure water the scattering is very small compared to the absorption. Dissolved salts absorb UV-B radiation. The presence of salts results in a weak absorption which slowly increases towards shorter wavelengths. Of more importance for the attenuation of UV-B are the particles and the Dissolved Organic Matter (DOM).

Dissolved organic matter absorbs radiation. In particular the so-called yellow substance (Gelbstoffe; KALLE, 1937) shows a very strong absorbance in the UV region (FOSTER, 1985). Yellow substance consists mainly of byproducts of organic decomposition. The material is present and produced in both freshwater and seawater bodies (KARABASHEV, 1977; HOJERSLEV, 1978; HARVEY *et al.*, 1983), in coastal waters the yellow matter is predominantly of terrestrial origin. The light attenuation by suspended particulate matter (SPM) is not highly selective. Most of the attenuation is probably due to scattering. Especially in turbid coastal waters, SPM exerts a great influence on the redistribution of light.

2.3.2. OPTICAL CLASSIFICATION OF MARINE WATERS

An optical classification of various types of seawater has been proposed by JERLOV (1976). It is based on amounts of suspended matter, productivity and yellow substance. Although only two optical types are used today in optical oceanographic literature (type 1: oceanic waters, type 2: coastal waters) the Jerlov classification appears to be useful for this study.

TABLE 1

Classification of open ocean (Roman figures) and coastal optical water types (Arabic figures), together with typical irradiance transmittance (in %/m) for two wavelengths (after Jerlov, 1976).

Water type	*SPM concentration*	*Yellow substance*	*Irradiance transmittance (% /m)*		
			310 nm	*350 nm*	
I	low	low	86	94)	
IA	|	|	83	92.5)	
IB	|	|	80	90.5)	oceanic waters
II	|	|	69	84)	
III	high	low	52	73)	
1	low	high	17	30)	
3	|	|	9	19)	
5	|	|	3	10)	coastal waters
7	|	|		5)	
9	high	higher		2)	

TABLE 2

Percentage of surface irradiance as a function of depth and water type, calculated for a wavelength of 310 nm (KE = absorption coefficient, Tr = transmission; from Zaneveld, 1975)

		Watertype							
		I	*IA*	*IB*	*II*	*III*	*1*	*3*	*5*
Depth (m)	0	100	100	100	100	100	100	100	100
	1	86	83	80	69	50	16	9	3
	2	74	69	64	48	25	2.6	0.8	0.1
	3	64	57	51	33	13	0.4	0.1	-
	4	55	47	41	23	6.3	0.1	-	-
	5	47	39	33	16	3.1	-	-	-
	10	22	16	11	2.4	0.1	-	-	-
	20	4.9	2.4	1.2	-	-	-	-	-
	30	1.0	0.4	0.1	-	-	-	-	-
KE in m^{-1}		0.15	0.19	0.22	0.37	0.69	1.83	2.41	3.50
Depth (m) at which radiation is 10% of surface value		15.4	12.1	10.5	6.2	3.3	1.26	0.96	0.66

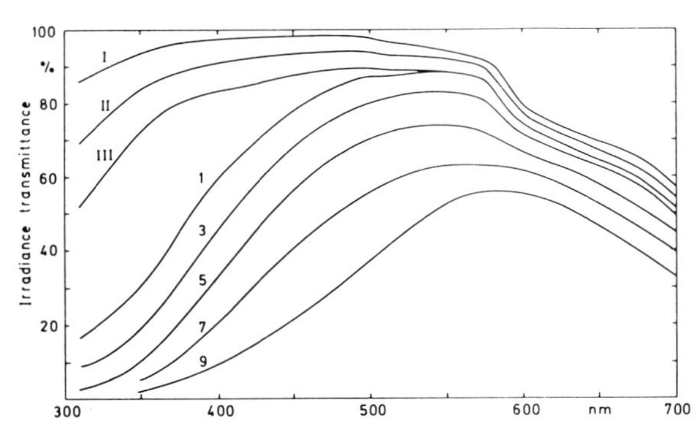

Fig. 4. Transmittance per meter of downward irradiance in the surface layer for optical water types. Oceanic types I, II, III, and coastal types 1, 3, 5, 7 and 9 (from: JERLOV, 1976)

Jerlov's scheme is divided into two major watertypes: open ocean waters (Roman numbers) and coastal waters (Arabic numbers). The optical water types are listed below (Table 1), and are a function of suspended matter concentration and absorbing organic matter. It will be clear from the table that low and high are relative measures within the oceanic and coastal water masses. The transmittance in the oceanic types generally depends only on the suspended matter content, while the coastal waters contain relatively high yellow substance concentrations. The transmittance curve for the clearest coastal watertype (1) coincides with that for oceanic type III between 500 and 700 nm (Fig. 4), but tends to a decreased penetration of shortwave light (the UV-B region) because of the absorption of the dissolved organic matter that marks these waters.

The percentage of surface irradiance that reaches a given depth for the different watertypes, calculated for a wavelength of 310 nm (UV-B) is given in Table 2.

The classification for the different optical watertypes is often used in optical oceanographic literature. Dutch coastal waters generally belong to water types higher than type 3, because of their specific concentrations of SPM and yellow substance.

2.4. EUPHOTIC ZONE

The euphotic zone is the upper water layer, which receives enough light to support photosynthesis by plants. Its thickness strongly depends on latitude, time of the year, optical properties of the seawater, and biological activity. The lower limit of the euphotic zone is usually taken to be the depth to which 1% of the total Photosynthetically Active Light (PAL) or Radiation (PAR) penetrates. Table 2 lists depths at which the radiation is 10% of the surface value, twice this value is the 1% level.

The photosynthetically active light does not exactly correspond to the component of visible light with maximum transmission. As natural waters are highly selective in attenuation, the light available for photosynthesis becomes, even at small water depth very close to the light at the wavelength of maximum transmission.

The UV-B/PAL ratio may be an important parameter (LORENZEN, 1975). The dependence of this ratio on watertype and depth is presented in Table 3. For the clearest ocean type waters (type I) these ratios are significant to more than 30 m water depth, for coastal waters only the upper few meters get a relatively higher UV-B irradiation. Thus increasing the surface UV-B irradiance would have an influence upon the upper layers of the eupothic zone, *i.e.* relatively more UV-B radiation will be available. ZANEVELD (1975) calculated the 'minimum zone of impact' (MZI) of UV-B radiation for several watertypes. This is the part of the (upper) watercolumn that receives a higher UV-B irradiance than the present surface level if ozone depth decreases. Of course, water below this minimum zone of impact would also receive increased UV-B irradiance, but not to a level that exists today at the surface. The results for a wavelength of 310 nm, and ozone depth of 0.16 and 0.24 cm (50% and 25% reduction) relative to the unperturbed ozone depth of 0.32 cm are summarized in Table 4.

It is seen that to up to 1% of a coastal waters euphotic zone may receive a larger amount of UV-B radiation than the current amount, if the ozone concentration in the stratosphere is decreased by 50%. Comparable results are reported by LORENZEN (1975). For coastal water type 5, and an increase in UV-B of 50%, a MZI of 0.11 m or 1% of the euphotic zone was calculated; the results for a 100% increase

TABLE 3

Ratio of UV-B irradiance (at 310 nm) and PAL irradiance (at wavelength of maximum transmission) as a function of water depth and coastal water type, with ratios normalized to the surface value (Zaneveld, 1975).

Depth (m) \ *Watertype*	*1*	*3*	*5*
0	1.00	1.00	1.00
1	0.18	0.11	0.041
2	0.033	0.012	0.002
3	0.006	0.002	
4	0.002		

in UV-B are 0.2 m and 1.8% of the euphotic zone, respectively.

2.5. CURRENTS, TURBULENCE

The above mentioned considerations are for a homogeneously mixed euphotic zone, without special attention for currents or turbulence. The calculated minimum impact zone for coastal waters, as mentioned above, does of course not mean that only the top 1% of the euphotic zone is affected by an increased UV-B irradiance. Mixing of the water mass by currents and turbulence will ensure that water from below the very upper water layers will be transported to the surface, and thus be subject to the UV-B irradiation (KULLENBERG, 1982).

It will be clear that the dose of UV-B radiation or radiant exposure will depend on the distance to the surface and the residence time. Both radiant exposure and residence time determine if effects upon organisms are to be expected. In a poorly mixed euphotic zone and/or stratification (*e.g.* due to a thermocline) the residence time of organisms in the upper layers of the sea and thus the effects of increased UV-B, may be increased.

3. BIOLOGICAL EFFECTS OF UV-B ON AQUATIC ORGANISMS

3.1. PRINCIPLES

The harmful effect of the different wavelengths of UV-B is not evenly distributed over the spectrum. In general, the actual amount of energy required to produce an effect varies with wavelength, generally the lower the wavelength the more effective. Biologically effective irradiance must therefore be based upon a weighting function ($\epsilon(\lambda)$) that takes into account the wavelength dependence of biological action. Ideally, the choise of a biological weighting function should

TABLE 4

Typical depth of the euphotic zone, water depth (in m) at which the UV irradiance is equal to the present ozone depth (0.32 cm) for different water types if the ozone depth decreases to 0.16 cm (-50%) and 0.24 cm (-25%), the "minimum zone of impact, MZI" (after Zaneveld, 1975).

	Watertype							
	I	*IA*	*IB*	*II*	*III*	*1*	*3*	*5*
Depth euphotic zone (m)	256	184	140	74	39	39	23	15
MZI (0.16 cm)								
depth (m)	2.27	2.58	2.33	1.33	0.712	0.268	0.204	0.140
% of euph.zone	1.3	1.4	1.6	1.8	1.8	0.69	0.89	0.94
MZI (0.24 cm)								
depth (m)	1.63	1.28	1.11	0.659	0.354	0.133	0.101	0.0697
% of euph.zone	0.64	0.70	0.79	0.89	0.91	0.34	0.44	0.46

be determined by the organism under investigation.

Several attempts have been made to formulate a generalized weighting function or action spectrum that can be used as an approximation for 'all' organisms (BAKER & SMITH, 1982A, CALDWELL *et al.*, 1986). Examples of weighting functions are:

- the 'photo-inhibition action spectrum, (ϵPI)' for photo-inhibition of chloroplasts (JONES & KOK, 1966).
- the 'plant response action spectrum, (ϵPR)' for higher plants (CALDWELL, 1971).
- the 'DNA action spectrum (ϵDNA) presented by SETLOW (1974).

In Fig. 5 two examples of these action spectra are graphically presented, the dotted curve represents Setlow's DNA action spectrum, the dashed curve Caldwell's plant response action spectrum. The solid

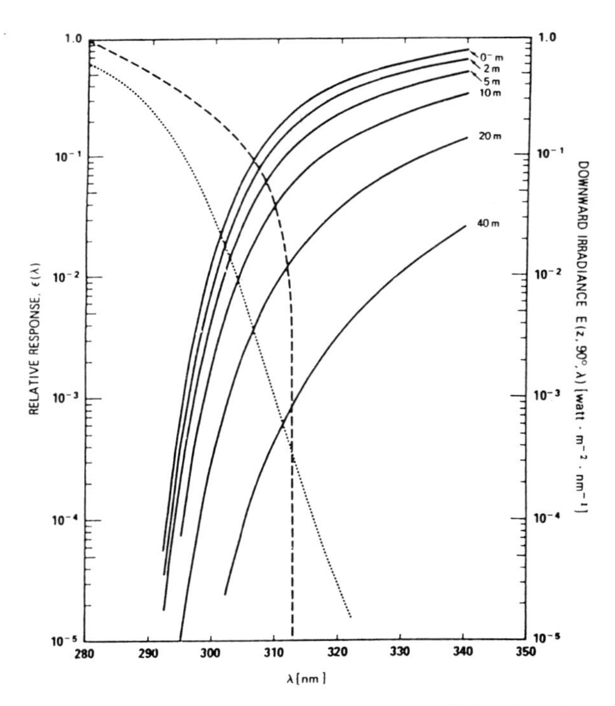

Fig. 5. Variation of relative biological efficiencies for SETLOW'S (1974) average DNA action spectrum (dotted curve) and for Caldwell's generalized plant response curve (dashed curve) with wavelength. The solid curves (corresponding to the right hand scale) give downward spectral irradiance at various depths (from: SMITH & BAKER, 1979)

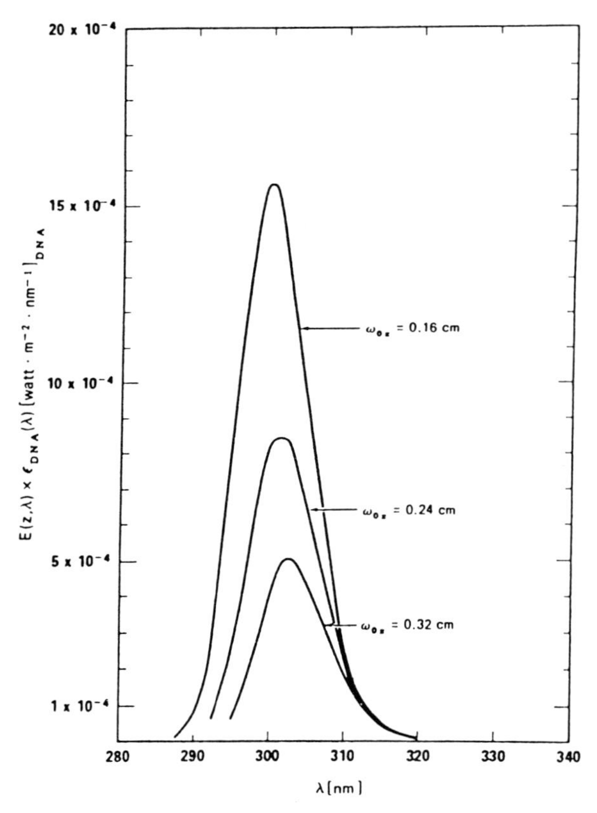

Fig. 6. Downward irradiance multiplied by Setlow's (1974) DNA action spectrum for ozone thicknesses of 0.16, 0.24 and 0.32 cm (from: SMITH & BAKER, 1979).

lines represent the spectral irradiance at various depths (SMITH & BAKER, 1979). The product of downward irradiance for UV-B and the 'DNA-action spectrum' is given in Fig. 6 for present and reduced ozone layer thickness (ozone depth = 0.32, 0.24 and 0.16 cm), based on Fig. 5 data.

ROBERTSON (1972) proposed the Robertson sensor that responds to UV-B in a manner weighted to simulate the biological effectiveness for the production of skin erythema by UV-B (erythema action spectrum, NACHTWEY & RUNDEL, 1982), the dose-unit often referred to is the 'sunburn unit, SU'.

In the literature the different action spectra are used by various authors. Comparisons have been made between the action spectra by CALKINS & BARCELO (1982). We should keep in mind that all action spectra are approximations, and care must be taken in comparing data from different weighting functions. However, quantitative comparisons can be made between biologically effective dose-rates, if the same specified wavelength normalizations are used (SMITH & BAKER, 1979). Dose-rate is the dose per unit of time, *e.g.* SU per hour.

3.2. NATURAL RESISTANCE TO UV-B

As discussed before, the UV-B irradiation largely depends on the geographic location. The daily expected dose in June, at 60°N is about 10 SU, and at the equator at the autumnal equinox about 33 SU (CALKINS & NACHTWEY, 1975). When organisms are translocated from high latitudes to the tropics, even current UV-B levels may be lethal to them (CALKINS & THORDARDOTTIR, 1980).

It seems therefore reasonable to suppose that organisms normally exposed to high levels of UV-B have developed a natural resistance mechanism to escape from the damaging effects of UV-B. KLUGH (1929) was among the first investigators to study the effects of the UV component of sunlight on certain marine organisms. He found a 'fairly close' relation between depth and susceptibility to shortwave radiation. The finding that tolerance and exposure to (present) solar UV are approximately equal (CALKINS & THORDARDOTTIR, 1980) implicates solar UV as a significant ecological factor. It suggests that no large reserve of resistance is available which could cope with increased UV exposure without requiring modification of physiology or behaviour. Several mechanisms have been found that diminish the harmful effects of UV-B: (vertical) migration, production of absorbing organic matter, development of repair capacity.

3.2.1. MIGRATION, AVOIDANCE

Zooplankton has apparently evolved mechanisms and behaviour (vertical migration) that enable it to adjust to current levels of UV irradiation (CALKINS, 1982B; DAMKAER & DEY, 1982; PENNINGTON & EMLET, 1986).

Shade-loving species of epifauna (sponges, tunicates, bryozoans) on coral reefs were found to suffer adverse effects when irradiated by full sunlight, except when wavelengths less than 400 nm were cut out (JOKIEL, 1980).

Horizontal and vertical movements of species of protozoans and crustaceans to sheltered areas suggest that they can detect and avoid UV-B (BARCELO & CALKINS, 1979). The aquatic protozoan *Coleps* in a lagoon system avoids light by descending to greater depths: when irradiance is high, it is usually found near the bottom, whereas when the irradiance is low it is more randomly distributed in the water column (BARCELO & CALKINS, 1980; BARCELO, 1982).

A positive correlation was found between sensitivity of the organisms and movement towards sheltered areas. BISHOP & HERRNKIND (1976) found burying effects of the pink shrimp *Penaeus duorarum* related to photo periods. TROCINE *et al.* (1981) found that epiphytes on leaf surfaces act as a shield and reduce the extent of photosynthetic inhibition from UV-B exposure in three species of seagrasses.

Thick epidermal cell-layers may inhibit effects induced by UV-B irradiation. Wavelength dependent refraction, reflection or absorption, and hence protection, is achieved by outer tissues (CHENG *et al.*, 1978).

3.2.2. UV ABSORBING COMPOUNDS, PIGMENTATION

Bathymetric adaptation of corals seems closely linked to the presence of UV-B absorbing compounds, such as mycosporine-like amino acids (DUNLAP *et al.*, 1986). SHIBATA (1969) observed that *Acropora* and a *Pocillopora* species contain large amounts of a water soluble, strongly absorbing (at 320 nm) organic compound (called S–320). The concentration of this compound in the coral *Porites lobata* is inversely proportional to the light intensity at the depth at which these corals grow (MARAGOS, 1972, cited by DUNLAP *et al.*, 1986). This is probably the reason why corals growing at depths of 1–2 m are more resistant to UV-exposure than their conspecifics from 5–6 m (SIEBECK, 1981).

The amount of pigmentation serves as a possible shield for UV-B. Several groups of organisms, comprising species both at tropical and at higher latitudes, show a darker appearance in the higher UV intensity regions (*e.g.* RINGELBERG *et al.*, 1984).

3.2.3. PHOTOREPAIR

The ability of organisms to repair the damage they suffer from UV-B exposure can be an important long-term mechanism to minimize the effects. There is some evidence that this ability can be induced by UV itself. The photorepair or photoreactivation of UV-B lesions was identified in *e.g.* embryonic anchovy larvae (KAUPP & HUNTER, 1981), and in larval shrimps (DAMKAER & DEY, 1983) and seagrasses (TROCINE *et al.*, 1981). JAGGER (1981) indicated that single-strand breaks in DNA are very readily repaired, and are not likely to be the major lethal lesion.

3.2.4. ADAPTATION

The ability of an organism to adapt to environmental change depends on the active or inducible systems it already possesses. Any of the above-mentioned mechanisms that decrease the effective amount of incident UV-B can serve as an adaptation mechanism.

However, each type of protection mechanism needs some time to become effective, a time that may be too long to match a rapid increase in UV-B intensity.

3.3. BIOLOGICAL EFFECT MECHANISMS

A number of physicochemical effects from UV-B can be distiguished: absorption by proteins and nucleic acids, introduction of photochemical reactions in organisms, tissue damage, genetic and/or chromosomal changes and phototoxicity. The general physical aspects of the biological effects have been discussed by NACHTWEY (1975a) and ZIGMAN (1982). Here only the possible effects as such will be treated; the effects on various organisms are dealt with below.

3.3.1. ABSORPTION

UV-B is readily absorbed by proteins and nucleic acids, and is effective in inducing photochemical reactions in plants and animals. As a result, many translucent (juvenile) organisms may suffer photochemical damage to enzymes, membranes and DNA and RNA (ZIGMAN, 1982; ZAFIRIOU *et al.*, 1984; MURPHEY, 1975; PEAK & PEAK, 1982).

Mutations and chromosomal aberrations are potential effects of absorbed UV-B interfering with chromosomes (NACHTWEY, 1975b).

3.3.2. TISSUE DAMAGE

UV-B can cause tissue damage in several ways. It may weaken the resistance of surface tissues (skin) of marine organisms to diseases and parasites, and to the action of toxic chemicals. These effects are under investigation, as are the pathological and parasitological effects of UV exposure (ROBBERTS & BULLOCK, 1981; BULLOCK, 1985). Skin and gill lesions attributable to UV action have been found in many species of fish (BULLOCK & COUTTS, 1985). The development of cancers of pigment cells in the skin (melanomas) and skin cancer (carcinomas) can be a direct effect of UV-B exposure (GEZONDHEIDSRAAD, 1986). PORTER (1975) has summarized the various effects on tissue.

Most tissue damaging effects observed so far have occurred in fish.

3.3.3. PHOTOTOXICITY

UV-B radiation can convert certain environmental contaminants into products (more) toxic to organisms. An evaluation of chemically induced phototoxicity to aquatic organisms has been presented by JOSHI & MISRA (1986).

Examples are the increased toxicity of benzo(a)pyrene to green algae exposed to UV-B (CODY *et al.*, 1984), and the effect of photo-oxidized crude oil on algae (KARYDIS, 1982). The possible toxic effects of photodegradation products of organic compounds are summarized in Section 4.

3.4. OBSERVED BIOLOGICAL EFFECTS ON MARINE ORGANISMS

The above-mentioned physicochemical processes can give rise to several ecological relevant biological effects upon organisms. The most important of these are:
- Decreased fecundity;
- Decreased growth and development;
- Inhibition of photosynthesis;
- Decreased survival rate;
- Decreased mobility;
- Inhibition of phototactic and photophobic response;
- Decreased photorepair ability;
- Change in species composition;
- Lethal response.

Depending on the local circumstances, synergistic factors (*e.g.* pollution) and the possible combination of these biological effects, largely determine the net result in an ecological and evolutionary way (CALKINS, 1982b). The structure of the marine ecosystem is very complex and dynamic. Diversity is associated with stability in ecosystems, allowing alternative routes and choices within food webs. Loss of species from an ecosystem results in loss of some of its natural resilience and flexibility. The marine environment contains a great number of species with different life stages and different trophic levels. Little is known about many of the fundamental processes affecting the abundance of marine resources in a quantitative sense. For these reasons, quantitative calculations of the ultimate effects of UV-B upon an ecosystem are not possible, even if direct effects upon single species are well known. An inventory of the effects of UV-B upon the different species may, however, result in a better understanding of the possible ecological effects and an estimate of their impact.

The observed effects on several groups of organisms will be discussed briefly.

3.4.1. BACTERIA

Bacteria were found to suffer from inhibition of amino acid uptake (BAILEY *et al.*, 1983) and growth delay (SIERACKI & SIEBURTH, 1986). Reduced activity of heterotrophs was found in the surface microlayer (CARLUCCI *et al.*, 1985) and in estuarine waters (MOEHRING *et al.*, 1984). Using heterotrophic populations in estuarine micro-ecosystems, THOMSON *et al.* (1980a) found reduced numbers of organisms but a relative increase of pigmented cells and heterotrophic respiration upon exposure to UV-B.

3.4.2. PHYTOPLANKTON

Reduction of growth, of activity or primary production have been studied by SMITH *et al.* (1980), WORREST *et al.* (1981a), DOEHLER (1984a,b), and others. CALKINS & THORDARDOTTIR (1980) estimated exposure and tolerance quantitatively, and found them remarkably similar. The survival of exposed diatoms was species- dependent, resulting in an altered community structure (DOEHLER, 1984a,b; WORREST *et al.*, 1981a).

Alteration of the natural phytoplankton community structure could result in a significant impact on successional pattern and primary producer-consumer trophodynamics. This change in species composition, rather than a decrease in net primary production (SMITH & BAKER, 1982), may be expected to be the result of increased UV-B irradiation (WORREST, 1986).

3.4.3. ZOOPLANKTON

Zooplankton in general comprises nearly all groups of aquatic animals, at least for some phase in their life history, *e.g.* Copepods *spp.* and Euphausiids *spp.*, but also drifting eggs and/or larvae. True drifters cannot avoid the upper water layers, and may therefore be exposed to elevated UV-B irradiation. A reduction in the number of eggs and nauplii, and a survival and fecundity as a result of UV-B exposure has, in fact, been observed (KARANAS *et al.*, 1979, 1981).

Adult Euphausiids possess a photorepair ability, which can keep pace with the damaging effects of UV-B radiation if the damage-rate is slow enough (DAMKAER *et al.*, 1980; DAMKAER & DEY, 1983).

Only about 50% of Anchovy larvae survived a UV-B dose of 1150 $J{\cdot}m^{-2}$ (DNA) in four days. The animals suffered retarded growth and development, lesions in brain and eye were induced (HUNTER *et al.*, 1979). KAUPP & HUNTER (1981) observed a photorepair mechanism and concluded that even at increased UV-B levels, sufficient photoreactive influence exists in the sea to ensure a maximum of photorepair of UV damage in Anchovy larvae.

3.4.4. CORALS

Mass bleaching of shallow-water scleractinian corals has been observed in the natural environment, and higher mortalities could be detected after sun bleaching (HARRIOT, 1985). It was postulated that bleaching may be the result of penetration of high levels of solar UV radiation. JOKIEL (1980) detected damaging effects of sunlight radiation only when wavelengths below 400 nm were not filtered out. VARESCHI & FRICKE (1986) also found a light response in corals.

3.4.5. FISH

In addition to the egg and larval stages of fish (HUNTER *et al.*, 1982), skin lesions have been observed in adult fish. Plaice (*Pleuronectes platessa*) exposed to 630 $J \cdot m^{-2}$ suffered skin lesions (ROBBERTS & BULLOCK, 1981); comparable experiments at high altitude in aquacultures of rainbow trout (*Salmo gairdneri*) confirm the UV action (BULLOCK & COUTTS, 1985). UV-B induced gill lesions were also observed in plaice (BULLOCK, 1982, 1985).

3.4.6. MARINE PLANTS

The only two references found for effects of UV-B upon marine plants concern seagrasses (TROCINE *et al.*, 1981) and benthic algae (POLNE & GIBOR, 1982). Differences in sensitivity between three different seagrass species were observed. A photorepair mechanism was only found in *Halodule spp*. All three species appeared to rely on epiphytic and detrital shielding, and on the shade provided by other seagrasses to reduce UV-B exposure. A study of *Porphyra, Ulva, Enteromorpha, Halimenia, Macrocystis* and other species revealed a damaging effect. Though able to adapt, deep ocean plants were at least twice as sensitive as high intertidal dwellers. Red algae were in general less sensitive to UV than green algae (POLNE & GIBOR, 1982).

4. UV INDUCED PHOTOCHEMISTRY

4.1. MARINE PHOTOCHEMISTRY

Photochemical reactions of various kinds may affect the euphotic zone. Direct or induced formation of new (organic) products may affect the biological system, resulting in positive (detoxification) or negative (formation of more toxic and/or resistant compounds) effects. Several reviews of the photochemistry of the aquatic environment (ZEPP, 1982; MILLER, 1983) and an extensive bibliography (ZAFIRIOU, 1984) are now available.

Various types of photochemical reactions and reaction chains may occur in the aquatic environment.

4.1.1. DIRECT AND INDIRECT PHOTOLYSIS

Direct photolytic reactions involve light absorbing entities called chromophores, which undergo chemical change as a direct result of absorbing photons. The primary products may participate in further, secondary reactions. Of only a few natural chromophores, the chemical structure is known. Examples are carbonyl compounds, riboflavin, tryptophane, thiamine and vitamin B12. Since pigments like chlorophyll and carotenoids are hydrophobic, they are readily adsorbed by particulate matter, and their photochemical reactions probably occur in the particulate phase (ZAFIRIOU *et al.*, 1984). Examples of these unknown chromophores include the yellow substance mentioned above, dissolved organic matter (DOM), humic- and fulvic acids, etc. In indirect photolysis a reaction is initiated through light absorption by a chromophore other than the substance itself. Sometimes the chromophore is regenerated, and thus plays an photocatalytic role. Known indirect photoreactions involve initial excitation of the chromophore, followed by energy transfer, or by transfer of electrons or hydrogen atoms to or from other components.

The energy transfer reaction is quantitatively the most important. Under environmental conditions, about 1–2% of the UV absorbing chromophores give rise to singlet oxygen. This only lives long enough in certain particular micro-environments or extremely pollutted areas to be chemically active. It reacts avidly with organic sulphides.

4.1.2. FREE RADICALS

Environment photoreactions often produce free radicals (ZAFIRIOU, 1987) and, among other compounds, oxidants. Important radicals are OH, O_2^-, ROO, RO, R and NO. Oxidants participate in non-radical reactions, and can also generate new radicals. They thus act as radical reservoirs. Examples of these are peroxides and hydroperoxides.

The number of organic compounds in the marine environment and the number of reactants (including radicals) and products from photochemical processes allow an enormous potential of newly formed compounds with different stability, half-life, toxicity, etc. No information is yet available on the nature and quantity of these compounds, formed by photosynthesis and interfering with the marine ecosystem.

4.1.3. PHOTODEGRADATION

Degradation or transformation by photochemical processes has been found for a relatively large number of organic and organometallic compounds.

For example, yellow substance strongly absorbs in the UV region. Humic substances have been shown to photosensitize transformations of several types of synthetic chemicals (FRIMMEL *et al.*, 1987). Photosensitized oxygenation and reactions involving energy transfer have been described for a number of organic model compounds in natural water media (ZEPP *et al.*, 1981). The relatively large amount of organic matter in the sea surface microlayer (GARRETT, 1967; MANINTYRE, 1974) and the high incidence of UV-B on the surface, leads to very special conditions with regard to photochemical processes in this micro-environment. No information on the photochemistry of this surface layer is, however, available.

Photodegradation and -transformation has been reported for several groups of xenobiotics and other pollutants. Some examples of photodegradation found for a series of pollutants are given in Table 5. The UV part of the spectrum was generally the most effective irradiation.

The fact that harmful (anthropogenic) substances are degraded or transformed by direct or indirect action of UV irradiation, certainly does not mean that the remaining fragments or the recombination products are less harmful. Newly formed products may become even more toxic, be more resistant to biological or physico-chemical breakdown or transformation, be more lipid soluble, etc.

Photochemical reaction products are (also) formed in the marine environment as a result of the action of UV-B. Increased UV-B irradiance would probably result in increased photochemical activity that would lead to increased concentrations of photolysis products (fragments and recombination products), probably including harmful substances. Thus indirectly, the UV-B might affect the marine ecosystem. No qualitative or quantitative data are, however, available.

5. SYNTHESIS OF POSSIBLE EFFECTS OF UV-B FOR DUTCH COASTAL WATERS

The effects of solar UV-B upon a given marine ecosystem can be assessed through four independently measurable factors:
- the amount of solar UV-B incident on the water surface;
- its attenuation upon reaching the position of various critical organisms;
- the UV-B sensitivity of the various species comprising the ecosystem;
- the ability of the irradiated organism to avoid the damaging radiation, to repair the injury and/or replace the components of the population that have been killed (Replacement Limiting Dose, RLD, CALKINS, 1975b).

5.1. AMOUNT OF UV-B REACHING THE SEA SURFACE

As described above ozone is not homogeneously distributed in the stratosphere. The seasonal changes in the ozone thickness (hole) have as yet only locally been observed, mainly near the polar regions. As a result of the solar angle at our latitude, large seasonal changes occur in the UV-B daily dose. No data were found for the Dutch coastal areas

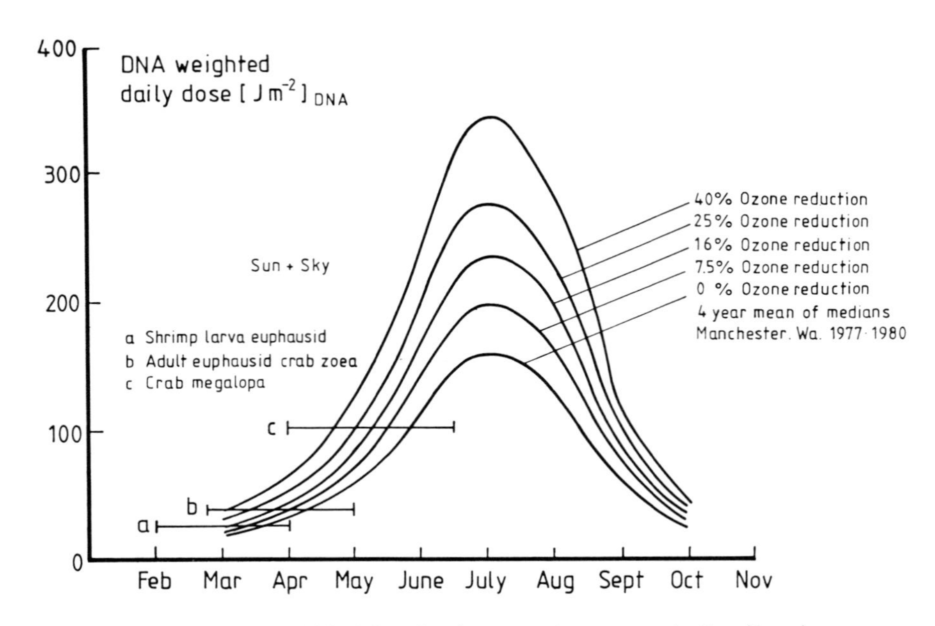

Fig. 7. Estimated effective UV-B daily solar dose at various ozone depths. Also shown are the thicknesses of 0.16, 0.24 and 0.32 cm approximate thresholds of UV-B daily dose for major experimental zooplankton groups (after: SMITH & BAKER, 1979) and in natural seasonal position (after: DAMKAER *et al.*, 1980).

(about 53°N), so data from places of similar latitude in the US (Manchester, WA, 47°30'N, DAMKAER *et al.*, 1981; Corvalis, Or., 44°30'N, NAGARAJA RAO *et al.*, 1984) will be used. In the Netherlands the overall mean dose will be slightly lower, and seasonal differences will be more pronounced. The US data show a daily dose of about 100 to 1100 kJ·m^{-2}·d^{-1} for December and July, respectively. Expressed as a DNA-weighted daily dose these figures correspond to 13 to 140 J·m^{-2} (DNA). A change in ozone depth results in an increased dose, with the major effects occurring in summer (Fig. 7).

Depending on the size of the increase, a shift will be observed towards earlier and later dates for a given dose. In particular the spring events might become critical, as in this period the eggs and larvae stages are abundant. For example, a daily dose of 100 J·m^{-2} (DNA) will be reached not at the end of May, but at the beginning of May for a 25% ozone reduction, or in the middle of April for an ozone reduction of 40% (Fig. 7). In the latter case the maximum daily dose could easily reach double the present maximum in summer.

5.2. ATTENUATION OF DUTCH COASTAL WATERS

The absorption of the water itself will of course not change, the effects of dissolved salts are present to some extent in the vicinity of fresh water outlets and rivers but will, however, be negligible. The major difference with the open sea waters is the amount of suspended particulate matter and the relatively high concentrations of yellow substance. In Dutch coastal waters the suspended particulate matter (SPM) content is subject to many changes in place and time. Typical concentrations are in order of 10–25 mg.dm^{-3} for surface waters; in bottom waters and during periods of high turbulence this may exceed 100 mg.dm^{-3} (POSTMA, 1954).

The distribution of yellow substance in the North Sea is shown in Fig. 8, where the relative unit 'millifluorescence' (mFL) of KALLE (1963) is used. In general the concentrations appear to be high in coastal waters, due to riverine input. The results of absorption by yellow substance and scattering by suspended particulate matter is a rather low transparency for UV-B radiation. ZANEVELD (1975) assumed an euphotic zone of 15 m for a type 5 coastal water and calculated a minimum zone of impact (MZI) of 14 and 7 cm (about 1% and 0.5% of the euphotic zone) for ozone depth of 0.16 and 0.24 cm respectively (Table 4). One may expect that Dutch coastal waters exceed often type 5 coastal waters in turbidity and dissolved organic matter concentration. Therefore the MZI might even be less, depending on the currents and turbulence conditions.

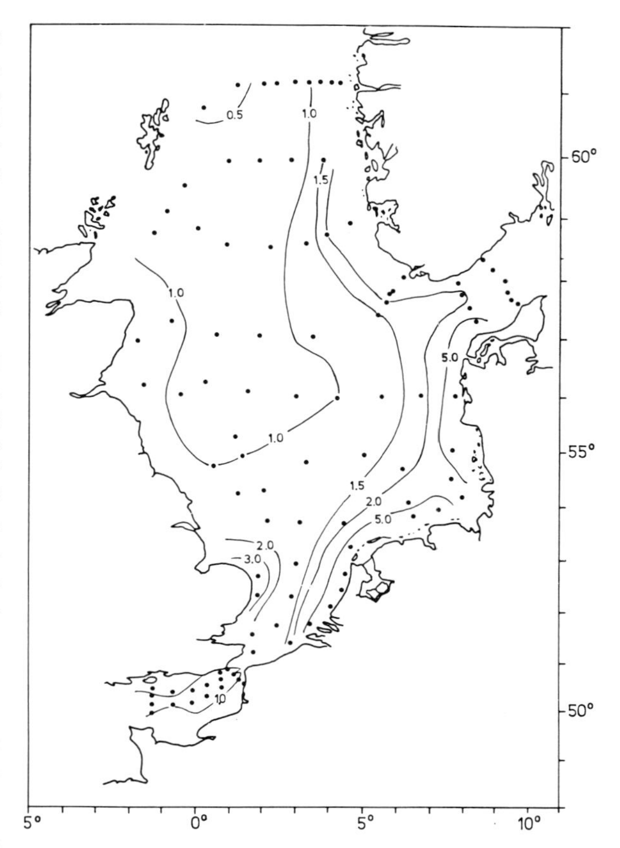

Fig. 8. Natural fluorscence (in mFl) in surface waters of the North Sea.

Special situations occur in areas where the sediments are above sea level at low tide. This occurs in relatively large areas in the Wadden Sea and in the delta formed by the Eastern and Western Scheldt. In these areas benthic organisms are directly exposed to the incident UV-irradiation.

Landlocked tidal pools or water remaining on the tidal flats at low tide will also contain nontypical benthic organisms. Size and depth of these water pools, and the absence of currents or turbulence results in deposition of the suspended matter. The resulting clear seawater (low turbidity) may show a relatively low UV-B attenuation compared to the parent coastal water, dissolved organic matter will not be decreased in concentration. Additionally, the shallow depth, usually less than 0.5 m, will enhance the changes of a relatively important MZI.

5.3. EXPECTED EFFECTS ON ORGANISMS

5.3.1. ORGANISMS IN THE WATER COLUMN

Due to the relatively high concentrations of dissolved

organic matter and suspended particulate matter only for organisms that remain constantly very close (maximum about 10–20 cm) to the surface of the water column, a serious damage is to be expected due to increased incident UV-B if ozone depletion occurs. In practice this means floating organisms or organisms which float during a certain lifestage and which dwell in coastal waters. Additionally, the floating lifestage must be present in the summer, when the elevation of the sun is high enough to ensure elevated UV-B irradiance. LORENZEN (1979) evaluated the present day level of UV-B upon phytoplankton. Although effects on the ^{14}C incorporation could be observed in a significant portion of the euphotic zone, the major effects were restricted to the upper portions of the zone. He concluded, that present day levels of UV-B have a minimal effect on (ocean) production (comparable in this respect to open North Sea waters), and that increased UV-B irradiation will only affect the upper euphotic zone.

For the Dutch coastal area this seems to be true as well. Even though the primary production is much more concentrated in the upper few metres, and in total much higher than in open sea or ocean waters, the optical properties of the water still prevent a substancial penetration of the UV-B. Here too, the Minimum Zone of Impact (MZI) is in the order of 1–2% of the population that might be affected (Table 4). It is assumed here that this will not interfere with the function of the ecosystem.

5.3.2. ORGANISMS ON TIDAL FLATS AND TIDAL POOLS

In the specific circumstances of tidal flats and small water pools the organisms that are already present will probably be adapted to the local situations. But increase of the UV-B irradiance might become an important ecological factor in these situations.

A review of most of the organisms present in the Wadden Sea has been edited by WOLFF (1983).

The present incident UV-B irradiation probably does not harm these organisms, due to adaptation and avoidance mechanisms. The finding that exposure to solar UV-B and tolerance are remarkably similar (CALKINS & THORDARDOTTIR, 1980) suggests however, that there may be no large reserve of resistance which can cope with increased UV-B irradiation. Assuming that this is the case, and no additional resistance mechanisms like pigmentation, repair capacity or avoidance can be developed in time, then the harmful effects mentioned in the preceding paragraphs might be expected.

Reduced growth, productivity and heterotrophic activity may be expected for bacteria. It is likely that benthic diatoms will show a negative growth response and inhibition of production. More important however, is the likely change in the species distribution that may occur (WORREST *et al.*, 1981). This change in community structure can have serious effects upon the ecosystem, as the change will almost certainly mean a decrease in diversity. The quantitative impact of these changes is not easy to calculate, because of the complex and dynamic structure of the ecosystem.

As no escape to deeper waters is possible in the shallow pools on the tidal flats, reduced fecundity and survival, and a reduction in the number of eggs and nauplii as found for copepods (KARANAS *et al.*, 1981) could result from increased UV-B irradiation upon zooplankton. On the other hand, photorepair mechanisms have been observed in larvae of fish and crustaceans, and this makes the net outcome unclear. Especially the effects of a change in UV-B dose with time of the year (see Fig. 7) should be considered, especially in relation to sensitive organisms or life stages.

Although present in large numbers on the tidal flats (WOLFF, 1983) Crustaceae, Polychaetae, Bivalvae etc. are usually buried in the sand, and thus probably unaffected by UV-B irradiation. Skin and gill lesions have been observed in a number of fish, including plaice. In the shallow pools on the tidal flats the young flat fish that use the Wadden Sea and the Delta as nursery, could be exposed for a sufficiently long period during low tide to result in damaging effects. These young lifestages are more translucent (thin skin) which increases the adverse effects. However, they also show burying habits which may (already?) be used to escape from direct solar irradiance. Only plaice (*Pleuronectes platessa*) and flounder (*Platichthys flesus*) can be expected to inhabit the tidal flat pools (in numbers up to several per m^2), and only for a limited period of the year. Plaice is present for about two months in the period February-May and flounder in the period April-May (VAN DER VEER, 1986). In the eastern parts of the Wadden Sea the immigration of the young fish is later in the year and damaging effects of UV-B could then be more marked, owing to the higher solar altitude.

Fortunately, most of the young fish are not retained in the tidal pools but escape to the tidal channels. Therefore, only a small effect is to be expected, but this is not likely to endanger the populations.

For the marine plants, only data on seagrasses and of some macroalgae are available. They indicate that the seagrasses do not suffer from intense radiation and have developed several protection mechanisms including a photorepair mechanism. The effects observed in the algae have been obtained at a UV irradiance ten times the natural level, and are therefore not realistic and difficult to interpret for the Dutch coastal waters.

TABLE 5

Examples of photodegradation or phototransformation of selected organic and organometallic compounds.

Group	*Reference*
Pesticides	Draper & Crosby,1984
herbicide	Wong & Crosby,1981
fungicide	Yumita & Yamamoto,1982
PAH	Mill *et al.*,1981
Oil compounds	
crude oil	Karydis,1982
	Tjessem *et al.*,1983
petroleum	Tjessem & Aaberg,1983
Organometallics	
tetraethyl lead	Charlou & Caprais,1982
organic mercury	Inoko,1981

No information is available on birds and mammals. Since both are subject to direct incident UV-B, at least for a considerable time, one can speculate on possible effects. Seals may suffer from sunburn, with possible formation of melanomas and skin cancers. The eyes of seals and birds may be damaged by increased UV-B irradiance.

5.4. EFFECTS OF SECONDARY PROCESSES

The change (degradation, transformation) in the organic matter composition in the marine environment caused or induced by UV-B irradiance is known for few identified and many unindentified compounds. The photochemical breakdown of pollutants has been studied in particular. Dutch coastal waters carry a rather high burden of pollutants, and photochemical degradation seems to be one advantage of increasing UV-B irradiation. However, no data are available on the nature and fate of the degradation products and their possible recombinations. Increased concentrations of biologically available fractions, or matter of a more lipophylic or a more toxic character might interfere with the marine coastal ecosystem in an as yet unknown way.

6. CONCLUSION

The waters of the Dutch coastal marine environment contain relatively high concentrations of suspended particulate matter and dissolved organic compounds, such as yellow substance. Extrapolation of data obtained under laboratory or semi-natural conditions reveals that transmission of UV-B will be limited to the uppermost layers of the water. The minimum zone of impact (MZI) for a 50% reduction of ozone concentration in the atmosphere will probably be less than 1% of the euphotic zone (<0.14 m). Assuming that the biota in this surface layer have by now been able to adapt to current levels of UV-irradiation, or are able to avoid it, we may expect that effects on this relatively small fraction of the euphotic zone are not likely to lead to drastic changes in the primary production part of the ecosystem. Tidal flats and tidal pools form a special situation, where decreased turbulence result in deposition of suspended matter. In such shallow pools, UV-B radiation can reach the bottom, where it may affect benthic organisms that cannot evade it, *e.g.* benthic diatoms. Other benthic organisms may escape it by burying themselves.

These conclusions have been reached after extrapolation of data derived from other (usually artificial) environments and from organisms other than those present in the Dutch coastal ecosystem. Experimental work in the natural environment or using semi-natural conditions in mesocosms will be necessary to verify these findings.

No qualitative and quantitative information is available on the nature of photodecomposition products and their toxicity to marine organisms. However, some examples of similar processes indicate that more toxic compounds can be formed, which may affect the biota.

ACKNOWLEDGEMENTS

This study was commissioned by the Netherlands Institute for Sea Research for the Ministry of Public Health, Physical Planning and Environment under project number 17545.

7. REFERENCES

BAILEY, C.A., R.A. NEIHOF & P.S. TABOR, 1983. Inhibitory effect of solar radiation on amino acid uptake in Chesapeake Bay bacteria.—Appl. Environ. Microbiol. **46:** 44-69.

BAINBRIDGE, R., G.C. EVANS & O. RACKHAM, 1966. Light as an ecological factor.—A symposium of the British Ecological Society Blackwell Sci. Publ. Oxford 452 pp.

BAKER, K.S. & R.C. SMITH, 1982. Bio-optical classification and model of natural waters, 2.—Limnol. Oceanogr. **27:** 500-509.

BAKER, K.S., R.C. SMITH & A.E.S. GREEN, 1980. Middle ultraviolet radiation reaching the ocean surface.—Photochem. Photobiol. **32:** 367-374.

——, 1982. Middle ultraviolet irradiance at the ocean surface: measurements and models. In: J. CALKINS. The role of solar ultraviolet radiation in marine ecosystems. Plenum, New York: 79-91.

BARCELO, J.A., 1982. Photomovement of aquatic organisms in response to solar UV. In: J. CALKINS. The role of solar ultraviolet radiation in marine ecosystems.—Plenum, New York: 407-409.

BARCELO, J.A. & J. CALKINS, 1979. Positioning of aquatic microorganisms in response to visible light and simulated solar shortwave UV-B irradiation.—Photochem. Photobiol. **29:** 75-84.

——, 1980. The relative importance of various environmental factors on the vertical distribution of the aquatic protozoan *Coleps spiralis*.—Photochem. Photobiol. **31:** 67-73.

BISHOP, J.M. & W.H. HERRNKIND, 1976. Burying and molting of pink shrimp, *Peneus duorarum* (Crustacea: peneidae), under selected photoperiods.—Biol. Bull. **150:** 163-182.

BULLOCK, A.M., 1982. The effect of UV-B irradiation on the integument of the marine flat fish *Pleuronectes platessa* L. In: J. CALKINS. The role of solar ultraviolet radiation in marine ecosystems. Plenum, New York: 499-508.

——, 1985. The effect of UV-B radiation upon the skin of plaice *Pleuronectes platessa* infested with the bodonid ectoparasite *Ichtyobodo necator*.—J. Fish. Dis. **8:** 547-550.

BULLOCK, A.M. & R.R. COUTTS, 1985. The impact of solar ultraviolet radiation upon the skin of rainbow trout, Salmo gairdneri Richardson, farmed at high altitude in Bolivia.—J. Fish. Dis. **8:** 263-272.

CALDWELL, M.M., 1971. Solar UV irradiation and the growth and development of higher plants. In: J.E. GIESE. Photophysiology. Vol. 6. Acad. Press New York: 131-177.

CALDWELL, M.M. & D.S. NACHTWEY, 1975. Introduction and overview. In: D.S. NACHTWEY, M.M. CALDWELL & R.H. BIGGS. Impacts of climatic change on the biosphere. CIAP mon. **5** (1): 1/3-33.

CALDWELL, M.M., L.B. CAMP, C.W. WARNER & S.D. FLINT, 1986. Action spectra and their key role in assessing biological consequences of solar UV-B radiation change. In: R.C. WORREST & M.M. CALDWELL. Stratospheric ozone reduction, solar ultraviolet radiation and plant life. Springer, Berlin: 87-111.

CALKINS, J., 1975a. Measurements of the penetration of solar UV-B into various natural waters. Impacts of climatic change on the biosphere. In: D.S. NACHTWEY, M.M. CALDWELL & R.H. BIGGS. Impacts of climatic change on the biosphere. CIAP mon. **5** (1): 267-296.

——, 1975b. Effects of real and simulated solar UV-B in a variety of aquatic microorganisms—possible implication for aquatic ecosystems. In: D.S. NACHTWEY, M.M. CALDWELL & R.H. BIGGS. Impacts of climatic change on the biosphere. CIAP mon. **5** (1): 5/31-72.

——, 1982a. The role of solar ultraviolet radiation in marine ecosystems. Plenum, New York: 724 pp.

——, 1982b. Modelling light loss versus UV-B increase for organisms which control their vertical position in the water column. In: J. CALKINS. The role of solar ultraviolet radiation in marine ecosystem. Plenum, New York: 539-542.

CALKINS, J. & J.A. BARCELO, 1982. Action spectra. In: J. CALKINS. The role of solar ultraviolet radiation in marine ecosystem. Plenum, New York: 143-150.

CALKINS, J. & D.S. NACHTWEY, 1975. Responses to UV radiation by bacteria, algae, protozoa, aquatic invertebrates, and insects. In: D.S. NACHTWEY, M.M. CALDWELL & R.H. BIGGS. Impact of climatic change on the biosphere. CIAP mon. **5** (1): 5/3-8.

CALKINS, J. & T. THORDARDOTTIR, 1980. The ecological significance of solar UV radiation on aquatic organisms.—Nature **283:** 563-566.

CARLUCCI, A.F., D.B. CRAVEN & S.M. HENRICHS, 1985. Surface film microheterotrophs amino acid metabolism and solar radiation effects on their activities.—Mar. Biol. **85:** 13-22.

CHARLOU, J.L. & M.P. CAPRAIS, 1982. Degradation of tetra ethyl lead in seawater.—Environm. Technol. Lett. **3:** 415- 424.

CHENG, L., M. DOUEK & D. GORING, 1978. UV absorption by gerrid cuticles.—Limnol. Oceanogr. **23:** 554-556.

CLYNE, M., 1976. Destruction of stratosheric ozone?—Nature **263:** 723-726.

CODY, T.E., M.J. RADIKE & D. WARSHAWSKY, 1984. The phototoxicity of Benzo(a)pyrene in the green alga *Selenastrum capricornutum*.—Environm. Res. **35:** 122-132.

DAMKAER, D.M. & D.B. DEY, 1982. Short-term responses of some planktonic crustacea exposed to enhanced UV-B radiation. In: J. CALKINS. The role of solar ultraviolet radiation in marine ecosystems. Plenum, New York: 417-427.

——, 1983. UV damage and photoreactivation potentials of larval shrimp, *Pandalus platyceros*, and adult euphausiids, *Thysanoessa rashii*.—Oecologia **60:** 169-175.

DAMKAER, D.M., D.B. DEY & G.A. HERON, 1981. Dose/dose-rate response of shrimp larvae to UV-B radiation.—Oecologia **48:** 178-182.

DAMKAER, D.M., D.B. DEY, G.A. HERON & E.F. PRENTICE, 1980. Effects of UV-B radiation on near-surface zooplankton of Puget Sound.—Oecologia **44:** 149-158.

DOEHLER, G., 1984a. Effect of UV-B radiation on the marine diatoms *Lauderia annulata* and *Thalassiosira rotula* grown in different salinities.—Mar. Biol. **83:** 247-254.

——, 1984b. Effect of UV-B radiation on biomass production pigmentation and protein content of marine diatoms.—Z. Naturforsch. Sect. C Biosci. **39:** 634-638.

DRAPER, W.M. & D.G. CROSBY, 1984. Solar photo oxidation of pesticides in dilute hydrogen peroxide.—J. Agric. Food Chem. **32:** 231-237.

DUNLAP, W.C., B.E. CHALKER & J.K. OLIVER, 1986. Bathymetric adaptations of reef building corals at Davies Reef, Great Barrier Reef, Australia. III. UV-B absorbing compounds.—J. Exp. Mar. Biol. Ecol. **104:** 239-248.

DÜTSCH, H.U., 1971. Photochemistry of atmospheric ozone.—Advan. in Geophys. **15:** 219-322.

FOSTER, P., 1985. Tracer applications of ultraviolet adsorption measurements in coastal waters.—Water Res. **19:** 701-706.

FRIMMEL, F.H., H. BAUER, J. PUTZIEN, P. MURASECCO & A.M. BRAUN, 1987. Laser flash photolysis of dissolved aquatic humic material and the sensitised production of singlet oxygen.—Environm. Sci. Technol. **21:** 541-545.

GARRETT, W.D., 1967. The organic chemical composition of the ocean surface.—Deep-Sea Res. **14:** 221-227.

GEZONDHEIDSRAAD, 1986. UV straling, blootstelling van de mens aan ultraviolette straling, Rijswijk.

GREEN, A., T. SAWADA & E. SHETTLE, 1974. The middle ultraviolet reaching the ground.—Photochem. Photobiol. **19:** 251- 259.

——, 1975. The middle ultraviolet reaching the ground. In: D.S. NACHTWEY, M.M. CALDWELL & R.H. BIGGS. Impacts of climatic change on the biosphere. CIAP mon. **5** (1): 2/29-37.

HARRIOTT, V.J., 1985. Mortality rates of scleractinian corals before and during a mass bleaching event.—Mar. Ecol. Prog. Ser. **21:** 81-88.

HARVEY, G.R., D.A. BORAN, L.A. CHESAL & J.M. TOKAR, 1983. The structure of marine fulvic acid and humic acids.—Mar. Chem. **12:** 119-132.

HIDALGO, H., 1975. Potential depletion of the ozone column from stratospheric flight. In: D.S. NACHTWEY, M.M. CALDWELL & R.H. BIGGS. Impacts of climatic change on the biosphere. CIAP mon. **5**(1): 2/3-26.

HOJERSLEV, N.K., 1978. Solar middle UV UV-B measurement in coastal waters rich in yellow substance.—Limnol. Oceanogr. **23:** 1076-1079.

HUNTER, J.R., J.H. TAYLOR & H.G. MOSER, 1979. Effect of ultraviolet irradiation on eggs and larvae of the northern anchovy, *Engraulis mordax*, and the Pacific mackerel, *Scomber japonicus*, during the embryotic stage.—Photochem. Photobiol. **29:** 325-338.

IONOKO, M., 1981. Photochemical decomposition of organo mercurials methyl mercury II chloride.—Environm. Pollut. Ser. B **2:** 3-10.

JAGGER, J., 1981. The expanding science of photobiology.—Nature **289:** 636-637.

JERLOV, N.G., 1976. Marine optics.—Elsevier, Amsterdam 231 pp.

JOKIEL, P.L., 1980. Solar ultraviolet radiation and coral reef epifauna.—Science **207:** 1069-1071.

JONES, L.W. & B. KOK, 1966. Photoinhibition of chloroplast reactions.—Plant Physiol. **41:** 1037-1043.

JOSHI, P.C. & R.B. MISRA, 1986. Evaluation of chemically-induced phototoxicity to aquatic organisms using paramecium as a model.—Biochem. Biophys. Res. Comm. **139:** 79-84.

KALLE, K, 1937. Nährstoff-Untersuchungen als hydrographisches Hilfsmittel zur Unterscheidung von Wasserkörpern.—Annln. Hydrogr. Berl. **65:** 1-18.

——, 1963. Uber das Verhalten und die Herkunft der himmelblauen Fluorescenz.—Dt. Hydrogr. Z. **16:** 153-166.

KARABASHEV, G.S., 1977. Influence of phytoplankton on the attenuation of short-wave solar radiation in the Baltic Sea.—Oceanol. **17:** 283-286.

KARANAS, J.J., H. VAN DYKE & R.C. WORREST,1979. Mid-ultraviolet (UV-B) sensitivity of *Acartia claussii* Giesbrecht (Copepoda).—Limnol. Oceanogr. **24:** 1104-1116.

KARANAS, J.J., R.C. WORREST & H. VAN DYKE, 1981. Impact of UV-B radiation on the fecundity of the copepod *Acartia claussii*.—Mar. Biol. 65: 125-134.

KARYDIS, M., 1982. Toxicity of a photo oxidized crude oil on 2 marine micro algae.—Bot. Mar. **25:** 25-30.

KAUPP, S.E. & J.R. HUNTER, 1981. Photorepair in larval anchovy, *Engraulis mordax*.—Photochem. Photobiol. **33:** 253- 256.

KLUGH, A.B., 1929. The effect of ultra-violet component of sunlight on certain marine organisms.—Can. J. Res. **1:** 100-109.

KRAMER, C.J.M., 1987. Effects of increased solar UV-B radiation on coastal marine ecosystems: a literature survey.—MT-TNO report R87/223, Delft, Netherlands.

KULLENBERG, G., 1982. Note on the role of vertical mixing in relation to effects of UV radiation on the marine environment. In: J. CALKINS. The role of solar ultraviolet radiation in marine ecosystems. Plenum, New York: 283-292.

LORENZEN, C.J., 1975. Phytoplankton responses to UV radiation and ecological implications of elevated UV irradiance. In: D.S. NACHTWEY, M.M. CALDWELL & R.H. BIGGS. CIAP mon. **5**(1): 5/81-91.

LORENZEN, C.J., 1979. Ultraviolet radiation and phytoplankton photosynthesis.—Limnol. Oceanogr. **24:** 1117-1120.

MACINTYRE, F., 1974. The top millimeter of the ocean.—Scient. Amer. **230:** 62-77.

MARAGOS, J.E., 1972. A study of the ecology of Hawaiian reef corals.—PhD. thesis Univ. Hawaii, Honolulu: 290 pp.

MILL, T., W.R. MABEY, B.Y. LAN & A. BARAZE, 1981. Photolysis of polycyclic aromatic hydrocarbons in water.—Chemosphere, **10:** 1281-1290.

MILLER, S., 1983. Photochemistry of natural water systems.—Environm. Sci. Technol. **17:** 568A-570A.

MOEHRING, M.P., R.C. WORREST & H. VAN DYKE, 1984. Influence of UV-B radiation on the heterotrophic activity of estuarine bacterio plankton.—Photochem. Photobiol. **39:** 65

MURPHEY, T.M., 1975. Effects of UV radiation on nucleic acids. In: D.S. NACHTWEY, M.M. CALDWELL & R.H. BIGGS. Impacts of climatic change on the biosphere. CIAP mon. **5**(1): 3/21-44.

NACHTWEY, D.S., 1975a. General UV irradiation physics and photo biological principles. In: D.S. NACHTWEY, M.M. CALDWELL & R.H. BIGGS. Impacts of climatic change on the biosphere. CIAP mon. **5**(1): 3/3-20.

——, 1975b. Genetic and chromosomal effects of UV radiation. In: D.S. NACHTWEY, M.M. CALDWELL & R.H. BIGGS. Impacts of climatic change on the biosphere. CIAP mon. 5 part 1: 3/45-49.

NACHTWEY, D.S. & R.D. RUNDEL, 1982. Ozone change: biological effects. In: F.A. BOWER & R.B. WARD. Stratospheric ozone and man. Vol. 2. CRC Press, West Palm Beach: 81-121.

NACHTWEY, D.S., M.M. CALDWELL & R.H. BIGGS, 1975. Impacts of climatic change on the biosphere. Part 1. Ultraviolet radiation effects. CIAP monograph 5: 704 pp.

NAGARAJA RAO, C.R., T. TAKASHIMA, W.A.I. BRADLEY & T.Y. LEE, 1984. Near ultraviolet radiation at the earth's surface: measurements and model comparisons.—Tellus **36B:** 286-293.

PEAK M.J. & J.G. PEAK, 1982. Lethal effects on biological systems caused by solar ultraviolet light: molecular considerations. In: J. CALKINS. The role of solar ultraviolet radiation in marine ecosystems. Plenum, New York: 325-336.

PENNINGTON, J.T. & R.B. EMLET, 1986. Ontogenetic and daily vertical migration of a planktonic echinoid larva, *Dendraster excentricus* (Eschscholtz)—occurence, causes, and probable consequences.—J. Exp. Mar. Biol. Ecol. **104:** 69-95.

POLNE, M. & A. GIBOR, 1982. The effect of high intensity UV radiation on benthic marine algae. In: J. CALKINS. The role of solar ultraviolet radiation in marine ecosystems. Plenum, New York: 573-579.

PORTER, W.P., 1975. Ultraviolet transmission properties of the vertebrate tissues. In: D.S. NACHTWEY, M.M. CALDWELL & R.H. BIGGS. Impact of climatic change on the biosphere. CIAP mon. 5 part 1: 6/3-14.

POSTMA, H., 1954. Hydrography of the Dutch Waddensea.—PhD Thesis, Univ. Groningen: 106 pp.

RINGELBERG, J., A. KEYSER & B. FLIK, 1984. The mortality effect of ultraviolet radiation in a translucent and in a red morph *Acanthodiaptomus denticornis* (Crustacea, Copepoda) and its possible ecological relevance.—Hydrobiol. **112:** 217- 222.

ROBERTS, R.J. & A.M. BULLOCK, 1981. Recent observations on the pathological effects of ultraviolet light on fish skin.—Fish. Pathol. Tokyo. **15:** 237-239.

ROBERTSON, D.F., 1972. Solar ultraviolet radiation in relation to human sunburn and skin cancer.—PhD Thesis, Univ. Queensland, Brisbane, Australia.

ROWLAND, F.S., 1982. Possible anthropogenic influences on stratospheric ozone. In: J. CALKINS. The role of solar ultraviolet radiation in marine ecosystems. Plenum, New York: 29-48.

SETLOW, R.B., 1974. The wavelength in sunlight effective in producing skin cancer: a theoretical analysis.—Proc. Natl. Acad. Sci. U.S. **71:** 3363-3366.

SHETTLE, W.P., M.L. NACK & A.E.S. GREEN, 1975. Multiple scattering and influence of clouds, haze and smog on the middel UV reaching the ground. In: D.S. NACHTWEY, M.M. CALDWELL & R.H. BIGGS. Impacts of climatic change on the biosphere. CIAP mon. **5**(1): 2/38-48.

SHIBATA, K., 1969. Pigments and a UV-adsorbing substate in corals and a blue green alga living in the Great Barrier Reef.—Plant Cell Physiol. **10:** 325-335.

SIEBECK, O., 1981. Photoreactivation and depth dependent UV tolerance in reef coral in the Great Barrier Reef/Australia.—Naturwissenschaften **68:** 426-428.

SIERACKI, M.E. & J.M. SIEBURTH, 1986. Sunlight-induced growth delay of planktonic marine-bacteria in filtered seawater.—Mar. Ecol. Prog. Ser. **33:** 19-27.

SMITH, R.C. & K.S. BAKER, 1979. Penetration of UV-B and biologically effective dose-rates in natural waters.—Photochem. Photobiol. **29:** 311-323.

——, 1982. Assessment of the influence of enhanced UV-B on marine primary productivity. In: J. CALKINS. The role of solar ultraviolet radiation in marine ecosystems. Plenum, New York: 509-537.

SMITH, R.C., K.S. BAKER, O. HOLM HANSEN & R. OLSON, 1980. Photo inhibition of photosynthesis in natural waters.—Photochem. Photobiol. **31:** 585-592.

THOMSON, B.E., H. VAN DYKE & R.C. WORREST, 1980a. Impact of UV-B radiation 290-320 nanometers upon estuarine bacteria.—Oecologia **47:** 56-60.

TITUS, J.G., 1986. Effects in stratospheric and global climate.—Proceedings of the Int. Conf. on health and environm effects of ozone modification and climate change. Arlington USA June 1986. UNEP/EPA, 1986.

TJESSEM, K. & A. AABERG, 1983. Photochemical transformation and degradation of petroleum residues in the marine environment.—Chemosphere **12:** 1373-1394.

TJESSEM, K., O. KOBBERSTAD & A. AABERG, 1983. Photochemically induced interactions in Ecofisk crude oil.—Chemosphere **12:** 1395-1406.

TROCINE, R.P., J.D. RICE & G.N. WELLS, 1981. Inhibition of seagrass photosynthesis by UV-B radiation.—Plant Physiol. **86:** 74-81.

VEER, H.W. VAN DER, 1986. Regulation of the population of O-group plaice (*Pleuronectes platessa* L.) in the Wadden Sea.—PhD. Thesis, Univ. Groningen: 91 pp.

VARESCHI, E. & H. FRICKE, 1986. Light response of a scleractinian coral (*Plerogyra sinuosa*).—Mar. Biol. **90:** 395-402.

WOLFF, W.J., 1983. Ecology of the Wadden Sea. (3 vols) Balkema, Rotterdam.

WONG, A.S. & D.G. CROSBY, 1981. Photo decomposition of pentachlorphenol in water.—J. Agric. Food Chem. **29:** 125-130.

WORREST, R.C., 1982. Review of literature concerning the impact of UV-B radiation upon marine organisms. In: J. CALKINS. The role of solar ultraviolet radiation in marine ecosystems. Plenum, New York: 429-457.

——, 1986. The effects of solar UV-B radiation on aquatic systems: an overview. In: J.G. TITUS. Effects of changes in stratospheric ozone and global climate. Vol.1 overview. UNEP/EPA: 175-191.

WORREST, R.C., K.U. WOLNIAKOWSKI, J.D. SCOTT, D.L. BROOKER, B.E. THOMSON & H. VAN DYKE, 1981. Sensitivity of marine phytoplankton to UV-B radiation impact upon a model ecosystem.—Photochem. Photobiol. **33:** 223-228.

YUMITA, T. & I. YAMAMOTO, 1982. Photodegradation of mepronil.—J. Pestic. Sci. **7:** 125-132.

ZAFIRIOU, O.C., 1984. A bibliography of references in natural water photochemistry.—Woods Hole Oceanographic Institution, Technical Memorandum 2-84.

——, 1987. Marine photochemistry: is seawater a radical solution?—Nature **325:** 481-482.

ZAFIRIOU, O.C., J. JOUSSIOT-DUBIEN, R.G. ZEPP & R.G. ZIKA, 1984. Photochemistry of natural waters.—Environm. Sci. Technol. **18:** 358A-371A.

ZANEVELD, J.R.V., 1975. Penetration of ultraviolet radiation into natural waters. In: D.S. NACHTWEY, M.M. CALDWELL & R.H. BIGGS. Impacts of climatic change on the biosphere. CIAP mon. **5**(1): 2/108-166.

ZEPP, R.G., 1982. Photochemical transformations induced by solar ultraviolet radiation in marine ecosystems.—NATO Conf. Ser. **4:** 293-307.

ZEPP, R.G., G.L. BAUGHMAN & P.F. SCHOLTZHAUER, 1981. Comparison of photochemical behaviour of various humic substances in water: I. Sunlight induced reactions of aquatic pollutants photosensitized by humic substances.—Chemosphere **10:** 109-117.

ZIGMAN, S., 1982. Mechanisms of actions of longwave-UV on marine organisms. In: J. CALKINS. The role of solar UV irradiation in marine ecosystems. Plenum, New York: 347-356.

EXPECTED CHANGES IN DUTCH COASTAL VEGETATION RESULTING FROM ENHANCED LEVELS OF SOLAR UV-B

J. VAN DE STAAIJ, J. ROZEMA and M. STROETENGA

Department of Ecology and Ecotoxicology, Vrije Universiteit Amsterdam, De Boelelaan 1087, 1081 HV Amsterdam, The Netherlands

ABSTRACT

Effects of increased UV-B radiation on plants from terrestric ecosystems will be negative, but large differences exist between species in their sensitivity to (enhanced) UV-B radiation. Field studies on the effects of enhanced UV-B are scant and should be extended to make general conclusions possible. Among two species of salt marsh plants species tested, the dicot *Aster tripolium* showed marked reduction of growth and photosynthesis. Growth and photosynthesis were less markedly inhibited by UV-B radiation in the monocot *Spartina anglica*. The prediction of the ecosystem response to increased UV-B radiation needs extension of experimental field studies. In further studies of the effects of increasing UV-B, other environmental factors forming part of the global climatic change should be included.

1. CLIMATIC CHANGE AND INCREASE OF SOLAR UV-B

The global change comprises man-made changes in the troposphere and stratosphere surrounding the earth. The increase of the atmospheric content of carbon dioxide due to the increasing use of fossil fuels leads to global warming. Also, the atmospheric enrichment with carbon dioxide directly affects all autotrophic organisms. Generally, an increase of photosynthesis and primary production of most ecosystems is to be expected. A differential response of plant groups of the terrestric biota may result in a shift in the vegetation zones on earth and in the competitive balance between plant species (see ROZEMA *et al.*, this issue). The so-called greenhouse effect has more direct and indirect consequences such as the rise of the sea level due to the melting of polar icecaps and thermal expansion of the ocean water, changes in weather and climate will occur as well, but are more difficult to predict. An increase of solar ultraviolet-B radiation as a result of the depletion of stratospheric ozone layer forms also part of the global change.

The release of chlorofluorocarbons, applied as aerosol propellants and in refrigeration systems, is related to the partial depletion of the stratospheric ozone layer (MOLINA & ROWLAND, 1974). The release of NOx-gases by the burning of fossil fuels also contributes to the photochemical destruction of ozone in the stratosphere. The stratospheric ozone column effectively absorbs all solar UV-C and part of the solar UV-B radiation. The filtering of solar radiation by the stratospheric ozone shield, makes that almost no solar radiation with a wavelength below 290 nm reaches the surface of the earth. Ozone (O_3) absorbs radiation with a wavelength below 330 nm. Maximum absorption by ozone is at 250 nm. With increasing wavelength the absorption by ozone diminishes and above 320 nm (UV-A) is hardly absorbed by ozone and reaches the surface of the earth for almost 100%. Reduction of the thickness of the stratospheric ozone column will inevitably lead to an increase of solar ultraviolet-B (UV-B) radiation on the surface of the earth. Although the increase of the lower part of the UV-B radiation waveband will be relatively small, the increase of the amount of biologically effective UV-B radiation at the earth surface will be disproportional. A 1% reduction of the stratospheric ozone concentration will lead to a 2% increase of biologically effective UV-B radiation (CALDWELL, 1977). A projection of a 5 to 9% decrease of the stratospheric ozone concentration implies a 19% increase in the biologically effective UV-B radiation (NAS, 1984). In 1987 international treaties have been made to reduce the emission of chlorofluorocarbons, according to the Protocol of Montreal. Although the emission of CFC is significantly being reduced in many of the western industrialized countries, a reduction of the thickness of the ozone layer with 10 to 15% during the next 40-50 years may be expected (RUNDEL, 1983). The Environmental Protection Agency of the USA considers a 6 to 8 % reduction of stratospheric ozone as a likely occurrence for the near future (TERAMURA, 1987).

J.J. Beukema et al. (eds.), Expected Effects of Climatic Change on Marine Coastal Ecosystems, 211–217.

2. DIFFERENCES IN NATURAL LEVELS OF SOLAR UV-B

There are large differences in the natural levels of UV-B radiation on different locations on earth. The total amount of UV-B reaching the surface of the earth at high altitudes in the tropics, is 1.6 times higher than the levels measured at the sea level in polar regions of the northern and southern hemisphere. The differences in the biological impact of the UV-B radiation levels are even larger. In tropical areas, radiation with shorter wavelength (290-300 nm) occurs, and this short wavelength UV-B is biologically very effective. The short wavelength UV-B radiation is lacking in polar regions (as long as no holes in the ozone layer occur). Therefore there is 7 times more biologically effective UV-B radiation in the tropics than in the polar regions (CALDWELL, 1980).

3. EFFECTS OF (ENHANCED) UV-B RADIATION ON PLANTS OF TERRESTRIAL ECOSYSTEMS

UV-B radiation generally has a negative effect on the growth of higher plants. This means that levels of solar UV-B now occurring at the earth surface have a suppressing effect on plant growth, which can be demonstrated when plants growing in the field are being excluded from UV-B radiation by installation of a UV-B absorbing screen above the plant canopy. Reduction of length growth is well known and mountain plants from high altitudes have often a prostrate occurrence with short stems and leaf stalks (CALDWELL, 1981). Depression of extension growth (HASHIMOTO & TAJIMA, 1980; LINDOO & CALDWELL, 1978; MURALI & TERAMURA, 1987), decrease of leaf area (KRIZEK, 1975; TEVINI *et al.*, 1983), and reduced fresh and dry weight (TERAMURA, 1983) have been reported in response to UV-B radiation. The extent of the plant response is dependent on the plant species and the dose of UV-B radiation applied. A summary of the growth response of a variety of crop species to UV-B radiation, both in greenhouse experiments and under field conditions is given in Table 1. The legume species Soybean and peas appear to be the most sensitive species. Most studies on the effect of UV-B radiation to plants refer to agricultural plant species (TERAMURA, 1987).

The negative effect of UV-B radiation on plant growth seems to be the result of a reduction of the rate of cell division, the process of cell extension seems not to be affected by UV-B radiation (DICKSON & CALDWELL, 1978). The reduced rate of cell division is the result of formation of pyrimidine dimers in the DNA-strains of irradiated plants. The dimers produced alter the spatial structure of the DNA-molecule and this makes the genetic code unreadable. This

TABLE 1

Effect of UV-B radiation on the dry weight of some crop plants in greenhouse and field experiments. More detailed information is available in TERAMURA (1983). The figures in this Table represent the numbers of studies in which the effect has been reported.

Crop plant	Greenhouse experiments			Field experiments		
	increase	decrease	no change	increase	decrease	no change
Rice, Oryza sativa	1	2	2	-	-	1
Corn, Zea mays	-	2	5	1	1	2
Sorghum, Sorghum vulgare	-	2	1	-	-	2
Millet, Panicum miliaceum	-	1	3	-	-	1
Soybean, Glycine max	-	10	-	-	1	2
Peas, Pisum sativum	1	5	-	-	-	-
Tomato, Lycopersicum esculentum	1	2	1	-	1	1

(UV-B effects on dry weights)

TABLE 2

Effect of UV-B radiation on the rate of photosynthesis in a number of plant species. The UV-B radiation level artificially supplied is expressed as the percentage decrease of thickness of the stratospheric ozone layer. The effect of UV-B reduction is expressed as the percentage reduction of the rate of photosynthesis as compared to the control treatment (no decrease of thickness ozone layer).

Plant species	UV-B level (% decrease thickness ozone layer)	Reduction rate of photosynthesis (%)	Reference
Oenonthera stricta	20	50	Robberecht & Caldwell, 1983
Rumex patienta	5 15 40	6 21 42	Tevini et al., 1983
Cucurbita pepo	10 40	20 35	Sisson, 1981
Aster tripolium	134 μ Watt UV-B.cm^{-2}	50	van de Staaij et al., 1990
Spartina tripolium	134 μ Watt UV-B.cm^{-2}	12	van de Staaij et al., 1990

negative effect of dimer formation in the DNA on growth has been demonstrated in *Lathyrus sativus* (SOYFER, 1983). A second cause of growth reduction by UV-B radiation may be the decrease of the photosynthetic activity of plants. A summary of the effect of an enhanced level of UV-B (expressed as the percentage reduction of the stratospheric concentration of ozone) on the rate of photosynthesis in some plant species is shown in Table 2. Considering UV-B levels corresponding to 5-20% depletion of stratospheric ozone, up to 50% reduction of the rate of photosynthesis may occur. The cause of the reduced rate of photosynthesis is considered to represent

an inactivation of the Photosystem II (PS-II) (IWANZIK *et al.*, 1983; TEVINI & PFISTER, 1985; TEVINI *et al.*, 1988).

4. UNCERTAINTIES IN THE KNOWLEDGE OF UV-B EFFECTS ON PLANTS OF TERRESTRIAL ECOSYSTEMS

The knowledge of effects of UV-B radiation on plants has increased in particular since the discovery of ozone depletion due to release of CFC's not more than two decades ago. Yet, this knowledge is rather general, based on a limited set of scientific data. Also the knowledge of effects of UV-B effects is mainly based on experiments with crop plants, mostly conducted in the greenhouse. It is doubtful whether the experience obtained for crop plants is representative for wild plant species, and secondly, more than with any other environmental factor: the results obtained from greenhouse experiments cannot simply be extrapolated to the field situation. There are several reasons for this. Usually greenhouses (**glasshouses!**) refer to constructions built of glass held with within a metal framework. Because glass absorbs UV-B radiation the greenhouse environment is UV-B free and UV-B radiation can only be obtained from UV-B tubelamps. Seedlings grown in a glasshouse experience a radiation environment without UV-B radiation. In field experiments there is always natural solar UV-B radiation. In addition to this, lamps can be used for supplementation of UV-B radiation according to one of the possible ozone depletion scenarios. The history of young plants may therefore be very important when considering the plant response to enhanced UV-B radiation. In field experiments plants that faced already natural solar UV-B may have developed flavonoid compounds as a protection against UV-B damage. Therefore it is understandable that the sensitivity of plants to UV-B radiation assessed in field experiments will be less than obtained in greenhouse experiments (Table 2). Also it has become clear that the effect of UV-B radiation to plants is dependent on the overall radiation level, which will usually be much higher in the field situation as compared to greenhouse conditions (TERAMURA, 1987). In addition to this, some early experimental work failed to use cut off filters in combination with UV-radiation tube lamps so that irradiation of plants was not only with UV-B but also with the rather destructive UV-C. Injury to plants ascribed to UV-B was in fact caused by UV-C radiation.

In a recent survey by TERAMURA (1987), it has been demonstated that of the 10 major terrestrial plant ecosystems UV-B studies are known of the tundra and alpine areas, the agricultural ecosystem and only very limited preliminary data are available for temperate forests and temperate grasslands, these ecosystems representing not more than 27% of the global net primary production.

It can be concluded that no reliable prediction can be made on expected effects of enhanced UV-B radiation due to stratospheric ozone depletion at the moment. This uncertainty becomes even larger, when other altering environmental factors as a result of anthropogenic activities are taken into consideration, such as air, soil and waterpollution, the rise of the atmospheric content of carbondioxide, global warming and the rise of the sealevel.

In the experiments described and reported here it was the aim to assess the effect of UV-B radiation on some species of the salt marsh ecosystem.

5. RESPONSE OF SALT MARSH PLANTS TO UV-B RADIATION

In a greenhouse experiment with UV-lamps covered with cut-off filter, in addition to Philips HPIT lamps, a level of UV-B radiation was reached representing about half the maximum level than can be measured outside in the open on a cloudless day in October, Amsterdam (Fig. 1). In the glasshouse itself there is no UV-B radiation present, unless supplied by special UV-lamps. During a growth period of four weeks, biomass growth of UV-B irradiated plants of *Aster tripolium* was about 25% reduced compared to plants without UV-B radiation. For the C_4- grass

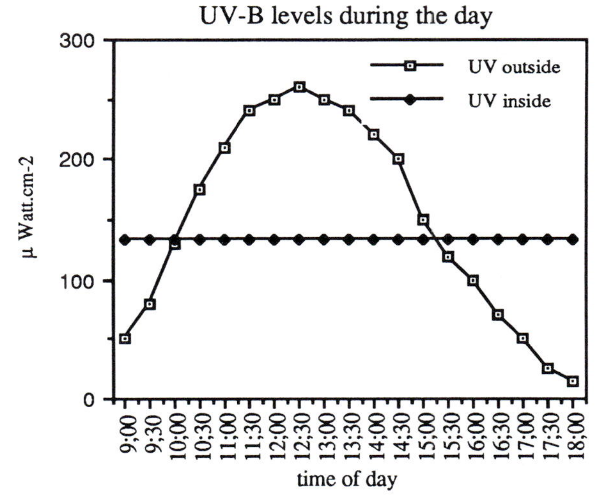

Fig. 1. Levels of UV-B radiation expressed as μwatt.cm^{-2} measured. Outdoors on a sunny cloudless day, October 10, 1988, Amsterdam, and inside a greenhouse with UV-lamps. UV-lamps covered with cellulose acetate foil. Values measured at half plant height. Measurements were done using a UVX radiometer with a UVX-31 sensor for the UV-B part of the spectrum.

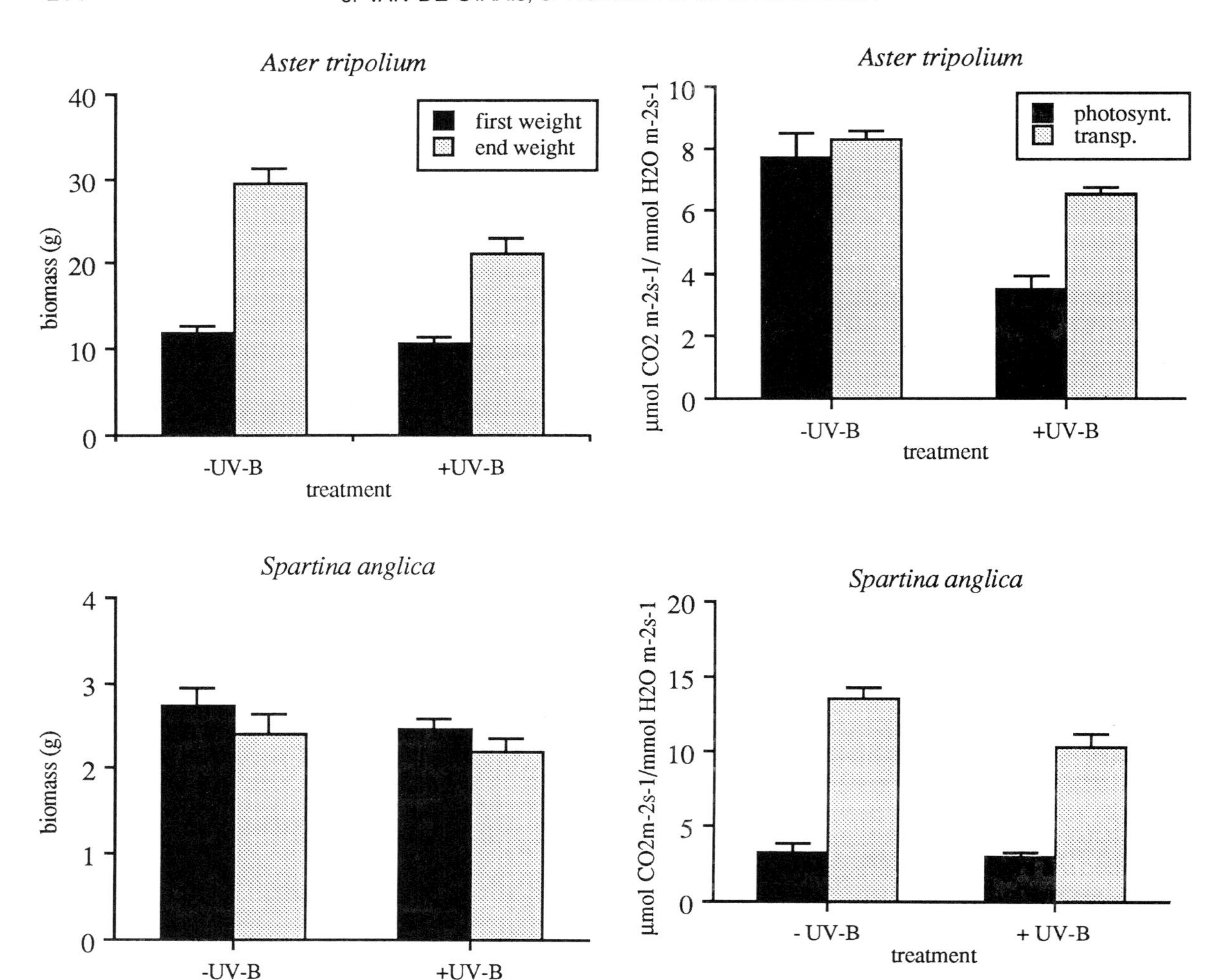

Fig. 2. Growth response of *Aster tripolium* and *Spartina anglica* to UV-B radiation (134 μwatt.cm^{-2}). Plants were grown in a greenhouse (20°C, 12 h light; 300 μE.m^{-2}.s^{-1} PAR from 400 W HPIT lamps measured with a Licor-Quantum Sensor Li-185 and a 12 h dark period, 15°C). Average values of 30 replications for *Aster tripolium* and of 35 replications for *Spartina anglica* with standard error of the mean. Black columns represent the biomass (*e.g.* fresh weight) at the start, dotted columns at the end of the 4 weeks growth period.

Fig. 3. Response of the rate of photosynthesis and transpiration to UV-B radiation in *Aster tripolium* and *Spartina anglica*. Measurements were done with an ADC-IRGA system in combination with a PLC leafchamber in the 4th week of the growth experiment. Black columns represent photosynthesis and dotted columns transpiration measurements. Average values of 10 replications with standard error of the mean.

species *Spartina anglica*, the reduction of biomass growth due to UV-B is much smaller and not statistically significant (Fig. 2). In an analysis of the carbon and water relations, the rate of photosynthesis of the leaves of *Aster tripolium* were significantly reduced with UV-B radiation, also there was a small depression of the rate of transpiration. For the leaves of *Spartina anglica* the rate of photosynthesis was not affected, the rate of transpiration was suppressed by UV-B radiation (Fig. 3).

In further experiments the above findings have been confirmed. The UV-B insensitive grass species *Spartina anglica* seems to contain flavonoids, a compound capable of absorbing UV-B, thus preventing destructive and disturbing activity of UV-B within the cell. One of the cell functions that can be disturbed by UV-B radiation concerns the processes of the Photosystem II, which may explain the reduced photosynthetic rate of *Aster tripolium* (Fig. 3). Apparently, reduction of the rate of photosynthesis in *Aster tripolium* is not due to stomatal closure, since (stomatal) transpiration seems to be unaffected by UV-B radiation (Fig. 3).

Studies of effects of UV-B on the level of the ecosystem and vegetation are scanty (TERAMURA, 1987). No such studies have been performed on the vegetation of coastal ecosystems (TERAMURA, 1987). Based on findings in the literature and the results of our group, some general remarks can be made. It seems to be a general trend that narrow leaved plant species are less sensitive to UV-B radiation than broader-leaved plant species (BASIOUNY & FOULAD, 1986). It is likely that this reflects a dose effect, the smaller leaf surface (perhaps combined with a different leaf angle) of the narrow leaved plant species receiving a smaller dose of UV-B radiation. Many studies show more in particular that grasses are less sensitive to UV-B radiation than dicotyledonous plant species. The work presented in this paper and other (unpublished) results on salt marsh plants confirm these conclusions. Both growth and photosynthesis of the dicot halophyte *Aster tripolium* were more reduced by UV-B radiation than of the monocot salt marsh *Spartina anglica*.

The growth of individual plant species might be reduced with increased solar UV-B due to ozone depletion. It is uncertain whether the primary production of the salt marsh ecosystem will be reduced by increased UV-B. Grasses dominate in the vegetation of Dutch salt marshes (ROZEMA *et al.*, 1988). Yet, the response of other grass species may differ from the response of *Spartina anglica* to UV-B. Based on the above findings, a shift in the competitive balance towards grasses might be expected. However, the number of studies of effects of enhanced UV-B on natural vegetation of the different ecosystems is very limited and apart from some types of vegetation from the temperate forest and grassland, tundra and alpine zone, no studies are available (TERAMURA, 1987). A recent experimental study on the effects of UV-B in the field wheat and wild oat (BARNES *et al.*, 1988) demonstrated no marked differences in the growth response to UV-B between these two grass species. Yet, a shift in the competitive balance between the two species occurred in response to UV-B enhancement. This effect could be explained in terms of differential morphological response of the two species that were exposed to enhanced UV-B. A trivial but important conclusion that may be reached is that not only greenhouse but also field studies of agricultural and natural ecosystems must be performed to make a reliable and realistic prediction possible.

6. INTERACTIONS OF UV-B RADIATION WITH OTHER ENVIRONMENTAL (CLIMATIC) FACTORS

Increase of UV-B radiation due to depletion of stratospheric ozone and climatic change will take place at the same time. Of these global changes taking place, the rise of the atmospheric content of carbon dioxide has been measured intensively. With preindustrial values of 280 ppm and 345 ppm today, it has been predicted that between the year 2075 and 2100 the atmospheric CO_2 concentration will reach 600 ppm (GAMMON *et al.*, 1985). In 1987, TERAMURA wrote 'At present we have no experimental evidence on the effects of enhanced levels of UV-B radiation under increased levels of atmospheric CO_2' (TERAMURA, 1987). However, from the literature and based on the responses of plants to UV-B and CO_2-enrichment, predictions can be made. Generally, it is found that plant growth and photosynthesis may be inhibited by UV-B radiation, as was reported here for the salt marsh halophyte *Aster tripolium*. It must be stressed however, that plant species differ in their sensitivity to UV-B radiation. Plants respond generally positively to CO_2 enrichment, C_3 plants in particular. Growth and photosynthesis are often stimulated by increased levels of CO_2, while transpiration is reduced as a result of (partial) stomatal closure. Since UV-B radiation primarily affects PS-II and thereby the rate of photosynthesis and the effect of CO_2 enrichment on photosynthesis is via the closure of stomata, one might expect that CO_2 enrichment and UV-B will not influence interactively photosynthesis of plants.

In 1988, a combined experiment of CO_2 enrichment and UV-B radiation was conducted in the greenhouses of the Department of Ecology and Ecotoxicology (ROZEMA *et al.*, this issue). The experiments were done with tomato and bean and the outcome was the following. Both plant species respond positively to CO_2 enrichment and negatively to UV-B radiation, be it that *Phaseolus vulgaris* is much more sensitive to UV-B than *Lycopersicon.* For *Phaseolus*, there was no interaction between the effect of CO_2 enrichment and UV-B radiation.

This means that the growth reduction by UV-B radiation is not alleviated by higher levels of atmospheric CO_2. On a global scale, these results indicate that increased UV-B radiation levels will govern the rate of primary production of plants in terrestrial ecosystems, independent of the CO_2 content of the atmosphere.

The rise of atmospheric CO_2 has not only a direct primary response to terrestrial autotrophic organisms; together with a number of other greenhouse gases such as nitrous oxide N_2O, the CFC's and methane, it causes an increase of the mean global temperature. This temperature increase causes through melting of polar ice caps and expansion of the volume of the oceans the rise of the sea level. The increase of the temperature of the earth's atmosphere may have important consequences, for the response of C_3 and C_4 plants to CO_2 enrichment.

Predictions of the global temperature increase range from 1.5 to 4.5°C for a doubling of the atmospheric CO_2 concentration (MACCRACKEN & LUTHER, 1985). IDSO *et al.* (1987) suggest an interaction between mean air temperature and the effects of atmospheric CO_2 enrichment. Mean air temperatures exceeding 18.5°C would increase plant response to elevated CO_2, while below 18.5°C the interaction effect of CO_2 enrichment might be negative and plant could be reduced with increased atmospheric CO_2. The consequences of these findings for coastal vegetation in Europe could be:

i. Absence or negative response to CO_2 enrichment in climate zones with a mean air temperature below 18.5°C. This would include major parts of the Scandinavian and Westeuropean and Atlantic coastline. To test this important hypothesis outdoor CO_2 enrichment studies at these prevailing temperatures are necessary. At the Department of Ecology and Ecotoxicology of the Free University of Amsterdam, field studies with open top chambers with enhanced CO_2 have started spring 1989.

ii. As has been presented in this and other papers (CURTIS *et al.*, 1989) C_3 plants respond positively to CO_2 enrichment, in contrast to C_4 plants and a shift in the competitive balance between C_3 and C_4 plants is to be expected accordingly. On the other hand, it is known that C_4 plants have a higher temperature optimum for photosynthesis and a low rate of photorespiration (EHLERINGER & BJØRKMAN, 1977). As a result, growth of C_4 plants will be increased with the predicted global warming. This will cause an effect on the competitive balance between C_3- and C_4-plants opposite to the effect of CO_2 enrichment.

Until sufficient well-designed experiments, both in the greenhouse and in the field, have been conducted in all the biomes, including the relevant environmental factors and modelling work, one can only speculate on the consequences of the global change.

ACKNOWLEDGEMENT

The authors are indebted to D. Hoonhout for typing the manuscript.

7. REFERENCES

BARNES, P.W., P.W. JORDAN, W.G. GOULD, S.D. FLINT & M.M. CALDWELL, 1988. Competition, morphology and canopy structure in wheat *(Triticum aestivum* L.) and wild oat *(Avena fatua* L.) exposed to enhanced ultraviolet-B-radiation.—Functional Ecology **2:** 319-330.

BASIOUNY, F.M. & M. FOULAD, 1986. Sensitivity of corn, oats, peanuts, rice, rye, sorghum, soybean and tobacco to UV-B radiation under growth chamber conditions.—J. Agr. Crop. Sci. **157:** 31-55.

CALDWELL, M.M., 1977. The effects of solar UV-B radiation (280-315 nm) on higher plants, implications of stratospheric ozone reduction. In: A. CASTELLANI. Research in photobiology. Plenum Press, New York: 597-607.

——, 1980. Light quality with special reference to UV at high altitudes. - NZFS FRI Technical paper no. 7.

——, 1981. Plant response to solar ultraviolet radiation. In: Encyclopedia of Plant Physiol. N.S. 12A.—Physiol. Plant Ecology **6:** 169-197.

CURTIS, P.S., B.G. DRAKE & D.F. WHIGHAM, 1989. Nitrogen and carbon dynamics in C_3 and C_4 estuarine plants grown under elevated CO_2 *in situ*.—Oecologia (in press).

DICKSON, J.G. & M.M. CALDWELL, 1978. Leaf development of *Rumex patienta* L. (Polygonaceae) exposed to UV-radiation.—Amer. J. Bot. **65:** 857-863.

EHLERINGER, J. & O. BJØRKMAN, 1977. Quantum yields for CO_2 uptake in C_3 and C_4 plants: dependence on temperature, CO_2 and O_2 concentration.—Plant Physiol. **59:** 86-90.

GAMMON, R.H., E.T. SUNDQUIST & P.J. FRASER, 1985. History of carbon dioxide in the atmosphere. In: J.R. TRABALKA. Atmospheric carbon dioxide and the global carbon cycle. US. Dept. Energy: 28-62.

GREEN, A.E.S., 1983. The penetration of ultraviolet radiation to the ground.—Physiol. Plant **58:** 351-359.

HASHIMOTO, T. & M. TAJIMA, 1980. Effects of ultraviolet irradiation on growth and pigmentation in seedlings. — Plant Cell Physiol. **21:** 1559-1571.

IDSO, S.B., B.A. KIMBALL, M.G. ANDERSON & J.R. MANNEY, 1987. Effects of atmospheric CO_2 enrichment on plant growth: the interactive role of air temperature.— Agricult. Ecosyst. and Environm. **20:** 1-10.

IWANZIK, W., M. TEVINI, G. DOHNT, M. VOSS, W. WEISS, P. GRABER & G. RENGER, 1983. Action of UV-B radiation on photosynthetic primary reactions in spinach chloroplasts.—Physiol. Plant **34:** 182-186.

KRIZEK, D.T., 1975. Influence of ultraviolet radiation on germination and early seedling growth.—Physiol. Plant **58:** 401-407.

LINDOO, S.J. & M.M. CALDWELL, 1978. Ultraviolet-B-radiation induced inhibition of leaf expansion and promotion of anthocyanin production.—Plant Physiol. **61:** 278-282.

MACCRACKEN, M.C. & F.M. LUTHER, 1985. Detecting the climatic effects of increasing carbon dioxide. US Dept. Energy; 1-198.

MOLINA, J.J. & F.S. ROWLAND, 1974. Stratospheric zink for chlorofluoromethanes; chlorine atom-catalysed destruction of ozone.—Nature **249:** 810-812.

MURALI, N.S. & A.H. TERAMURA, 1987. Insensitivity of soybean photosynthesis to UV-B radiation under phosphorus deficiency.—J. Plant. Nutr. **10:** 501-506.

NAS, NATIONAL ACADEMY OF SCIENCES, 1984. Causes and effects of stratospheric ozone reduction: an update. National Academy Press, Washington D.C.

ROBBERECHT, R. & M.M. CALDWELL, 1983. Protective mechanisms and acclimation to solar ultraviolet-B-radiation in *Oenothera stricta*.—Plant, Cell and Environment **6:** 477-485.

ROZEMA, J., M.C.T. SCHOLTEN, P.A. BLAAUW & J. VAN DIGGELEN, 1988. Distribution limits and physiological tolerances, with particular reference to the salt marsh environment. In: A.J. DAVY, M.J. HUTCHINGS & A.R. WATKINSON. Plant population ecology. Blackwell, Oxford: 137-164.

ROZEMA, J., G.M. LENSSEN, W.J. ARP & J.W.M. VAN DE STAAIJ, 1990. Global change, the impact of increased UV-B radiation and the greenhouse effect on terrestrial plants. In: J. ROZEMA & J.A.C. VERKLEY. Ecological responses to environmental stresses. Junk Publishers, Dordrecht (in press).

RUNDEL, R.D., 1983. Action spectra and estimation of biological effective UV-radiation.—Physiol. Plant **58:** 360-366.

SISSON, W.B., 1981. Photosynthesis, growth and ultraviolet irradiance absorbance of *Cucurbita pepo* L. exposed to ultraviolet irradiance (288-315 nm) simulating a reduced atmospheric ozone column.—Plant Physiol. **58:** 563-565.

SOYFER, V.N., 1983. Influence of physiological conditions on DNA-repair and mutagenesis in higher plants.—Physiol. Plant **58:** 373-380.

TERAMURA, A.H., 1983. Effects of ultraviolet-B-radiation on the growth and yield of crop plants.—Physiol. Plant **58:** 415-427.

——, 1987. Assessing the risks of trace gases that can modify the stratosphere. - Vol. VIII. Technical support documentation ozone depletion and plants. EPA report: 1-117.

TEVINI, M. & K. PFISTER, 1985. Inhibition of Photosystem 2 by UV-B radiation.—Zeitschr. Naturforsch. **40c:** 129-133.

TEVINI, M., U. THOMA & W. IWANZIK, 1983. Effects of enhanced UV-B radiation on germination, seedling growth, leaf anatomy and pigments of some crop plants.—Zeitschr. Pflanzen Physiol. **109:** 435-448.

TEVINI, M., P. GRUSEMAN & G. FIESER, 1988. Assessment of UV-B stress by chlorophyll fluorescence analysis. In: H.K. LICHTENTHALER. Applications of chlorophyll fluorescence. Kluwer Academic Publishers, Dordrecht: 229-238.

INDEX

Abra, 85-91
Acanthis, 192
Acropora, 201
Agropyron, 148, 149
Agrostis, 148
Alauda, 192
Anaitides, 89
Anas, 148, 150, 191, 192
Angulus, 85-91, 93-96
Anthus, 192
Antinoella, 85-91
Arenaria, 192
Arenicola, 86-89, 100, 146
Armerium, 176
Artemisia, 176
Aster, 49-53, 168, 171, 176, 177, 211-215
atmosphere
  chemistry, 6
Atriplex, 176
Bathyporeia, 89
benthos
  distribution - present and past, 81
  Gironde estuary, 80
  latitudinal differences, 80
  Seine estuary, 80
  sensivity to high temperatures, 83, 93-96
  sensivity to low temperatures, 83-91, 93-96
  Wadden Sea, 77-81, 83-91
birds, 189-193
boundaries, 55-64, 69-76, 90
Branta, 148, 191
Calidris, 150, 192
Callophyllis, 70-75
Capitella, 100
carbon dioxide
  increase, 7
  influence on plant transpiration, 49-53
  photosynthesis, 7, 23-31, 33-38, 41-46, 49-53
  seasonal course, 42
Carcinus, 85-91, 100
Cardium, see *Cerastoderma*
Carex, 148
Centroceras, 60, 61
Cerastoderma, 85-91, 143
CFC, see halocarbons
CH4, see methane
Charadrius, 192
Chorda, 70-75
Cladophora, 62-64, 70-75
climate parameters
  annual course, 20, 25
  latitudinal gradients, 55-64, 72
  quantitative relations, 21
climatic change
  causes, 5-9
  changes in productivity, 37, 38, 215, 216
  impacts, 10, 17-22, 28-31
  national responses, 14
  prevention, 11, 15
  regional scenarios, 18-20, 23-31
  sea-level rise, 123, 133-138
CO_2, see carbon dioxide
conferences, 1, 13, 15
Corophium, 89, 100, 146
Coturnix, 148, 149
Crangon, 85-91, 100, 149
Cucurbita, 212
Denmark, 113-121
Desmarestia, 55-64
Dictyosphaeria, 55-64
dinitrous oxide, 8
distribution patterns
  benthos, 80, 81, 90
  geographic boundaries, 55-64, 69-76, 90
  past ranges, 72, 73
  seaweed species, 55-64, 69-76
Dumontia, 55-64
Elymus, 176, 177
Elytrigia, 168
Emberiza, 192
Engraulis, 101
Enteromorpha, 203
erosion, 111, 123-130, 137, 173-186
Eteone, 89
Festuca, 41-46, 148, 149, 176, 177
fishes, 99-102, 203, 206
GCM, see General Circulation Model
General Circulation Model, 17, 18, 23-31
Gironde estuary
  benthic fauna, 80
  temperatures, 80
Glaux, 148, 176
Glycine, 212
Gobius, 101
Gracilaria, 70-75
greenhouse gases
  relative efficiency, 7

greenhouse warming (see also temperature change)
  preventive strategies, 10, 11
  scenarios, 10, 11
  sea-level rise, 105-111, 123
growth rates
  benthos, 95
  fishes, 100, 101
  plants, 33, 41-46, 49-53
Haematopus, 150, 192
Halimenia, 203
Halimione, 148, 149, 176, 177
halocarbons, 6, 9, 211
Halodule, 203
Harmothoe, 85-91
Heteromastus, 89
Holocene deposits, 133-138, 139-151
Hydrobia, 89, 134
Juncus, 148, 176, 177
Laminaria, 74-76
Lanice, 85-91, 100, 146
Larus, 191
Lathyrus, 212
latitudinal gradients
  benthos, 77-81, 90
  climate, 55-64, 69-76, 77-81
  seaweeds, 55-64, 69-76
Limonium, 168, 176, 177
Littorina, 89
Lolium, 41-46
Lomentaria, 70-75
Lycopersicon, 215
Lycopersicum, 212
Macoma, 83-91, 146, 191
Macrocystis, 203
Magelona, 89
Merlangus, 149, 191
methane
  concentrations, 8
  emission, 8
Mira estuary, 99-102
Mississippi Delta, 155-163
models
  characteristics, 24
  crop prediction, 29-31
  general circulation models, 17, 18, 23-31
  hierarchy, 28
  kinds, 24
  model-ecosystems, 41-46
  primary productivity, 34-37
Mont-Saint-Michel Bay, 139-151
Mya, 86-89
Mysella, 85-91
Mytilus, 86-89, 192
Nephtys, 85-91, 100
Nereis, 88, 89, 100
N_2O, see dinitrous oxide
Oenothera, 212
Oosterschelde
  salt marshes, 167-171
temperatures, 78, 79
Oryza, 212
ozone, 6, 195, 196, 211
Panicum, 212
Pectinaria, 100
Penaeus, 200
Phalacrocorax, 191
Phaseolus, 215
Philomachus, 150
photosynthesis, see primary production
Phragmites, 134, 148
Pisum, 212
plant production, see primary production
Platichthys, 99-102, 206
Plantago, 168, 176
Pleuronectes, 100, 149, 203, 206
Pluvialis, 192
Pocillopora, 201
Pomatoschistus, 100, 101
Porites, 201
Porphyra, 203
primary production, 33-38, 41-46, 49-53, 211-216
Puccinellia, 33-38, 147, 149, 150, 168, 171, 175-179
Raja, 149
Retusa, 89
Rumex, 212
Salicornia, 147, 168, 171, 175-177, 182
salinity
  effect on plant growth, 49-53
  increase, 12, 158
Salmo, 203
salt marshes
  birds, 192
  evolution, 139-151, 178-183
  flooding frequency, 167-171
  Oosterschelde, 167-171
  plant production, 33-38, 49-53, 211-216
  species composition, 167-171
  Wadden Sea, 173-186
  zonation vegetation, 167-171, 175
Scirpus, 148
Scrobicularia, 89
Scolelepis, 89
Scoloplos, 89
sea-level rise
  costs for the society, 12, 13
  effects on birds, 189-193
  endangered coasts, 12, 13
  history, 133-138, 139-151
  land loss, 155-163
  long-term trends, 107, 124
  marsh vegetation, 171, 176, 177
  Mississippi Delta, 155-163
  Mont-Saint-Michel Bay, 139-151
  the Netherlands, 12, 13, 133-138

salinity increase, 155-163
sedimentation, 105-111, 123-130, 133-138, 139-151, 173-186
vulnerable areas, 13
Wadden Sea morphology, 123-130, 173-186
sea water
alkalinity, 6, 7
pH, 6, 7
temperatures, 55-64, 69-76, 77-81, 84, 93-96, 100
seaweed species, 55-64, 69-76
sedimentation, 105-111, 123-130, 133-138, 139-151, 158, 173-186
sediment transport, 113-121
Seine estuary
benthic fauna, 80
temperatures, 80
Sepia, 149
shore lines, 113-121
solar radiation, 34
Solea, 100-102
Somateria, 191
Sorghum, 212
Spartina, 33-38, 147, 157, 168, 175-182, 211-215
Spergularia, 49-53
Sphoeroides, 102
Sterna, 192
Suaeda, 147, 171, 176, 177
Syngnathus, 101
Tadorna, 191
Tagus estuary, 99-102
Tellina, see *Angulus*
temperature
annual course, 79, 100
effect on fish growth, 101
effect on fish spawning, 101
effect on plant growth, 33-38
effect on respiration, 95, 96
influence on life cycle, 69-76
latitudinal differences, 80
lethal boundaries, 69-76
past periodes, 81
seasonal course, 79
seaweed species, 55-64, 69-76
summer maximum, 79
winter minimum, 79, 80, 84
temperature change
Dutch coastal waters, 77-80
effects on benthos, 77-81
global trend, 78
global trend expectation, 9, 17
past periods, 27, 28
predictions for Europe, 25
regional scenarios, 18, 19, 23-27
regional trends, 78
seaweed distribution, 55-64, 69-76
tidal amplitude
increase, 107-111
influence on vegetation, 167
tidal flats
benthos, 83-91
birds, 191
fishes, 101
UV-threats, 206
trace gases
lifetime, 5
listed, 6
reductions, 11
trends
carbon dioxide, 7
sea level, 107, 124, 140-142, 156
temperture, 9, 78
Trifolium, 41-46
Triglochin, 168, 177
Trisopterus, 149
Ulva, 203
UV-B radiation
effects on aquatic organisms, 199-207
effects on terrestrial vegetation, 211-216
levels, 195-200, 205, 212, 213
transmission in water, 195-199, 205
Vanellus, 192
Venerupis, 100
Wadden Sea
benthos, 77-81, 83-91
erosion, 111, 123-130, 173-186
fishes, 99-102
fish nursery, 101, 102
morphology, 105-111, 123-130, 173-186
salt marshes, 173-186
temperatures, 78-80, 84
tidal basins, 123-130
tidal-flat heights, 105-111, 123-130
wind regime, 113-121
winter temperatures
data, 79, 84
effects on benthos, 83-91
workshops, see conferences
yellow substance, 197, 205
Zea, 212
Zostera, 99-102, 168